Barriers and Transport in Unsteady Flows

A Melnikov Approach

Mathematical Modeling and Computation

About the Series

The SIAM series on Mathematical Modeling and Computation draws attention to the wide range of important problems in the physical and life sciences and engineering that are addressed by mathematical modeling and computation; promotes the interdisciplinary culture required to meet these large-scale challenges; and encourages the education of the next generation of applied and computational mathematicians, physical and life scientists, and engineers.

The books cover analytical and computational techniques, describe significant mathematical developments, and introduce modern scientific and engineering applications. The series will publish lecture notes and texts for advanced undergraduate- or graduate-level courses in physical applied mathematics, biomathematics, and mathematical modeling, and volumes of interest to a wide segment of the community of applied mathematicians, computational scientists, and engineers.

Appropriate subject areas for future books in the series include fluids, dynamical systems and chaos, mathematical biology, neuroscience, mathematical physiology, epidemiology, morphogenesis, biomedical engineering, reaction-diffusion in chemistry, nonlinear science, interfacial problems, solidification, combustion, transport theory, solid mechanics, nonlinear vibrations, electromagnetic theory, nonlinear optics, wave propagation, coherent structures, scattering theory, earth science, solid-state physics, and plasma physics.

Sanjeeva Balasuriya, *Barriers and Transport in Unsteady Flows: A Melnikov Approach*

A. J. Roberts, *Model Emergent Dynamics in Complex Systems*

John B. Drake, *Climate Modeling for Scientists and Engineers*

Erik M. Bollt and Naratip Santitissadeekorn, *Applied and Computational Measurable Dynamics*

Daniela Calvetti and Erkki Somersalo, *Computational Mathematical Modeling: An Integrated Approach Across Scales*

Jianke Yang, *Nonlinear Waves in Integrable and Nonintegrable Systems*

A. J. Roberts, *Elementary Calculus of Financial Mathematics*

James D. Meiss, *Differential Dynamical Systems*

E. van Groesen and Jaap Molenaar, *Continuum Modeling in the Physical Sciences*

Gerda de Vries, Thomas Hillen, Mark Lewis, Johannes Müller, and Birgitt Schönfisch, *A Course in Mathematical Biology: Quantitative Modeling with Mathematical and Computational Methods*

Ivan Markovsky, Jan C. Willems, Sabine Van Huffel, and Bart De Moor, *Exact and Approximate Modeling of Linear Systems: A Behavioral Approach*

R. M. M. Mattheij, S. W. Rienstra, and J. H. M. ten Thije Boonkkamp, *Partial Differential Equations: Modeling, Analysis, Computation*

Johnny T. Ottesen, Mette S. Olufsen, and Jesper K. Larsen, *Applied Mathematical Models in Human Physiology*

Ingemar Kaj, *Stochastic Modeling in Broadband Communications Systems*

Peter Salamon, Paolo Sibani, and Richard Frost, *Facts, Conjectures, and Improvements for Simulated Annealing*

Lyn C. Thomas, David B. Edelman, and Jonathan N. Crook, *Credit Scoring and Its Applications*

Frank Natterer and Frank Wübbeling, *Mathematical Methods in Image Reconstruction*

Per Christian Hansen, *Rank-Deficient and Discrete Ill-Posed Problems: Numerical Aspects of Linear Inversion*

Michael Griebel, Thomas Dornseifer, and Tilman Neunhoeffer, *Numerical Simulation in Fluid Dynamics: A Practical Introduction*

Khosrow Chadan, David Colton, Lassi Päivärinta, and William Rundell, *An Introduction to Inverse Scattering and Inverse Spectral Problems*

Charles K. Chui, *Wavelets: A Mathematical Tool for Signal Analysis*

Barriers and Transport in Unsteady Flows

A Melnikov Approach

Sanjeeva Balasuriya

University of Adelaide
Adelaide, South Australia
Australia

siam.

Society for Industrial and Applied Mathematics
Philadelphia

10 9 8 7 6 5 4 3 2 1

Publisher	David Marshall
Acquisitions Editor	Elizabeth Greenspan
Developmental Editor	Gina Rinelli Harris
Managing Editor	Kelly Thomas
Production Editor	Ann Manning Allen
Copy Editor	Bruce Owens
Production Manager	Donna Witzleben
Production Coordinator	Cally Shrader
Compositor	Cheryl Hufnagle
Graphic Designer	Lois Sellers

Library of Congress Cataloging-in-Publication Data

Names: Balasuriya, Sanjeeva.
Title: Barriers and transport in unsteady flows: a Melnikov approach / Sanjeeva Balasuriya, University of Adelaide, Adelaide, South Australia, Australia.
Description: Philadelphia : Society for Industrial and Applied Mathematics, [2016] | Series: Mathematical modeling and computation ; 21 | Includes bibliographical references and index.
Identifiers: LCCN 2016035018 (print) | LCCN 2016038029 (ebook) | ISBN 9781611974577 | ISBN 9781611974584
Subjects: LCSH: Unsteady flow (Fluid dynamics) | Fluid dynamics.
Classification: LCC TA357.5.U57 B35 2016 (print) | LCC TA357.5.U57 (ebook) | DDC 620.1/064–dc23
LC record available at https://lccn.loc.gov/2016035018

Contents

List of Figures

Preface

Over the past several years, there has been an explosion of interest in analyzing flow barriers in fluids. This has been driven by a variety of reasons. One is a concern for our environment: the spreading of pollutants (such as the Deepwater Horizon oil spill in the Gulf of Mexico in 2010) is known to be greatly influenced by the location and movement of such flow barriers, which impede the transport of pollutants. Another is an increasingly urgent desire to understand geophysical processes in an era of rapid climate change. For example, the transport of ozone across the Antarctic Circumpolar Vortex boundary, or of heat across oceanic or atmospheric vortex boundaries, has a profound influence on our planet. How these boundaries (i.e., barriers) evolve with time, and the amount of transport which occurs across such evolving boundaries, has impact on the earth's weather systems. A third reason for flow barriers to become important recently is the ongoing biotechnological revolution, in which fluidic devices at the micro- and nanoscales are being developed for a variety of applications, including DNA synthesis, drug delivery and monitoring, lab-on-a-chip technology, and biofluidic cells. The tiny dimensions of these devices ensure that turbulent mixing is suppressed, and in cases in which the efficient mixing of fluids is required—to expedite a chemical reaction between a sample and a reagent, for example—improving transport across a fluid interface (i.e., a barrier once again) is needed.

Flow barriers and the transport associated with them are currently being studied in many ways, including observationally at the geophysical scale, experimentally at the laboratory scale, computationally from either numerically generated velocity fields or from observational data, or theoretically through the development of analytical methods to define, understand, and quantify flow barriers. This book is a contribution to the last of these approaches, based on insights I have built up over many years of working in this area. While much of the development is theoretical in nature, the inspiration for the relevant theory arises directly from applied questions, which I have made an effort to motivate throughout. With this in mind, I have chosen to begin the book with an introductory chapter which positions the material within the large research area of fluid barriers, associated with oceanography, atmospheric science, engineering, physics, and of course mathematics. Chapters 2 and 3, which respectively develop the Melnikov approach to characterizing flow barriers as stable/unstable manifolds for nonautonomous flows and assess the transport across such flow barriers, are at the heart of the book. In the next chapters, these methods are modified and applied toward optimizing transport across barriers (Chapter 4) and controlling the barriers (Chapter 5), inspired primarily by applications in microfluidics.

Parts of this book are fairly technical, and it is not necessary for a reader to digest every section to be able to proceed to the next. I provide a brief picture of the chapter dependencies in Figure 1 to enable a reader to choose a pathway through the book. For example, a reader seeking to access the optimization chapter needs only Sections

3.1–3.4 from Chapter 3. To assist the reader, I have identified sections which I call "fundamental" as being Sections 1.1–1.6, 2.1–2.4, and 3.1–3.4. The "later" sections in each chapter are often self-contained in the sense that while they do depend on the fundamental earlier sections, they are independent of the (nonfundamental) other sections. Readers interested in microfluids may find Chapters 4 and 5 as the main goals to reach, while those interested in geophysics might be interested in the transport determinations in Sections 3.4 and 3.8. Optimizing and controlling geophysical flows is of course highly challenging, but perhaps aspects of Chapters 4 and 5 will be useful to geophysics in the future.

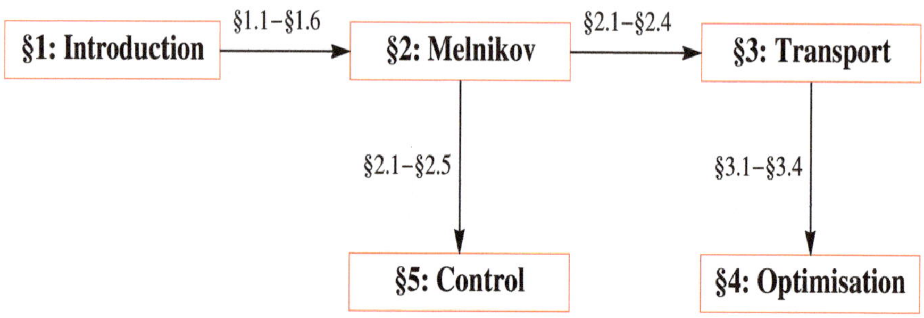

Figure 1. *Chapter and section dependencies. The "fundamental" sections are Sections 1.1–1.6, 2.1–2.4, and 3.1–3.4, which are needed to access the remaining sections of the book.*

The many theoretical and applied areas which are touched on by the material in this book necessitate my providing an extensive and current bibliography, which I hope will benefit the reader wishing to track down more information on a particular aspect of interest. While dealing comprehensively with the details of the theory, I have also strived to provide geometric intuition for the results in all situations. The theory I discuss is motivated by applied problems primarily in fluid mechanics, and many standard examples, such as the double-gyre, the Duffing oscillator, the Taylor-Green flow, the Kelvin-Stuart cat's-eyes flow, and the Hadamard-Rybczynski droplet flow, are used in a recurring fashion to illustrate the theory developed, with many other examples also provided. As such, I trust that this book will provide an applications-inspired theoretical approach to dynamical systems methods applied to fluid dynamics in which the geometrical intuition is emphasized throughout. The connections I provide to many branches of research will I hope inspire additional extensions in the future.

SIAM's publications staff and several anonymous reviewers of the original manuscript provided invaluable advice in expanding the book's scope, detecting errors, and improving readability. Finally, I wish to thank my wife, Rasika, for her unending support, critical perusal of the manuscript, and good humor while I was writing this book, which, inevitably, turned out to take much more time than anticipated. This book is dedicated to her.

Sanjeeva Balasuriya
Adelaide, 2016

Chapter 1

Unsteady (nonautonomous) flows

Let reality be reality. Let things flow naturally forward in whatever way they like.
—Lao-Tzu

1.1 ▪ Unsteady flows versus steady flows

Transport occurring as a result of time-varying fluid velocities is of paramount importance in geophysics. The transport of heat from the Gulf of Mexico toward Europe by Gulf Stream waters, for example, has profoundly influenced the climate in Europe. Ozone movement due to wind velocities of the Antarctic Circumpolar Vortex impacts ultraviolet radiation and the global heat balance. The capture of plankton within oceanic eddies mobilizes oceanic communities toward the eddies, resulting in marine ecosystems which last over meso-timescales of the order of months. The spread of pollutants—such as arising from the Fukushima Daiichi or Deepwater Horizon disasters—is governed strongly by fluid velocities. When a sample and a reagant come together in a microfluidic device, how quickly and well they mix together is affected by the velocities within the devices.

In these and in a multitude of similar examples, a basic question is to determine how a quantity of interest (heat, contaminant, energy) is moved around by the background fluid (sea water, air) in which it flows. Coupled with this is the more basic question of how the *fluid itself* moves. Along with these transport questions sometimes comes the issue of what *mixing* is achieved as a result of the transport.

Traditionally in fluids problems, the major information one seeks is the fluid velocity $u(x,t)$ as a function of space x and time t. To accomplish this, one often requires information about the dynamical evolution of u, typically from the Navier-Stokes equations (subject to complications and/or simplifications). From the perspective of the transport problem, however, determining u is only the first step. Since the rate of change of position is the velocity, the evolution of fluid particles is then governed by

$$\dot{x} = u(x,t), \tag{1.1}$$

where the overdot denotes differentiation with respect to time t, which is an ordinary differential equation. In typical problems, the spatial variable x (and, correspondingly, u) will be in $\mathbb{R}^2$ or $\mathbb{R}^3$. For a particle at a location x_0 at time 0, (1.1) would need to be solved subject to the initial condition $x(0) = x_0$ to determine the fate of that *one* particle. Understanding global transport therefore requires knowledge of solutions to (1.1) for *all* initial conditions. If one requires information on the evolution of a scalar quantity (such as heat, pollutant concentration, plankton density, etc.), the simplest approach would be to assume that these are moved "along with the flow," and therefore this problem reduces to the question of understanding the trajectories of (1.1). (To a next-order approach, one might instead impose some other dynamically/biologically/chemically inspired rule—such as an advection-diffusion equation—on how the scalar quantity interacts with the flow.) In typical situations, examining all trajectories of (1.1) would reveal *coherent structures* where fluid parcels in these structures tend to remain together. The boundaries of these will be *flow barriers*. As shall be seen, these simple intuitive statements carry within them inherent difficulties, the resolution of which *in some circumstances* is a goal of this book. For the moment, let us continue with this intuitive but imprecise understanding that coherent structures comprise fluid parcels which clump together, following similar motions. If so, passively advected scalars would tend to get organized according to these coherent structures. Thus, understanding coherent structures in (1.1) is an important first step in identifying how flow-advected quantities evolve over time. Small but nonzero diffusion would "smear out" information that one would get from these structures. On the other hand, the structures will remain robust on the nondiffusive timescale. Therefore, examining the nondiffusing situation—i.e., effectively (1.1)—is crucial to understanding the coherent structures.

If $u(x,t)$ does not depend on time t, the velocity field is deemed *steady*, and the corresponding dynamical system (1.1) is *autonomous*. While there is this subtlety in difference of meaning, in keeping with common usage, the terms *steady* and *autonomous* will be used interchangably in this book. Consequently, (1.1) is *unsteady* or *nonautonomous* since there is an explicit t-dependence in the velocity field. Under the *steady* condition, fluid particles will follow trajectories of a steady system

$$\dot{x} = u(x). \tag{1.2}$$

In this case, particularly in two dimensions, coherent structures are easy to identify using standard nonlinear dynamical systems methods. *Fixed points*, or *critical points*, are first identified by seeking x-locations at which $u(x) = 0$. In the language of fluid mechanics, such a point is also called a *stagnation point* in view of the fact that fluid particles which are at that location have zero velocity for all time and hence do not move. Flow near a fixed point can be determined by considering the locally linearized velocity. If a is a fixed point, then utilizing an expansion $x(t) = a + y(t)$ and retaining terms only to order $|y|$ leads to the equation

$$\dot{y} = Du(a)\,y, \tag{1.3}$$

in which $Du(a)$ is the (Jacobian) matrix derivative of the steady velocity field u, evaluated at $x = a$. This is now a standard scenario from which easy implications on the behavior near a can be imputed. For example, suppose $Du(a)$ contains a full-dimensional set of eigenvalues which are real, with some of them being positive and the remainder negative. The space spanned by the eigenvectors associated with the positive eigenvalues corresponds to solutions which expand away from $y = 0$ as t evolves. This space determines the *local unstable manifold* of a in the system (1.2). The remaining space de-

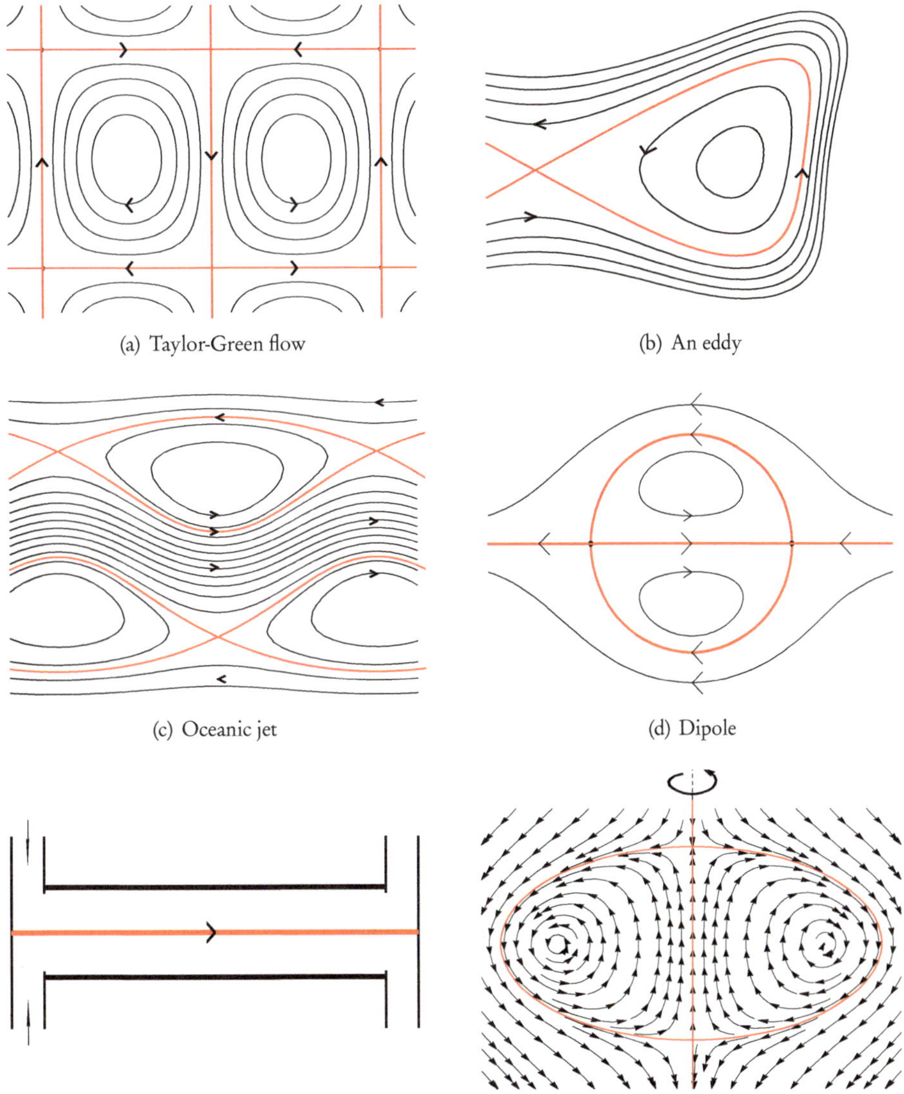

(a) Taylor-Green flow (b) An eddy (c) Oceanic jet (d) Dipole (e) Microchannel (f) Hill's spherical vortex

Figure 1.1. *Some phase portraits associated with the steady flow (1.2).*

termines the *local stable manifold*, comprising trajectories of (1.2) which get attracted exponentially to a. The simplest example of this is if $x \in \mathbb{R}^2$; if $Du(a)$ has one positive and one negative eigenvalue, corresponding to these are one-dimensional local unstable and stable manifolds, and hence a is a *saddle point*. In general, if $Du(a)$ has eigenvalues which do not fall on the imaginary axis, the point a is called a *hyperbolic point*. Technically speaking, this could mean that the real part of all the eigenvalues takes the same sign, though in more common usage hyperbolic points are often thought of as analogues of saddle points in that there are eigenvalues with both positive and negative real parts. Since there are advantages to both viewpoints, a particular decision on this definition will not be attempted; the usage will be made clear in context.

So how are these entities associated with the coherent structures and flow barriers? To answer this question, the *phase portraits* of (1.2) are shown for several different choices of $u(x)$ in Figure 1.1. A phase portrait for a steady flow (1.2) is a diagram in the relevant x-space—called the *phase space*—with arrows indicating the directions of the vector field $u(x)$. A wonderful connection here is that for steady fluid flows, *the phase space is identical to the physical space* in which the fluid is flowing. The phase portraits shown in Figure 1.1 relate to idealized (toy) velocity fields which have been suggested for modeling fluid flows ranging in scales from geophysical to microfluidic. In all cases, the curves shown are locally tangential to the velocity field and are thus *streamlines*. Flow barriers between *regions which have different characteristics* are shown in red.

Figure 1.1(a) is a model associated with Taylor-Green flow [5, 27, 1, 26, 54, 69, e.g.], which has also been used (with suitable modifications) to model an oceanic double-gyre [364, 54, 422, 316]. Adjacent gyres have opposite directions of rotation (different signs for the vorticity), and the bounding line between them is one such flow barrier. An eddy is shown in Figure 1.1(b), which has counterclockwise rotation; this can be thought of as a kinematic model for an oceanic eddy/ring or an atmospheric cyclone/hurricane/vortex [216, 226, 193, 346, 24]. (The term *kinematic* or *kinematical* is applied to models which are constructed to display some desired behavior but which are not necessarily based on physical rules. In contrast, a *dynamical* or *dynamically consistent* model is one in which the governing equations of motion, say, the Navier-Stokes equations, are specifically used in deriving the model.) If attempting to answer the question "Where does the eddy end?" one might seek the outermost streamline beyond which the streamlines no longer exhibit a closed pattern, and indeed the red curve is precisely that. This represents a *change in topology* of the streamline pattern, going from closed curves to something different. Figure 1.1(c) is a kinematic model associated with an eastward-flowing oceanic jet [323, 314, 48, 355, 125, 352], and here the boundaries between the oceanic jet and its outer eddies are indicated by the red curves. If examining each of these flow barrier curves carefully, since these are associated with a change in topology in a smoothly varying streamline pattern, there *must* be at least one point at which the streamlines intersect. Indeed, the eddies here serve to show, as a generalization of Figure 1.1(b), that there may be more than one streamline intersection point on the boundary of an eddy. Now, a fluid particle which is on a streamline must follow the streamline as time progresses; i.e., each streamline is a trajectory of the steady flow (1.2). A point of intersection of streamlines leads to a contradiction: Which direction would a particle which finds itself at the intersection point go? Which of the two streamline directions will it follow? Since (1.2) is an ordinary differential equation and the uniqueness theorem applies for sufficiently smooth u, this ambiguity is not permissible. Thus, an intersection point between streamlines—a necessary part of a flow barrier for a two-dimensional steady flow—is not permitted, *unless that intersection point corresponds to a zero velocity*. If $u = 0$ at such a point, a fluid particle located at that point does not receive conflicting instructions on which direction to go; since the velocity is zero, it just sits there. Thus, flow barriers (in the sense of representing a change in topology) in steady two-dimensional flows must contain a fixed point. Since the remainder of the barrier is connected to the fixed point, these curves must of necessity be stable and unstable manifolds. In particular, the fixed point must be saddle-like in this situation.

A particular feature of the flow barriers is their relationship to *stable* and *unstable manifolds*. A stable manifold of a fixed point is the set of points in the phase space which approach it in forward time. Since the velocity is zero at the fixed point, such an approach can only happen in infinite time. In examining the flow barriers illustrated in Figure 1.1 discussed so far, it is clear that the red curves—the flow barriers—are stable manifolds of fixed points. These are one-dimensional curves in this two-dimensional flow. Indeed, they comprise *trajectories* of a system (1.2) which approach the relevant fixed point. Now, an unstable manifold is the set of points in phase space which approach a fixed point in backward time. It is clear from Figure 1.1 that the flow barriers consist of situations in which *stable and unstable manifolds coincide*. In general these entities are called *heteroclinic manifolds*. For the instance in which they are the stable and unstable manifold of the *same* saddle fixed point, as in Figure 1.1(b), the term *homoclinic manifold* is used. In general, the idea obtained from these steady two-dimensional pictures is that *flow barriers consist of heteroclinic manifolds*.

Figure 1.1(c) is also inspired by oceanography: this flow is associated with idealized eastward currents (such as the Gulf Stream) with adjacent cat's-eyes eddies flanking the main oceanic jet. Kinematical and dynamically consistent models illustrating the structure of Figure 1.1(c) are common in the literature [125, 323, 314, 417, 225, 48, 355, 175]. The barriers between the jet and the eddies, or between the eddies and the outer flow, all consist of heteroclinic manifolds.

The same theme is continued in Figure 1.1(d), which shows an idealized dipole, or a dual-core eddy, or alternatively a two-cell microdroplet in a background flow; once again, the flow barriers demarcating regions of distinct flow are stable and unstable manifolds which are associated with saddle fixed points. Variants of this kinematical structure are commonly studied in relation to microdroplets [149, 378, 375, 90, 176, 403, 116, 248, 403, 363, 78, 386, 381, 97, 177, 281, 38, 102]. The situation of Figure 1.1(e) is slightly different in that the flow barrier is not necessarily a fluid trajectory which separates topologically distinct regions but rather one which separates two different fluids. In a standard microfluidic situation, two different fluids enter the microchannel via a T-junction at the left, but as they flow along the channel they tend not to mix across the centerline of the channel [383, 420, 308, 6, 425, 12, 412, 118, 376, 307, 236, 289, 287, 288, 390, 79, 25, 42, 237]. However, here too one can envisage this flow barrier as a connection between two saddle stagnation points, which in this case are located on the left and right ends of the main channel.

In contrast to all the figures discussed so far, Figure 1.1(f) shows a three-dimensional flow. The entity displayed here is kinematically equivalent to *Hill's spherical vortex*, or a *bubble vortex* from classical inviscid fluid theory [234, 3, 50, 238, 201, 274, 260, 195, 233, 370, 20, 196, 241, 345, 356, 249, 11, 87, 94, 335], or the Hadamard-Rybczynski solution from highly viscous (Stokes) flow [184, 347, 381, 38, 229, 426, 380, 176, 38, 51]. This entity is cylindrically symmetric, and Figure 1.1(f) shows the behavior in a plane going through the axis of symmetry. The full three-dimensional picture is obtained by rotating about the axis of symmetry shown by a dotted line. The main separating flow barrier here is shown by the red ellipse; this corresponds to an ellipsoid in the full picture. The surfaces inside the ellipsoid are nested torii, whereas those outside are concentric cylinders. Thus, the ellipsoid represents a change in topology of the flow. This is achieved through the presence of saddle fixed points at the "north pole" and "south pole" of the ellipsoid, which are connected together by a straight line along the axis of the ellipsoid. The larger torii get vanishingly close to both this line and the ellipsoid's surface. This line is in fact simultaneously a branch of the one-dimensional unstable manifold of the north pole and a branch of the one-dimensional stable manifold of the south pole, once again forming a heteroclinic manifold. Despite the importance of this, it is a one-dimensional entity in the background three-dimensional space. Therefore, it cannot *separate* regions and is not a flow barrier. On the other hand, the surface of the ellipsoid is simultaneously the two-dimensional unstable manifold of

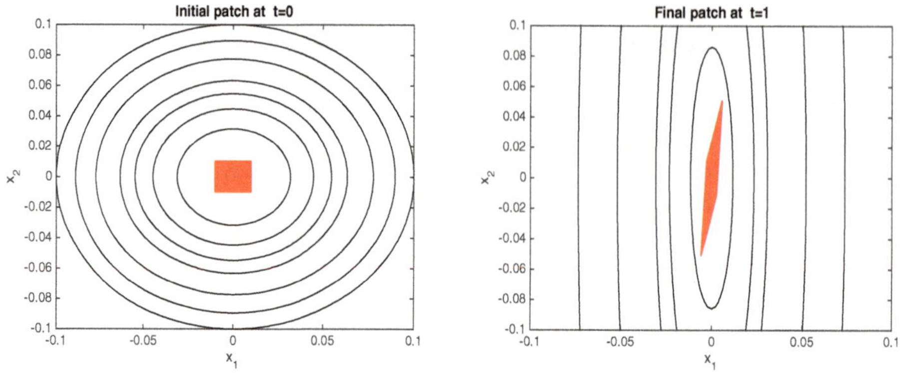

(a) $t=0$ (b) $t=1$

Figure 1.2. *Evolution of a patch of initial conditions (red) from time* 0 *to* 1*, according to the velocity (1.4) of Example 1.1 with* $c=2$.

the north pole and the two-dimensional stable manifold of the south pole. This two-dimensional heteroclinic manifold does indeed separate regions where distinct behavior occurs; there are invariant torii inside of this flow barrier, whereas there are invariant cylinders outside of it. In a slightly different variation of Figure 1.1(f) which is sometimes considered, an additional swirl is introduced around the axis [50]; once again, the ellipsoid is the flow barrier between flow on torii and flow on cylinders.

The particular structure of Figure 1.1(f) shows integrable flow trajectories which do not display any complicated motion. While this is necessarily so in two-dimensional steady flows, it need not be the case that in three dimensions one gets regular flow structures as in Figure 1.1(f); examples to the contrary are the ABC flow [162, 338] and of course Lorenz's classical equations [250]. Figure 1.1(f), however, indicates a situation in three-dimensional steady flow in which regular motion, with easily identifiable regular flow barriers, occurs. Regularity arises here because there are additional symmetries and/or dynamical constraints. *General* three-dimensional steady flows would usually have flow barriers which are less easy to identify and indeed to define. Similarly, *unsteady* two-dimensional flows, in contrast to Figures 1.1(a)–(e), will not in general have unambiguous flow barriers. One difficulty which leads to this is the fact that fixed points do not generically exist in unsteady flows since flow velocities are changing with time. Sketching pictures such as those in Figures 1.1(a)–(e) is no longer reasonable because of continual time-variation. The difficulties arising in unsteady situations are illustrated below via two simple examples.

Example 1.1. Consider the system

$$\dot{x}_1 = -2x_2 e^{-ct}, \quad \dot{x}_2 = 2x_1 e^{ct}, \tag{1.4}$$

where c is a parameter. Figure 1.2 displays the evolution of an initial patch (red) at time 0 under (1.4) until time $t=1$, with $c=2$. The direction fields associated with (1.4) at each of the times are shown by the arrows, and it is true that at each time, the streamlines (curves which are instantaneously tangential to these) are given by contours of the streamfunction

$$\psi(x_1, x_2, t) = x_1^2 e^{ct} + x_2^2 e^{-ct}.$$

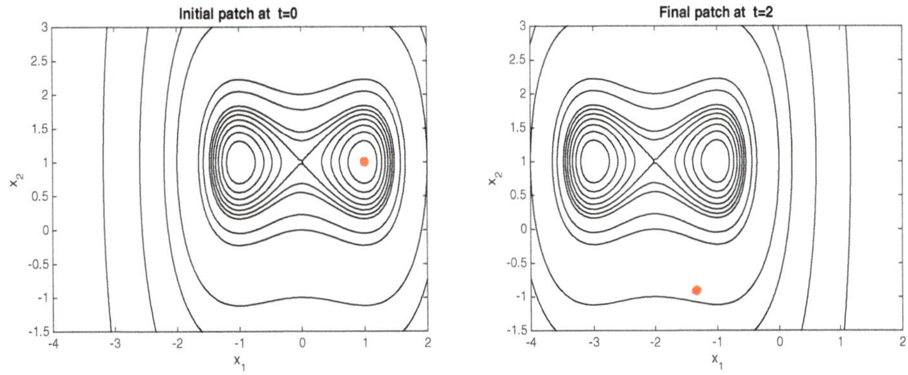

(a) $t=0$ (b) $t=2$

Figure 1.3. *Evolution of a patch of initial conditions (red) from time* 0 *to* 2*, according to the velocity (1.5) of Example 1.2 with* $c=1$.

These are ellipses at *all* instances in time, giving the impression that the motion is that of pure rotation. However, as indicated in Figure 1.2(b), an initial fluid element experiences stretching. The presence of $c \neq 0$ in (1.4) makes this flow an unsteady one, and the Eulerian understanding (via instantaneous streamlines always being circles) clearly gives an invalid intuition as to the Lagrangian impact of the velocity field. ■

Example 1.2 ("moving" Duffing oscillator). A different unsteady example is

$$\dot{x}_1 = x_2 - c, \quad \dot{x}_2 = (x_1 + ct) - (x_1 + ct)^3, \tag{1.5}$$

where c is a parameter. In this case, the velocity field (1.5) is everywhere tangential to the contours of the function

$$\psi(x_1, x_2, t) = x_2^2 - 2cx_2 - (x_1 + ct)^2 + \frac{1}{2}(x_1 + ct)^4 . \tag{1.6}$$

The contours of $\psi(\cdot, \cdot, 0)$ possess a "figure-8" structure as shown in Figure 1.3(a). The form of (1.6) indicates that any structure present at $t = 0$ simply translates to the left at speed c. (This flow is in fact obtained by applying a simple translation to the steady Duffing oscillator which shall be analyzed in considerably more detail in Examples 2.11, 2.21, 2.33, 3.16, 3.34, and 5.4 in this book.) In comparison with the pictures shown in Figure 1.1, the indication from Figure 1.3(a) might be that the figure-8 forms a flow barrier, with particles inside the two loops of the figure-8 exhibiting rotational motion. Given that the time-evolution of the streamlines indicates that this structure simply moves to the left at speed c, one might believe, then, that the red patch of particles within one of these "rotational regions" as indicated in Figure 1.3(a) would continue to remain inside the structure. However, Figure 1.3(b) shows the fate of this patch by time $t = 2$, where $c = 1$ has been used for the velocity in (1.5). The patch is clearly outside the ostensible flow barrier formed by the figure-8. Moreover, it has exhibited excursions in the x_2 direction, which seems contradictory to the fact that the streamline pattern only moves in the x_1 direction as time evolves. This example highlights that tracking the time-evolution of Eulerian entities does *not* in general provide information on the Lagrangian nature of the flow. The ostensible flow barriers identified in correspondence with Figure 1.1 are *not* flow barriers at all. ■

The unsteadiness present in the above two examples is actually trivial. The first example corresponds to applying a time-dependent modification to a steady flow, whereas the second is a constant translation of the reference frame in one direction. Even such minor introductions of unsteadiness affect our intuition on flow barriers and transport in nonobvious ways. While both the examples—generated as they are via affine transformations applied to a steady flow (see also [33] for more such examples)—in reality have no complicated flow patterns, generic unsteady flows typically display significantly more complex Lagrangian trajectories. That is, a smooth and regular fluid velocity $u(x,t)$ in a nonautomous flow (1.1) does *not* by itself guarantee that the motion of fluid particles is smoothly distributed. Simple counterexamples are the forced Duffing system [203] and the double-gyre [364] in two dimensions and the Lorenz [250] and ABC flows [162, 338] in three. In all these cases, the right-hand side of (1.1) is very smooth, but the resulting trajectories are *chaotic*. This fact was highlighted in the fluids dynamics literature by Aref [14], who pointed out that the "classical" fluids approach of thinking of $u(x,t)$ as either *laminar* (regular) or *turbulent* concealed an intermediate stage. The classical viewpoint evaluates the smoothness of $u(x,t)$ (usually from a solution to the Navier-Stokes equation) in terms of its dependent variables (x,t) and classifies the flow as laminar (smooth dependence) or turbulent (nonsmooth dependence). This attitude of thinking of the velocity in terms of (x,t) is called the *Eulerian* approach to fluids. In contrast, assessing the flow in terms of how fluid particles evolve with time is the *Lagrangian* approach. The point highlighted by Aref [14] was that it was possible for a flow to be regular in the Eulerian sense but irregular in the Lagrangian sense. Put another way, (1.1) can generate complicated fluid trajectories despite the velocity u being very smooth. Since (1.1) is precisely the correct formulation for addressing particle trajectories, its implications are often termed "Lagrangian." Thus, the coherent structures resulting from (1.1) are usually called *Lagrangian Coherent Structures* (LCSs) in the literature [193, 189, 309]. Somewhat strangely, the *boundaries* of LCSs are now commonly referred to as LCSs; in this book these boundaries shall be thought of as the *flow barriers* which demarcate distinct coherent structures, as illustrated by the red curves in the steady situation in Figure 1.1.

In steady flows, coherent structures are well understood in the sense of pictures such as those in Figure 1.1. These show *streamlines*, which are the curves to which the velocity field is locally tangential at a fixed instance in time. Since the flow is steady, the streamline patterns do not change. Another class of "lines" (these are actually *curves*) which are often referred to in the fluids dynamics literature are *pathlines*. As is clear from the context, these are the paths followed by fluid particles. In other words, these are fluid trajectories associated with the advection equation (1.1). Since streamlines are associated with a velocity pattern at a certain instance in time while pathlines demonstrate the eventual trajectories as a result of the fluid velocity, these two entities are, respectively, key players in the Eulerian and Lagrangian viewpoints. Now for a steady flow, fluid particles are pushed by the local velocity along streamlines, and since streamlines remain stationary, *pathlines are identical to streamlines*. Fluid dynamics terminology identifies a third class of "lines": *streaklines*. These are curves followed by fluid particles which are expelled from a *fixed* location. For example, if dye or smoke is continually released at a certain point in the flow, the way that the dye distributes itself is the streakline at that instance in time. Streaklines evolve (even in steady flows) since the dye spreads over time; indeed, understanding this evolution is crucial, for example, in determining how oil from a spill at a certain location would pollute the surroundings. However, in a steady flow, if streaklines are allowed to spread over a large period of time, since all particles which go through the same point undergo the same velocities as they evolve, streaklines will fill out streamlines. In this sense, it is reasonable to say that *streamlines, pathlines, and streaklines coincide for a steady flow.*

When the flow is unsteady, the velocity field in (1.1) is nonautonomous. The fact that the velocity is changing with time means that fluid particles going through a specific location at different times are not pushed in the same direction by the velocity field. Their behavior (i.e., the corresponding pathlines and streaklines) is no longer identical to the streamlines at that instance in time. From the transport perspective, however, it is pathlines and streaklines which are important. The key observation here is that one *cannot* impute transport implications purely by examining how a streamline pattern changes with time (i.e., by viewing an evolution of an Eulerian quantity); solutions of the dynamical system (1.1) *must* be considered. Examples 1.1 and 1.2 illustrate this clearly.

A main goal of this book is to address the issue of how the straightforward discussion of flow barriers for *steady* (autonomous) flows extends to *unsteady* (nonautonomous) flows. For *unsteady* flows, how can one

1. *Define* flow barriers?
2. *Characterize* their time variation?
3. *Quantify* transport across them?
4. *Optimize* such transport?
5. *Control* the flow barriers?

These questions are difficult and continue to attract considerable attention. This book will cover a range of techniques which can help answer each and every one of the above questions under certain hypotheses. While the main emphasis shall be on two-dimensional, compressible, nearly autonomous situations, excursions into other aspects, such as higher dimensions (Section 5.2), finite times (Sections 2.8 and 5.3), impulsive nonautonomousness (Sections 2.7 and 3.6), and nonheteroclinic flow barriers (Sections 3.7 and 4.4), shall also be made.

1.2 ▪ Importance in geophysical flows

The previous section discussed the importance of *coherent structures*, giving examples in autonomous flows. *Flow barriers* are the *boundaries* of such coherent structures, and thus understanding their motion in time-varying flows effectively quantifies the motion of coherent structures. This section outlines why understanding coherent structures is particularly important in the geophysical context.

First, a brief description of the historical connections established between geophysics and dynamical systems is in order. A substantial literature on why dynamical systems methods are relevant to oceanic and atmospheric transport began emerging in the late 1980s and early 1990s. Early evidence for the necessity for using Lagrangian data in assessing transport was furnished by authors such as Bower and Rossby [83], Davis [122], and Bowman [84], and many studies and models for assessing Lagrangian transport [81, 252, 82, 120, 136, 137, 143, 352, 8, 273, 314, 125, 417, 225] also appeared during this time. The theoretical connection between oceanic transport, chaotic advection, and intersections of stable and unstable manifolds was developed, with excellent reviews available [422, 354, 421, 353, 227]. The present monograph is in the same spirit, but with the intention to move toward *general* time-dependence. There continues to be a range of studies which analyze geophysical transport from this perspective, using observational or computational fluid dynamics data [216, 275, 350, 277, 276,

316, 339, 321, 318, 351] or synthetic data from models [351, 369, 314, 125, 324, 212, 134, 322, 348]. A study on the implications of the geophysically relevant constraint of potential vorticity conservation by Brown and Samelson [91] highlighted the necessity of using dynamically consistent velocity models in transport assessment; they concluded that complicated transport is impossible if the potential vorticity gradient was never zero. A weakening of this assumption to the absence of constant patches of potential vorticity at time zero was later obtained by Balasuriya [23]. Including dynamic consistency, i.e., dynamical constraints—such as the conservation of potential vorticity—which the fluid velocity field must obey, would seemingly be important to consider in evaluated models for geophysical transport. Accordingly, there have been many dynamically consistent (as opposed to simply kinematically reasonable) models which have been studied from the perspective of geophysical transport assessment [197, 339, 277, 67, 47, 48, 275, 350, 277, 276, 316, 339, 321, 318, 351, e.g.].

How is geophysical transport connected to the ideas of unsteady flows as presented in the previous section? Figures 1.1(a)–(c) show toy kinematic models for a double-gyre [364, 54, 422, 316]; an eddy, cyclone, or vortex [216, 226, 193, 346, 24, 285]; and an oceanic jet [81, 324, 314, 48, 355, 125, 352, 322, 273, 354], respectively. One dominant feature appearing in all of these models is the regions of rotation; these are variously thought of as *eddies*, *gyres*, *rings*, or *vortices*, with the appropriate term depending on the atmospheric or oceanic context and/or size. Such entities have a profound influence on global weather, climate, and the environment since in particular they remain *coherent*. One such entity is the Antarctic Circumpolar (Stratospheric) Vortex, commonly known as the ozone hole [216, 226]. Jupiter's Great Red Spot is an extraterrestrial example. For a more detailed description of the influence of such entities, imagine that Figure 1.1(c) is the Gulf Stream with warmer waters to the south and cooler waters to the north. Should the jet develop a large meander which curls back on itself, thereby causing one of the flanking eddies to pinch off, then the end result would be a warm-core eddy in cooler waters, or vice versa. Such eddy detachment occurs often in the Gulf Stream, and it is not unusual for these eddies to exist as *mesoscale* eddies (of the order of tens to hundreds of kilometers in diameter), over time scales spanning months. The eddy detachment process transports heat across the main stream, and the time scale over which eddies remain has an impact on the global heat budget. Indeed, the coupling process between the ocean and the atmosphere implies that such oceanic entities influence atmospheric circulation patterns and consequently the weather. For example, there is evidence of cyclone intensification due to the interaction of these rotating structures of anomalous hotspots. Marine ecology is also strongly influenced by eddies; it is well established that the patchiness of plankton, and the propensity for marine predators to congregate, is well correlated to such coherent structures [119, 124, 131, 211, 217].

Examples of oceanic jets include the Gulf Stream and the Agulhas, Kuroshio, Humboldt, and Antarctic Circumpolar Currents. These too are coherent structures, but in a different sense from eddies in that they are not composed of a system of fluid particles *rotating* together but rather *translating* together in a way which is anomalous with surrounding waters. The toy model of Figure 1.1(c) demonstrates that in the steady situation, identifying the boundary of such a current once again topologically requires the presence of saddle-like fixed points. In Figure 1.1(c), the boundary of the current is composed of red curves which separate the current from the flanking eddies. This is once again a flow barrier, which is identifiable with heteroclinic manifolds. Any flux (of water, salinity, heat, nutrients, etc.) occurring between the jet and its surroundings requires transport across such a flow barrier. Given the dominance of oceanic jets in transport in our oceans, understanding the "leakages" across their boundaries is important. From the geophysical perspective, it is necessary not only to identify how jet barriers change with time but also to comprehend the fluxes across these barriers.

The problem with oceanic and atmospheric flows is that the velocities are unsteady. Thus, the simple characterization of flow barriers as in Figure 1.1, and the implied understanding that there is *zero* flux across them, does not hold water (and may be a lot of hot air). As discussed in Section 1.1, the difficulty here is in the difference between the Eulerian and Lagrangian viewpoints. Most oceanic velocity data that is collected, for example, is Eulerian in nature. This is typically done by taking satellite images of sea-surface height (SSH), which are then plotted as contours or color maps. Since this is performed usually at each instance in time, the observational data consists of Eulerian snapshots of SSH (also referred to as *altimetry*). The *geostrophic* assumption [310] is that the gradient of the SSH, rotated by $\pi/2$, is proportional to the velocity, thereby providing an indirect method for computing the sea-surface velocities.[1] Dominant oceanic features are distinguishable clearly in SSH maps; for example, the meandering nature of the Gulf Stream, and eddies which have pinched off from it, are strongly visible in SSH field maps. When viewing how the SSH field varies with time, the changes tend to be slow and subtle, giving the indication that fluid transport occurs in a predictable way since the basic understanding from physical oceanography was that fluid parcels would tend to move along contours of the SSH map [310]. However, when floats were released and tracked in various locations in the ocean, their trajectories were observed to display seemingly erratic behavior overlaid on a more regular SSH-contour following tendency [83, 437, 251, 295, 68]. Indeed, the complicated ways in which collections of such floats behaved led to the term *spaghetti plots* being applied to float trajectory data [83, 82]. Given the observed regularity of how SSH contours evolved, this was initially surprising. Today, of course, this is well understood as the difference between and Eulerian and Lagrangian viewpoints [354, 189, cf.]; regular Eulerian velocity fields can certainly engender complicated Lagrangian motion in unsteady situations.

SSH/altimetry data has been a strong driver of how we understand our ocean currents. The resolution of SSH measurements is set to increase substantially (to better than 10 km) with the implementation of NASA's Surface Water and Ocean Topography (SWOT) program around 2020. This will allow for a better understanding of submesoscale (of the order of 1 to 10 kilometers) transport in the oceans. While SSH altimetry is one Eulerian data source available from satellite observations, there is an ongoing effort to improve satellite obtained data on quantities such as sea-surface temperature and sea color. The latter quantity is a proxy for the chlorophyll concentration and thus represents plankton availability. Since plankton patches tend to passively flow with the fluid [211], their patchiness correlates with coherent structures and, as mentioned previously, has been shown to link with the presence of higher trophic levels in the food chain. In addition to improving satellite-obtained data, there is considerable ongoing data collection which is directly Lagrangian in nature, through an increased disposition and tracking of floats and drifters [432, 340, 311, 312, 129]. Since these floats have inbuilt sensors, data on the time-variation of the local position, velocity, temperature, pressure, density, salinity, etc., can be collected. Under the assumption that the floats passively follow the flow, all this data is Lagrangian in nature.

Assuming that various data is available, quesions one might ask are first how one could identify an *unsteady* flow barrier and second how one could quantify cross-

[1]The SSH is therefore effectively a streamfunction for the sea-surface velocity.

barrier transport. These questions seem conflicting in the sense that if it is a flow *barrier*, then how could there be any transport? Unlike in steady flows, unfortunately, this apparent contradiction is unavoidable. In unsteady flows, there are generically no *absolute* flow barriers, but there could be entities which are dominantly barriers in that they separate blobs of fluid which are "mostly" coherent, with a mild leakage across the "barriers." This of course leads to the difficult question of how one might *define* the barriers. Definitions which are universally accepted are not available; instead, a rich array of methods (to be outlined in some detail in Sections 1.4 and 1.5) has evolved. Whatever definitions one uses, there has been considerable recent interest in applying such definitions to characterize oceanic flow barriers and the associated transport across them. A point to be emphasized is that *using purely Eulerian definitions is well known to be inadequate in identifying flow barriers* (despite their continued usage by some communities) since transport of fluid is by definition *Lagrangian* in nature. Within this correct understanding, the earlier work in this area tended to be on purely kinematical models for geophysical features [324, 314, 417, 225, 125, 352], but later work takes into account the dynamics the velocity field needs to obey [48, 355, 23, 47, 364] or, alternatively, observational data. Observations, coupled with a choice of definitions, have led, for example, to identifying and tracking Agulhas rings [152], determining the cross-shelf transport barrier off the west coast of Florida [295], forecasting oil movement due to the Deepwater Horizon oil spill [275, 296], iron supply (for plankton blooming) in the Kerguelen plateau [129], atmospheric transport of plant pathogens [360, 392], identifying garbage patches where global garbage congregates in the oceans [160, 406], and analyzing the variability of the boundary of the Antarctic (stratospheric) Circumpolar Vortex [348, 226].

The large variations in the *definitions* used for flow barriers and associated transport in the articles mentioned above mean that while a considerable understanding of oceanographic and atmospheric phenomena is developing using ideas of coherent structures, there are no *definitive* answers to the five questions which culminate the previous section. It turns out, however, that the hypotheses which underscore much of this monograph—namely, nearly autonomous two-dimensional, potentially compressible flows—are quite reasonable in describing a range of geophysical entities. One interesting aspect, for example, is the interpretation of the autonomous part as the mesoscale process, which has at a gross level the information about the larger coherent structures, with an added nonautonomous perturbation to account for submesoscale processes. Two specific applications in this regard, in relation to wavenumber dependence of flux in Rossby eddies and the impact of viscosity on oceanic eddies, will be presented in Sections 3.4 and 3.8. These are two examples which are part of a much broader current interest in understanding the influence of submesoscale processes at mesoscale level models, as a method for parametrizing the smaller scales in moderate-scale oceanic models. Indeed, a nonautonomous perturbation can be viewed as a particular realization of small-scale stochasticity. Stochasticity at the small scale is sometimes also viewed in terms of *eddy diffusivity* [204, 164], or *hyperdiffisivity* [24, 205]; methods for additionally modeling this effect and estimating its relevance is a topic which is under considerable investigation today [76, 256, 218, 2, e.g.]. In this vein, could one characterize submesoscale-inspired transport across mesoscale coherent structures? (This is the *stochastic parametrization* problem.) The methods that are developed in this monograph offer promise for new approaches to problems such as these, which are becoming increasingly pertinent in ocean/climate modeling.

1.3 ▪ Importance in micro/nanofluidic flows

Microfuidic devices are small enough to contain/carry microliters of fluid and are increasingly finding use in a variety of applications including in in-situ drug delivery and/or monitoring, single-cell analysis, and DNA synthesis. As the biotechnological revolution continues, microfluidic (and even nanofluidic) devices are continually being developed, designed, and fabricated with increasingly complex applications in mind. From the dynamical systems perspective, an important aspect of microfluidic devices is that they are inevitably of low Reynolds number [389, 14, 15, 302, 420]; their lengthscale L and typical speed U are small, leading to a smallness in the Reynolds number $\mathrm{Re} := UL/\nu$, whatever the value of the kinematic viscosity ν. The *Stokes limit* of infinite Re is therefore appropriate for modeling such flows [171, 97, 381, 38, 177, 229, 372, 103, 102, 51]. An important consequence of the Reynolds number being small is that the flow is akin to that of highly viscous fluids like treacle; it is *laminar* and never *turbulent*. The absence of turbulence means that fluids in microfluidic devices do not tend to mix well, whereas in many applications such mixing is exactly what is desired. For example, one often needs to mix together a sample and a reagant quickly in order to uniformly impact a chemical reaction which has a fast reaction rate, in an experiment in which a protein is being analyzed. The random motion of fluid particles, i.e., *diffusion*, can contribute to better mixing in these devices, but it is often the case that in and of itself, diffusion is an inefficient mechanism for achieving good mixing in microfluidic devices [420, 389]. The diffusive timescale is too long for most applications.

The fact that the flow in microfluidic devices is very regular is actually better phrased by the statement that the *velocity* is regular. This is the Eulerian velocity $u(x,t)$, which expresses the fluid velocity u at a position x and time t, and the claim is that u is a smooth function of (x,t). As observed by Aref [14], this by itself does *not* preclude good mixing within the flow. The reason is that the trajectories of fluid particles are governed by (1.1), and it is possible that solutions to (1.1) are irregular even when $u(x,t)$ is regular. This is easily illustrated by the idea of chaotic transport, in which seemingly irregular motions arise from regular dynamics [250, 183]. There are some restrictions on the possibility of engendering complicated dynamics. First, if $x \in \mathbb{R}^2$ and the velocity u is steady (independent of t, autonomous), then the Poincaré-Bendixson Theorem [17, 183] precludes chaotic mixing. All bounded trajectories are forced to approach a fixed point, a periodic orbit, or a heteroclinic cycle (a closed curve which connects together trajectories between fixed points) as $t \to \infty$ [183, 9]. Second, for $x \in \mathbb{R}^3$, if the velocity and the vorticity (the curl of the velocity) are linearly independent, and the flow is inviscid, Arnold has shown that complicated trajectories are impossible for (1.1) [16, 50]. If alternatively a three-dimensional steady, or a two-dimensional unsteady, flow had an extra conserved quantity (such as potential vorticity [91, 23]), the effective reduction to two dimensions implies regular trajectories. From the microfluidic perspective (in which the flow is the opposite of inviscid, and there is no obvious conserved quantity), one might achieve good mixing in two-dimensional *unsteady* (nonautonomous) flows, or in three-dimensional flows.

Two typical illustrations of the regularity of microfluidic flows are shown in Figures 1.1(e) and (f). The first of these indicates a typical situation in which two fluids—coming in from the two inlets on the left—are to be mixed together thoroughly. Nevertheless, the laminar nature of the flow tends to promote these two fluids to flow in a parallel fashion along the microchannel, forming a flow barrier (shown in red) across which they tend not to mix. This is a typical scenario on microfluidics [383, 420, 308, 6, 425, 12, 412, 118, 376, 307, 236, 289, 287, 288, 390, 79, 25, 42, 237, 409].

Figure 1.1(f) shows the streamfunction pattern within a microdroplet; this is a three-dimensional picture which is symmetric about the vertical axis. The flow lies on nested torii which wrap around the vertical axis, until reaching the red surface, which is topologically a sphere. Outside of this sphere the flow is on nested cylinders. This flow structure is called the Hadamard-Rybczynski solution [184, 347, 381, 38, 229, 426, 380, 176, 38, 51], obtained from the steady Stokes (highly viscous) flow equations. The kinematics here are topologically equivalent to the classical Hill's spherical vortex [234, 3, 50, 238, 201, 274, 260, 195, 233, 370, 20, 196, 241, 345, 356, 249, 11, 87, 94, 335], but the dynamical constraints are different in that this flow is thought of as a Stokes flow, in contrast to the (opposite) zero viscosity limit associated with the Hill's vortex. In the microfluidic context the interior of the sphere in Figure 1.1(f) represents a microdroplet of one fluid, traveling within an exterior, immiscible carrier fluid (e.g., oil), and the picture is drawn in the frame of reference of the microdroplet [149, 378, 375, 90, 176, 403, 116, 248]. A two-dimensional version of Figure 1.1(f) is also often considered in microfluidics; in this case, the figure is to be thought of as is (with no rotation), and here the interior of the microdroplet consists of two cells of different fluids [149, 378, 375, 90, 176, 403, 116, 248]. This scenario occurs, for example, when two fluids are injected together into a microchannel carrying an immiscible carrier fluid. This picture is exactly what is shown in Figure 1.1(d) (but it is rotated by $\pi/2$ in comparison to (f)). One often seeks to achieve good mixing within the microdroplet, and versions of this picture have been studied both experimentally [375, 90, 149, 116, 248, 403, 363, 78, 386] and theoretically [381, 97, 177, 176, 281, 38, 102, 38].

Figures 1.1(e) and (f) are for steady flows, and the relevant transport/mixing is to be achieved by introducing velocity agitations which destroy the flow barriers (in red). In practice, one might use ad hoc methods such as pumping fluid back and forth in the transverse direction to the channel flow in Figure 1.1(e) by using syringes or cross-channels [438, 334, 123, 411, 79, 390, 237, 279, 424, 198, 409], vibrating boundary membranes to generate an additional interior flow velocity [123, 411], or applying external electromagnetic fields to influence charged particles within the fluid [242, 376, 288, 289, 437, 62, 4, 206]. Since these supply energy to the system, they are called *active* methods. In constrast, *passive* methods take advantage of physical/chemical phenomena such as gravity, capillary action, surface tension, and osmosis. A particularly simple method would be to introduce grooves or curves within the microchannels [6, 308, 425, 12, 107, 375, 102, 381, 38].

If some velocity agitation, i.e., advective mechanism, is used, one typically likes to move small blobs of one fluid into a region which mainly comprises the other fluid. This means that fluid blobs are to be transported across the flow barriers as shown in red in Figure 1.1. If the blobs are sufficiently small and elongated, *then* diffusion will begin to be effective, and genuine mixing will occur. Thus, mixing within the microfluidic devices can be achieved using an *advection-driven diffusion* process [286], where the mixing effect of the advective mechanism could, for example, be modeled using an advection-diffusion equation [174, 400, 165, 171]. It is the *advective* mechanism which will be the main focus here. How exactly could one *optimize* the design of the velocity agitation in order to achieve good transport?

The "practical" approach to this would be to fabricate many devices with slightly different configurations and agitations, *experimentally* test the mixing in each, and choose the design which gives the best mixing. This would require ultra-high resolutions for the measurement method (e.g., *particle image velocimetry*, or PIV), and realistically, only a small number of devices can be fabricated because of the considerable time and cost of the process. Most reports in the literature [237, 12, 107, 289, 288, 308, 376, 438, 293, 375, 90, 149, 116, 248, 403, 363, 78, 386] indeed are based on *one* device at a time. Comparison between these is moreover difficult since the mixing measure used by each study is usually *different* [248, 102, 90, 403, 381, 283, 372, 116]. To deal with the difficulty of using an experimental method for optimizing the device design, many investigators adopt a *direct numerical simulation* (DNS) of a particular design, in which the Navier-Stokes equations (or suitable simplifications such as the Stokes equations which characterize extremely low Reynolds number flows) are numerically solved using computational fluid dynamics (CFD). This too is not without issues since in particular implementing the boundary conditions on moving interfaces between fluids is difficult. Therefore, most microfluidics-inspired DNSs limit themselves to highly simplified geometries or idealized interface/boundary conditions [6, 107, 118, 376, 425, 103, 102, 358, 357, 283, 128]. As in the experimental literature, computational results are also usually reported for one specific design at a time [6, 107, 118, 376, 425, 103, 102, 358, 357, 283, 128], making decisions on optimal designs difficult.

Dynamical systems offers a method—albeit highly idealized—for dealing with these difficulties, using *toy models*. Basically, one expresses the fluid velocities in the device based on some kinematically or dynamically plausible hypothesis [177, 97, 38, 229, 426, 380, 176, 381, 51]. Since the velocity field is then *specified*, one can use tools of dynamical systems to attempt to quantify and optimize transport across flow barriers. Any insight gleaned from this process can later be used to inform decisions on what configurations or parameter regimes might be important to test in a more realistic DNS or experimental investigation. Within the context of the idealized models, several different types of questions might be posed from the perspective of dynamical systems, and the interested reader is referred to the review article [36] for an extensive discussion on these. Here, a few possibilities will be mentioned. If sloshing fluid back and forth using some transverse cross-channels in Figure 1.1(e), there is experimental [237, 293, 244, 161, 336, 367, 414, 242, 413, 376, 387, 289, 287], computational [288, 244, 161, 336, 414, 294], and theoretical [342] evidence that there is an optimal frequency at which to do this sloshing. Is there a method for determining this frequency [31]? See Section 4.2. Alternatively, is there any insight as to *where* in the microchannel to position the transverse channels [27, 25, 42]? This is addressed in Section 4.3.

1.4 ▪ Unsteady flow barriers: introduction

The previous sections have outlined the usefulness of the idea of flow barriers in geophysical and microfluidic applications. One obvious issue immediately emerges: how does one *identify* a flow barrier in a realistic flow?

Figure 1.1 seems to offer a resolution to this question. The flow barriers are those which demarcate regions of distinct motion. As argued in Section 1.1, in this case flow barriers can be identified as being heteroclinic manifolds. For the two-dimensional situations in Figures 1.1(a)–(e), these are indeed heteroclinic trajectories since the manifold is a one-dimensional entity, which consists of a solution trajectory to (1.1), in which u does not depend on t. For Figure 1.1(f), the ellipsoidal heteroclinic manifold consists of a collection of heteroclinic trajectories, each of which begins at the saddle point at the north pole and progresses along constant longitudes to the saddle point at the south pole. In steady flows, therefore, flow barriers are associated with stable and unstable manifolds.

Figure 1.1 is associated with a *steady* or *autonomous* flow, in which the velocity field does not change with time. Thus, all arrows which are drawn indicate the direction of the velocity at all times. Indeed, all curves drawn are simultaneously streamlines, pathlines, and streaklines. Streamlines, being instantaneous curves drawn to be tangential to the velocity field, are an Eulerian entity. Since for steady flows these do not change with time, particle trajectories will follow streamlines; i.e., the streamlines are therefore also pathlines, which are Lagrangian by nature. The pleasing Eulerian-Lagrangian coincidence breaks down immediately if the velocity field is *unsteady* or *nonautonomous*. Now, for assessing fluid transport, it is clearly pathlines which are important (Section 3.7 addresses a scenario in which *streaklines* are the relevant quantity, but this will be disregarded in the discussion for the moment). If the velocities are changing with time, then one cannot immediately know the fate of particles by examining instantaneous streamline pictures such as in Figure 1.1. (Lagrangian) particles will move according to a continuously varying velocity field, and any flow barriers (if they exist) need to take into account this variation. How exactly do the concepts of stable and unstable manifolds—the definers of flow barriers in the autonomous situation—generalize to nonautonomous flows?

Saddle points were the entities to which flow on stable and unstable manifolds decayed as $t \to \infty$ (resp., $t \to -\infty$), and thus the first step might be to discuss their analogues in the nonautonomous (unsteady) case. An obvious guess might be *instantaneous fixed points*: points at which the velocity is at that instance in time zero. As an example, consider the flow

$$\dot{x}_1 = x_1 + ct - c, \quad \dot{x}_2 = -2x_2, \tag{1.7}$$

where $(x_1, x_2) \in \mathbb{R}^2$, the dot denotes the t-derivative, and $c > 0$ is a constant. Here, the velocity field is nonautonomous and is given by $(x_1 + ct - c, -2x_2)^\top$. There is only one instantaneous fixed point at each time t, given by $(x_1, x_2) = (c - ct, 0)$. As time increases, this point moves to the left in the x_1x_2-plane, along the x_1 axis. An immediate caution as to the relevance of this point is the fact that its evolution does not constitute a solution to the differential equation (1.7). Now, in the transformed (ξ_1, ξ_2) coordinate system defined by $\xi_1 = x_1 + ct$, $\xi_2 = x_2$, the flow satisfies $\dot{\xi}_1 = \dot{x}_1 + c = (x_1 + ct - c) + c = \xi_1$ and $\dot{\xi}_2 = -2\xi_2$. Thus, the system is trivial in a coordinate system which simply translates at speed c. Clearly, there is a saddle fixed point at $(\xi_1, \xi_2) = (0, 0)$ in this steady flow, to which is attached stable and unstable manifolds. This then is the "central" point which separates the flow into four quadrants. In the x_1x_2-plane, this transforms to $(x_1, x_2) = (-ct, 0)$, which is emphatically *not* the instantaneous stagnation point. As shall be seen later, the time-varying point $(x_1, x_2) = (-ct, 0)$ for (1.7) is what is called a *hyperbolic trajectory*, which generalizes the concept of a saddle fixed point to a nonautonomous situation. (This shall be defined formally in Section 1.6.) The conclusion from this example is that *instantaneous fixed points do not in general identify the entities to which are connected stable and unstable manifolds*. Clearly, determining instantaneous eigenvalues and eigenvectors[2] at these points would be similarly meaningless when trying to identify important flow barriers.

The above example shows that even introducing quite trivial time-dependence in the vector field makes it difficult to exactly *define* what a flow barrier is. Even having

[2]The appropriate theoretical analogues to the eigenvectors would be Oseledets spaces [299, 150, 153, 240, 170, 39], which are beyond the scope of this monograph; however, the time-varying local tangent vectors to stable and unstable manifolds will be seen in Sections 2.5 and 5.3 to provide the "correct" geometric analogues to eigenvectors.

full knowledge of the vector field, to arbitrary resolution in space and time, and over all $t \in \mathbb{R}$, does not remove this difficulty. A second fact often exacerbates the issue in realistic flows: velocity field data is often only available in some restrictive fashion, often on a discrete spatial grid, at discrete times, and over a finite time duration. This *finite-time* issue causes difficulties even in steady flow situations because stable and unstable manifolds are defined in terms of exponential decays as $t \to \pm\infty$ to the saddle points. If time is restricted, say $t \in [-T, T]$ for some finite $T > 0$, then *every* function on $[-T, T]$ can be bounded by an exponential function in the form $Ke^{\alpha t}$ with constants K, α chosen uniformly. The characterization of trajectories $x(t)$ on the stable manifold decaying to a saddle fixed point a in the form $|x(t) - a| \leq Ke^{\alpha t}$ for $t \in \mathbb{R}$ can therefore no longer be used for finite time. The formal characterization of this exponential decay is in terms of *exponential dichotomies*, which shall be addressed in Section 1.6; the issue for finite times is that modifications to the exponential dichotomy conditions must be pursued [127, 219, 135, 66]. Basically, stable and unstable manifolds, in the classical sense of dynamical systems, are no longer well defined for finite-time vector fields.

These issues have led to the development of a variety of "definitions" of flow barriers for finite-time nonautonomous systems of the form

$$\dot{x} = u(x, t), \quad x \in \Omega, t \in [-T, T] \tag{1.8}$$

where often $\Omega \subset \mathbb{R}^2$ or $\mathbb{R}^3$ for fluid applications. The methods for identifying flow barriers of (1.8) shall be called *diagnostics*, in the sense that each definition offers a diagnostic of what one intuitively thinks of as a flow barrier but is not necessarily defined in terms of *being* a flow barrier. An important point is that there is *no* universally accepted definition for these entities in nonautonomous finite-time flows. The diagnostics instead seek entities which obey *different* properties from one another, with each such property being chosen from among the characteristics of stable and unstable manifolds. In all cases, these purported flow barriers will be of lower dimension than the background space; for example, if $\Omega \subset \mathbb{R}^2$, this would mean determining *curves* which have whatever property is being used as the finite-time definition. Some geometric properties used for defining these curves are listed below, with additional detail given for the more commonly used ones. In all cases, the understanding is that the barriers are to be assessed according to a flow beginning at time t_i and ending at time t_f, with $-T \leq t_i < t_f \leq T$.

1. *Finite-Time Lyapunov Exponents (FTLEs):* This is possibly the most commonly used diagnostic for flow barriers and is valid in any dimension. The intuition behind FTLEs comes from the fact that trajectories on stable/unstable manifolds decay to the corresponding hyperbolic fixed points *exponentially* in steady flows. To illustrate the intuition for FTLEs via example, consider the flow in $\mathbb{R}^2$ given by

$$\frac{dx}{dt} = \begin{pmatrix} -3 & 0 \\ 0 & 2 \end{pmatrix} x + \mathcal{O}\left(|x|^2\right). \tag{1.9}$$

When the $\mathcal{O}\left(|x|^2\right)$ terms are discarded, (1.9) is easily seen to have saddle fixed point at the origin. Nearby motion is therefore as shown in Figure 1.4, with the stable manifold being in the direction $(1, 0)^\top$ with compression rate -3 and the unstable manifold in the direction of $(0, 1)^\top$ with expansion rate 2. If $|x| = \mathcal{O}(\varepsilon)$, within a $\mathcal{O}(\varepsilon)$-neighborhood B_ε, the higher-order term could be neglected.

Within B_ε, trajectories to (1.9) would have the form

$$x(t)=\begin{pmatrix} x_1(t_i)e^{-3(t-t_i)} \\ x_2(t_i)e^{2(t-t_i)} \end{pmatrix}, \tag{1.10}$$

when thinking of t_i as an initial time, and $x=(x_1,x_2)$. Let $x^*(t)$ be a trajectory with nearby initial condition $x^*(t_i)$, whose evolution can be represented exactly as in (1.10) with a star superscript. Now let $\delta x(t) := x^*(t)-x(t)$ be the difference between the two trajectories, with the notation $\delta x = (\delta x_1, \delta x_2)$ representing δx in component form. The explicit expression in (1.10) enables the distance between the trajectories' evolution with time t to be expressed as

$$d(x(t_i),x^*(t_i),t)=\sqrt{\delta x_1(t_i)^2 e^{-6(t-t_i)}+\delta x_2(t_i)^2 e^{4(t-t_i)}} \tag{1.11}$$

as long as they both remain within B_ε. Consider the issue of maximising d at fixed (but fairly large) $t=t_f>t_i$ and fixed initial condition $x(t_i)$, by choosing the direction of the initial condition of $x^*(t_i)$. That is, suppose the direction of $\delta x(t_i)$ can be varied, subject to a fixed $|\delta x(t_i)|$. From (1.11), it is clear that one can gain more by putting more into the second component of $\delta x(t_i)$, indeed, by choosing $\delta x_1(t_i)=0$. If so,

$$d\left(x(t_i),x^*(t_i),t_f\right)\sim |\delta x_2(t_i)|\, e^{2(t_f-t_i)}. \tag{1.12}$$

Two observations can be made. First, the rate of exponential decay of this "maximal" choice is 2, which is the positive eigenvalue associated with the linearized flow of (1.9) near the fixed point at the origin. One might indeed say from (1.12) that

$$2=\frac{1}{t_f-t_i}\ln \sup_{\delta x(t_i)} \frac{d\left(x(t_1),x^*(t_i),t_f\right)}{|\delta x(t_i)|} \tag{1.13}$$

since it is by maximizing the direction of $\delta x(t_i)$ that (1.12) emerges. Second, the direction of the nearby point which exhibits the most distance increase is in the direction of $(0,1)^\top$, which is the eigenvector associated with this eigenvalue. Now if one poses the same maximization of distance problem in *backwards* time, i.e., with $t_f<t_i$, then one would get $d(x(t_i),x^*(t_i),t_f)\sim|\delta x_1(t_i)|\,e^{-3(t_f-t_i)}$, with the eigenvalue -3 and eigenvector $(1,0)^\top$ playing central roles and a formula such as (1.13) also applying. Now, (1.9) is a special situation in which the stable and unstable manifolds are orthogonal to each other, arising from the fact that the linearization matrix is diagonal. In a more general situation where the fixed point at the origin is still hyperbolic (with a positive and a negative eigenvalue for the linearization) but the eigenvectors are not orthogonal, the argument leading to, say, (1.13) would still apply in the relevant direction.

The above example represents *local* intuition regarding *steady* flows. The analogous entity for an *unsteady* flow would be a situation in which the picture of Figure 1.4 is moving around in time. The fixed point at the origin is no longer a fixed point, but a hyperbolic trajectory whose defining characteristic is that it has attached stable and unstable manifolds. These manifolds are themselves moving around with time. Nevertheless, once one moves into the frame of reference of the hyperbolic trajectory, the above arguments are still legitimate. An official change of coordinates in performing such a move is, however, not necessary

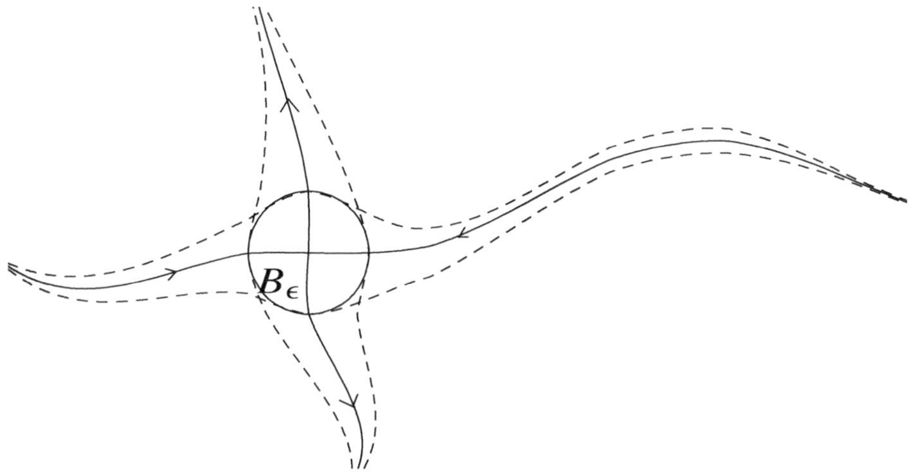

Figure 1.4. *Phase portrait near the origin for (1.9), motivating the concept of FTLEs; the dashed envelopes indicate regions which are eventually influenced by the linearized flow in B_ε.*

since equations such as (1.13) take into account the motion of nearby trajectories *relative* to $x(t)$. Therefore, a direct extension of (1.13) to an *unsteady* flow (1.8) can be made, and the FTLE field shall be *defined* as

$$\lambda(x,t_i,t_f) := \frac{1}{\left|t_f - t_i\right|} \ln \sup_{\delta x(t_i)} \frac{d\left(x,x^*,t_f\right)}{|\delta x(t_i)|}, \tag{1.14}$$

where $x^*(t)$ is the trajectory of (1.8) such that $x^*(t_i) = x(t_i)+\delta x$. Equation (1.14) is usually thought of as a scalar field for $x \in \Omega$, and depending on whether $t_i \lesseqgtr t_f$, it is associated with capturing the stable or the unstable manifold's local direction and expansion. In typical applications, it is common to think of the FTLE as a time-varying field with the t_i representing the time, and with t_f chosen such that $t_f - t_i$ is constant. That is, the trajectories from a beginning time t_i are flowed for a fixed time interval $|t_f - t_i|$ in *all* cases; this quantity is the "finite-time" in the definition of FTLEs. In an alternative computational method, one flows until d reaches some specified threshold value, and t_f would be this value which would vary for each trajectory. This approach, which still utilizes the equation (1.14), is called the *Finite-Size Lyapunov Exponent* (FSLE) method, which is also in considerable use [130, 216, 70].

Having computed the field λ in (1.14), with the usual understanding that t_i is the time of interest and that $|t_f - t_i|$ is fixed, "flow barriers" are then found for general unsteady flows by looking for the *FTLE ridges*, i.e., sharp transitions in the FTLE field. A simple (but not recommended) method might be to simply threshold the FTLE values, but more sophisticated methods continue to be developed [362, 220]. What one obtains from this process are analogues for the stable manifold (obtained when $t_f > t_i$, i.e., from the *forward-time FTLE* field) and for the unstable manifold (if choosing $t_f < t_i$, i.e., when considering the *backward-time FTLE* field). For example, if Ω is two-dimensional, these would appear as curves in each time-slice t_i.

An important sidenote, regarding a situation in which there is apparently considerable confusion among users of FTLEs, must be made. If (1.9) is truly linear, with no additional higher-order terms, then *all* trajectories will satisfy (1.13). That is, when the quantity on the right-hand side of (1.13) is evaluated, the value 2 would emerge for any choice of initial condition $x(t_i)$. In other words, it is not just trajectories on the stable manifold which will exhibit an FTLE value of 2; *all* trajectories will. There is no FTLE *ridge*! Thus, computing the FTLE will *not* enable a method for distinguishing the stable manifold, which consists of trajectories starting on the x_1-axis. This is indeed the case: *FTLEs do not provide a method for distinguishing the stable and unstable manifolds in linear flows.* Their utility arises in *nonlinear flows*, in which there are higher-order terms in (1.9). Then, it is only within B_ε that the argument leading to (1.13) works, and thus FTLE values of 2 will appear only for trajectories which are eventually within B_ε. (For additional analysis of the role of the neighborhood B_ε, the reader is referred to [49].) The location of such trajectories is indicated by the dashed envelope around the stable manifold displayed in Figure 1.4. It is for $x(t_i)$ values within this strip that a value of 2 will appear when the FTLE is calculated. Outside the strip, the computed FTLE values will typically be substantially less than 2 since the exponential influence of the hyperbolic point will not be felt by those trajectories. Notice that this strip becomes thinner when going away from the fixed point at the origin because of the expansion occurring in the unstable manifold direction. Thus, when computing FTLEs and seeking its "high" values, one would typically recover an *FTLE ridge* which is essentially this envelope. The ridge would be fatter near the hyperbolic point but better defined away from it. In backwards time, the associated ridges would be formed from the envelope around the unstable manifold in Figure 1.4. The computational resolution will be strongly connected to the size of B_ε, which itself is intimately connected with the fatness of the ridge. However, the numerical output will not show the neighborhood B_ε, and since one usually refines the ridge using some process, the envelope will also usually not be visible. On the other hand, it is the presence of B_ε which will allow for the presence of a distinguished ridge in the nonlinear flow; such will not be apparent for *linear* flows. This steady argument will, of course, carry over to the more general unsteady case, with the understanding that the envelopes and hyperbolic point in Figure 1.4 are moving with time, and so Figure 1.4 would give the FTLE ridges at an instance in time. Theoretical [189, 187, 188, 220, 247, 290, 359, 88, 49] developments in FTLEs continue to occur, as do numerical methods for more efficient computations [364, 113, 92, 247]. The usage of FTLEs/FSLEs is pervasive in applications in fluids, for example in laboratory [329, 278, 303, 96, 410, 269, 223], geophysical [315, 130, 364, 240, 328, 87, 395, 104, 321, e.g.], and microfluidic [237, 289, 287, 307] contexts. For example, FTLE/FSLE ridges have been shown to line up with gradients of sea-surface temperature [168, 167] or plankton density [131, 132] observations. To highlight its role in (geophysical or other) fluid-mechanical situations, Samelson [353] suggests calling the FTLE a finite-time Lagrangian strain instead. This is because the $DF_{t_i}^{t_f}(x)$ is exactly that: the strain (i.e., rate of change with respect to space) of the Lagrangian flow map $F_{t_i}^{t_f}(x)$. This interpretation may resonate with readers seeking to connect physical interpretation to the dynamical systems idea of attempting to capture exponential stretching. Building on the interpretation of Samelson [353], using the phrase *Finite-Time Lagrangian*

Stretching may be an even more descriptive terminology for FTLEs, avoiding the ambiguity of what the term "strain" means in different communities. How and whether FTLEs do indeed capture "flow barriers" is a continuing endeavor, with many interesting examples and counterexamples presented in [188, 189, 364, 49].

Example 1.3 ("moving" Duffing oscillator (cont.)). Given the high usage of FTLEs in the applied literature, a quick illustration of FTLEs and their relationships to stable and unstable manifolds is in order. The forward- and backward-time FTLEs associated with the system (1.5) with $c = 1$ are shown in Figure 1.5. In all cases, t_i is taken as 0, and the reader is directed to Figure 1.3(a) to compare with an instantaneous (Eulerian) picture at the same time. It turns out that the flow barriers *are* a figure-8 structure, but their location *cannot* be determined using the Eulerian picture. The FTLE field, a Lagrangian diagnostic, is shown in Figure 1.5. Initial conditions chosen at $t_i = 0$ are evolved until the indicated t_f value, and the FTLE field (1.14) is shown. It turns out that the hyperbolic trajectory is at the origin at $t = 0$, and not at $(1, 0)$ as might be guessed from Figure 1.3(a). (The point $(1, 0)$ in Figure 1.3(a) is an instantaneous stagnation point with instantaneously a saddle-like structure of the adjacent streamlines; this Eulerian observation turns out to have nothing to do with the Lagrangian flow barriers. It has already been established in Example 1.2 that the Eulerian curves do not form flow barriers.) The stable manifold crosses through the origin from the second to the fourth quadrants, and it is this which the forward-time FTLE field is attempting to identify in the left panels of Figure 1.5. When $t_f = 1$, the "ridge" in the forward-time FTLE field is incorrect, but as t_f passes through 3 to 6, it becomes better refined. The stable manifold is identified by the forward-time FTLE since, as shown in Figure 1.4, points near to the stable manifold enter B_ε and are strongly pulled apart by the *unstable* manifold, thereby exhibiting very large stretching rates in forward time. Similarly, the unstable manifold of the hyperbolic trajectory which is instantaneously at the origin crosses the origin between the first and the third quadrants, and its refinement using backward-time FTLEs appears in the right panels of Figure 1.5.

This simple example illustrates several facts about FTLEs. First, the ridge detection is messy; it is not always clear how one would identify a ridge. How sharp should it be in comparison to nearby values? How long should it be? Second, the FTLE values depend strongly on the time-interval over which the integration is done. What is the "right" value to use? Third, the identification of a hyperbolic trajectory may be imputed as the point where the stable and unstable manifolds intersect, and thus the diagnostic for this would be where the forward- and backward-time FTLE ridges intersect. Notwithstanding the uncertainty in defining what a ridge is, there are possibly several such intersections. How does one identify the correct one? Indeed, the system (1.5) is an integrable one (since it is obtained by an affine transformation applied to a steady system), and thus the stable and unstable manifolds do not intersect transversely except at the hyperbolic trajectory location, and so the pictures in Figure 1.5 are actually simpler than what might be expected in a genuinely nonautonomous system, in which multiple interestions are typical. ■

2. *Trajectory complexity:* This diagostic too applies for any dimension. Let $x(t, x_i)$ be the trajectory of (1.8) with initial condition at x_i. Consider any function $f(x, t)$ defined on $\Omega \times [-T, T]$; a common choice is the fluid speed, i.e.,

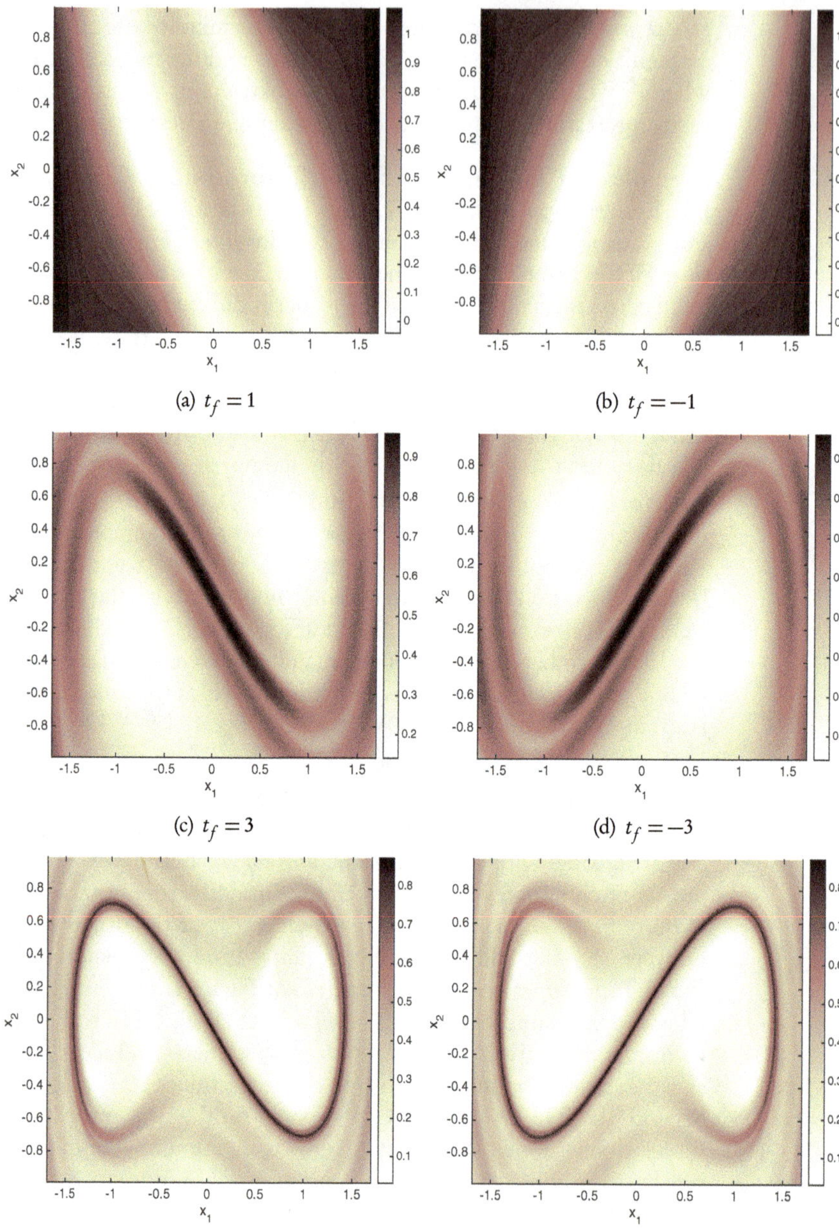

(a) $t_f = 1$ (b) $t_f = -1$

(c) $t_f = 3$ (d) $t_f = -3$

(e) $t_f = 6$ (f) $t_f = -6$

Figure 1.5. *Forward-time (left) and backward-time (right) FTLE fields computed for (1.5) with $c = 1$, at $t_i = 0$.*

$f(x,t) = |u(x,t)|$. Then, consider the quantity

$$a(x_i, t_i, t_f) = \frac{1}{t_f - t_i} \int_{t_i}^{t_f} f(x(t, x_i), t)\, dt\,, \tag{1.15}$$

which (for fixed t_i, t_f) is a scalar field for $x_i \in \Omega$. The quantity a represents an *average along trajectories*, of an observable f, and might be thought of as characterising a complexity of each trajectory, indexed by x_i. Moreover, the definition (1.15) has a connection to the Koopman operator K from ergodic theory; $K \circ f$ is given by the integrand in (1.15) [93, 94]. Now, by plotting the values of a on Ω (say, as a color plot indicating levels of a) and seeking codimension-1 "ridges" of this, one can identify "flow barriers." Initially suggested by Poje et al. [317], this method has popularity in oceanographic applications [266, 265, 275, 351, 263]. A basic intuition regarding this is that should a cluster of trajectories have similar values of the complexity a, then they can be considered "coherent." An abrupt change in the value of a, that is, a "ridge" of the scalar field, constitutes a barrier between such coherent structures.

3. *Perron-Frobenius or transfer operator methods:* This is yet another method which works for any dimension of Ω, but the computational cost depends on the dimension. Here, the approach is to *directly* go after a *transport* occurring between sets of Ω. To describe the simplest implementation think of partitioning Ω into boxes B_j, $j = 1, 2, 3, \ldots, m$, and seeding many particles within each box. Then, the particles are followed by the flow (1.8), and depending on which box they end up in, one finds an approximate probability P_{jk} which indicates the probability of transition from box B_j to B_k when following the flow (1.8) from time t_i to t_f (each being thought of as fixed). The probability transition matrix P with elements P_{jk} expresses how transport occurs and is usually thought of as (a spatial discretization of) the Perron-Frobenius operator. Now, however, one considers a "diffusive perturbation" P_ε to P, which in practice may happen, for example, using the numerical discretization procedure used in computing the matrix elements of P. Then, the singular vector V corresponding to the second-largest singular value of P_ε (i.e., the eigenvector/eigenvalue corresponding to the matrix $P_\varepsilon^* P_\varepsilon$) is of importance.[3] Since the process is numerical on a spatial grid, let V_j be the value of V in the box B_j. Now, by plotting V_j as colors related to some color-scale across Ω, one obtains a visual representation of the transport. At some gross level, examining the colors of V_j will partition Ω into two disjoint sets, the boundary between which there is an abrupt change. This is the flow barrier, and this process enables the breaking up of Ω at time t_i into two sets, each of which consists of particles which move "almost coherently." Now, there are many intricacies associated with this process; for a theoretical description see [150], with a range of aspects and applications also covered in [157, 159, 154]. Meanwhile, it turns out that the Perron-Frobenius operator described here is the adjoint of the Koopman operator mentioned under "trajectory complexity," forming a connection between these methods whose long-term implications are still being analyzed [93, 94].

4. *Curves/surfaces of extremal attraction or repulsion (hyperbolic Lagrangian Coherent Structures):* If Ω is two-dimensional, this diagnostic identifies curves in the time-slice t_i to which there is the most attraction (or repulsion) when the flow from t_i to t_f is considered [140, 74, 399, 189]. A curve of maximum attraction would be a curve such that in comparison to "nearby" curves (i.e., any slight deformation of the curve), nearby points are attracted more strongly towards it by the flow. If Ω were three-dimensional, the extremal problem would relate to surfaces. The original idea of Lagrangian Coherent Structures (LCSs) pro-

[3]The largest singular value is 1, with corresponding singular vector being the invariant density associated with the flow from t_i to t_f.

posed by Haller and Yuan [193] indeed thinks of LCSs in this extremal sense. (From the perspective adopted here, one might instead think of these entities as the *boundaries*—i.e., flow barriers—between coherent structures.) Since then, Haller has expanded his definition for LCSs, and curves/surfaces of extreme attraction/repulsion are now termed as *hyperbolic LCSs* [189, 140, 74]. The reader should beware that the term LCS is now commonly used to denote not this particular definition but *any* entity which might be construed as a flow barrier. That is, the term LCS is often used for entities uncovered by any of the various diagnostic methods which are being described in this section. To avoid this confusion, the term LCS will be avoided when a more generic flow barrier is intended. For curves/surfaces of extremal attraction/repulsion associated with a flow field (1.8), one needs to first compute $F_{t_i}^{t_f}(x)$, which maps points in Ω at time t_i to where the flow (1.8) takes them by time t_f. Then, the Cauchy-Green strain tensor

$$C_{t_i}^{t_f}(x) := \left[DF_{t_i}^{t_f}(x)\right]^{\top} DF_{t_i}^{t_f}(x), \tag{1.16}$$

a symmetric positive definite matrix,[4] is defined on $\Omega \subseteq \mathbb{R}^n$, where $DF_{t_1}^{t_f}$ is the spatial (Jacobian) derivative matrix. This, therefore, has positive eigenvalues λ_i with orthogonal eigenvectors ξ_i satisfying $C\xi_i = \lambda_i \xi_i$ for $i = 1, 2, \ldots, n$, and generically $0 < \lambda_1 < 1 < \lambda_n$ for incompressible flows since the product of the eigenvalues must equal unity. It is shown by Haller and collaborators [140, 74] that the hyperbolic LCSs are orthogonal to $\xi_1(x)$ if locally most attracting and to $\xi_n(x)$ if locally most repulsive; this representation leads to a computational algorithm for identifying hyperbolic LCSs [297]. These methods have been used in environmental applications [85, 396, 397, 296, 68].

5. *Other methods:* There are a variety of other diagnostics which are proposed, each of which deals with (1.8) and extracts its own entities from it. These include, but are not limited to, extremal length deformation [188, 190], extremal curvature deformation [257], topological entropy [10], ergodic theory [94, 155], minimum flux curves [44, 43], residence times [49, 322], clustering [158, 185], and vorticity-averaging [191].

The large number of diagnostic methods proposed for identifying flow barriers in nonautonomous flows is a result of the difficulties in agreeing to a unified definition as to what a flow barrier in an unsteady flow is. Consequently, different investigators tend to use and promote their own definition. Exploring *relationships* between the different methods has become a recent focus in the community (see [189, 34]). It is pertinent that *all* the above methods attempt to utilize properties which are analogous to stable and unstable manifolds. While some of the approaches are inspired by these entities, the diagnostics do not always obey all the properties that stable and unstable manifolds do. Indeed, there are often problems which arise when the diagnostics are used in a thoughtless way. Among the issues are the following:

1. *Preserving time-dependence:* In using the many finite-time diagnostics above, the usual attitude is to think of t_i and t_f as fixed. So, for example, one might seek a set of curves in Ω in the time-slice t_i whose length deforms the least after flowing according to the finite-time vector field until time t_f. With this viewpoint, one

[4] It turns out that the argument inside the natural logarithm of the FTLE definition (1.14) can also be couched in terms of the Cauchy-Green strain tensor, which forms another connection between the various diagnostics.

is essentially simply applying a flow map to Ω, which takes points in Ω to the points they flow to by time t_f. Unless the method is able to think of t_i or t_f as *variable*, this is in some senses a fairly limited perspective. For a *fixed* map, one might be able to calculate the relevant entities, but in genuinely nonautonomous flows, understanding the nonautonomous nature of the flow barriers is essential. Is the method able to allow, for example, for t_f to be a variable easily? If so, are calculations for each t_f computationally expensive?

2. *Preventing implied time-periodicity:* As discussed above, many diagnostics are based on a one-step flow map $F_{t_i}^{t_f}$ with fixed $t_{i,f}$ from Ω to itself. Consider, for example, a fixed point a of $F_{t_i}^{t_f}$ such that $DF_{t_i}^{t_f}(a)$ has eigenvalues lying within the unit circle. Classical discrete dynamical systems theory would claim that a is a stable and attracting fixed point, but in this situation, this *is incorrect.* The reason is that the conclusion for attraction relies on *repeated applications* of the map—implying time-periodicity in the flow—whereas under this viewpoint, one can make sense of *exactly one* application of the flow map and nothing else. Another situation in which time-periodicity might be implicit is if using Fourier expansions in time for the map. While this might be a useful strategy in some situations, care should be taken to ensure that its effect of artificially extending the time-interval from $[t_i, t_f]$ to $\mathbb{R}$ by periodic extensions does not interfere with whatever conclusions one is trying to make.
3. *Lagrangian nature of flow barriers:* Some have argued that any flow barriers identified via a diagnostic method must be Lagrangian in nature, in the sense that those entities travel with the flow (1.8). For example, if one identifies a curve in $\mathbb{R}^2$ as a flow barrier, when this is seeded with points and advected according to (1.8), does the resulting curve at a new time satisfy the diagnostic flow barrier condition as well? If it does, this condition is satisfied. However, in many methods, there is no automatic guarantee of this, since some procedure will be followed for each fixed times t_i and t_f, and it is not clear that by varying t_f, the entity one gets is one which flows according to (1.8).
4. *Frame-independence (objectivity):* Borrowing from solid mechanics, the concept of objectivity is that if some entity is identified via a procedure, then the *same* entity would be identified if using the procedure after using an affine transformation in space from $x \in \Omega \subset \mathbb{R}^n$ to $z \in \Omega'_t \subset \mathbb{R}^n$ given by

$$z = A(t)x + b(t), \tag{1.17}$$

where $A(t)$ is a time-varying orthogonal $n \times n$ matrix and $b(t)$ is a time-dependent vector [189]. The former encodes rotation, reflection, and axis permutation, whereas the latter captures the possibility for translation of the frame of reference. It is important to note that these are permitted to be time-varying. An alternative statement of this is to say that the method is frame-independent. Thus, the question here is that if a diagnostic identifies a flow barrier with respect to Ω, would it identify the image of that flow barrier if applied in Ω'_t? (The t-dependence in the space Ω'_t means that care should be taken when understanding exactly what this means.) In other words, are (diagnostically identified) flow barriers invariant under the transformation (1.17)?

Not all users of the various diagnostic methods for identifying flow barriers agree that the above issues are important. Some diagnostics can be shown to clearly *not* obey one or the other of the above conditions. There is one which obeys them all, without any issues: stable and unstable manifolds which are defined for $t \in \mathbb{R}$ and then clipped in time for $t \in [-T, T]$ as required by the finite-time advection (1.8). It is therefore stable and unstable manifolds which are the focus here. In relation to the above four points:

1. Stable and unstable manifolds of nonautonomous flows quite naturally have within them a time-variation. Viewing these in fixed time-slices will give different pictures; for example, if $\Omega = \mathbb{R}^2$ and the stable/unstable manifolds are one-dimensional curves, then a separate curve would result at each time step.
2. It is important to ensure that time-periodicity is not implied. Allowing for general time variation of the velocity field is desirable. This will mean that classical techniques such as Poincaré maps will be inapplicable, and any theory would need to be developed in a genuinely time-varying way.
3. Stable and unstable manifolds are obviously Lagrangian in the sense that they move with the flow. This is in fact one of their defining characteristics.
4. Stable and unstable manifolds of nonautonomous systems can be defined in terms of exponential decay rates to hyperbolic trajectories, as shall be described in Section 1.6. These can be shown to be invariant under objective transformations [33, Appendix B].

The time-variation of these invariant manifolds will indicate how the flow barriers are moving with time. In contrast with the steady pictures in Figures 1.1(a)–(e) in which the stable and unstable manifolds fall on top of each other, there is however a substantial difficulty when the flow is nonautonomous. When pictured in a time-slice, a given one-dimensional stable manifold may intersect with a one-dimensional unstable manifold at points which are not fixed points. Indeed, two such manifolds may intersect once, finitely many times, infinitely many times, or not at all! Unlike for steady flows, such intersections offer no contradictions to uniqueness of trajectories, since a particle at an intersection point will get an instruction to travel in the direction of the local velocity *at that instance in time*, and this velocity does not need to be in either of the directions of the two manifolds which intersect at that point. More specifically, these manifolds are *not* instantaneously tangential to the velocity vector, which is a major reason for the difficulties that arise in this case. Eulerian methods (instantaneous velocity) cannot help identify these Lagrangian flow barriers. Now, if a stable and an unstable manifold intersect in some complicated way, "the" flow barrier created by this is ambiguous; rather than being a well-defined barrier impervious to fluid flow as in Figures 1.1(a)–(e), "the" barrier is a *region* consisting of two intermingling curves. This region evolves with time, to complicate the issue further. The intersections (or lack of such) between a stable and unstable manifold also implies that there is some sort of transport occurring across them, which begs the question as to how one might *define* transport across an entity which is defined to be two intermingling curves. These are the questions that will be addressed here—in particular, in Chapter 3—since the nonautonomous stable and unstable manifolds continue to be the dominant flow structures which govern how global transport is organized.

1.5 ▪ Quantifying unsteady transport: introduction

Understanding the transport of heat, nutrients, pollutants, biological organisms, and chemicals in a background fluid flow has clear implications across many disciplines.

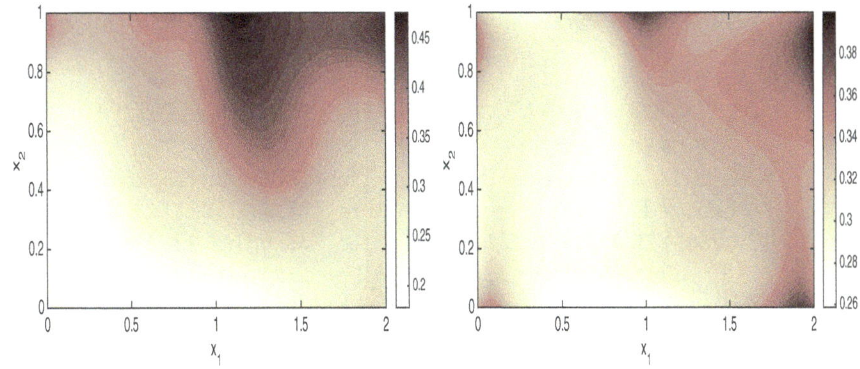

(a) $t = 1$ (b) $t = 2$

Figure 1.6. *A passive tracer concentration c at two different times, simulated according to (1.18) with $\delta = 0.1$, using the double-gyre velocity field, to be introduced in Example 2.25.*

Quantifying such transport would therefore be of obvious importance. A regularly used method for this is the idea of a *concentration variance*. Suppose the flow is (1.1), in which the velocity field $u(x,t)$ is known (in applications, it may be known in terms of observed data on a discrete spatial and temporal grid, say). Assume that $\Omega \ni x$ is a bounded set. Let $c(x,t)$ be a time-varying concentration of a quantity (temperature, chemical concentration, pollutant concentration) which evolves according to the flow (1.1) from an initial distribution $c(x,t_i)$ at time t_i. This evolution may be specified in a particular way depending on the application; for example, each particle (each initial condition) might conserve its initial value of c, or c might evolve according to the advection-diffusion equation [400, 171, 174, 165]

$$\frac{\partial c}{\partial t} + u(x,t) \cdot \nabla c(x,t) = \delta \nabla^2 c(x,t), \tag{1.18}$$

where $\delta \geq 0$ is the diffusivity parameter. If $\delta = 0$, (1.18) states that particles which follow the flow conserve their value of c, but in typical situations, $0 < \delta \ll 1$ might be a plausible hypothesis. Diffusion as represented by the Laplacian operator on the right-hand side of (1.18) is the averaged effect of small-scale random (Gaussian) particle motions. Figure 1.6 is an illustration from a simulation of a tracer concentration c evolving according to (1.18), where in this case $\delta = 0.1$ and the velocity field $u(x,t)$ used is that of the double-gyre, which shall be analyzed in considerable detail in many examples later on. Alternatively, c might be *measured* from observations, without necessarily having knowledge of a physical mechanism (such as (1.18) above). A popular recent method is to use optical flow methods, in which c would be a *pixel intensity* [237, 107, 288, 307, 293]. Indeed, (1.18) may be thought of as the modeling equivalent of a standard scenario used in microfluidic experiments: taking optical measurements to capture the spread of a dye [90, 375, 177, 109, 248]. While the above examples have c as a *passive scalar* being advected by the flow, c in (1.18) may instead be an *active scalar* which is functionally dependent on the fluid velocity u. As an example, c may be the *potential vorticity* of a two-dimensional oceanic flow with the components of $x = (x_1, x_2)$ being eastward and northward and with $c(x,t) = [\nabla \times u] \cdot \hat{x}_3 + \beta x_2$, where β is the β-plane parameter associated with barotropic flow [310, 48, 47, 355, e.g.].

In any case, suppose that an active or passive scalar distribution is obtained numerically or observationally at a final time t_f. A measure for how well-mixed c is at time

t_f could be the *concentration variance*

$$\sigma(t_i,t_f)=\frac{1}{S(\Omega)}\int_\Omega\left|c(y,t_f)-\frac{1}{S(\Omega)}\int_\Omega c(x,t_f)\,dx\right|^2 dy, \tag{1.19}$$

where $S(\Omega)$ is the measure (size) of Ω. This measure (or trivial modifications of which) is often used in the fluid mixing literature [121, 383, 12, 237, 6, 118, 236, 107, 288, 376, 109, 248, 403, 102, 381, 128, 283, 372, 163, 236, 6, 237, 12, 236, 308].

Thiffeault [400] (see also [267, 36]) presents an argument as to why using (1.19) may present a problem unless a spatial resolution of order better than $\delta^{1/2}$ is available. An alternative is the so-called *mix-norm*, which was introduced by Mathew et al. [268, 267], which defines a variance-like quantity in which the smaller length-scales are suppressed. The mix-norm is essentially equivalent to the Sobolev $H^{-1/2}$ norm, and indeed some researchers use slightly different Sobolev-type norms [246, 400], which have theoretical utility in expressing how to optimize global mixing via global velocity specifications [246, 267, 268, 400, 117].

An alternative method for estimating the global mixing is the FTLE, which was described in Section 1.4 in some detail. The FTLE is a time-dependent quantity λ as given in (1.14), but with fixed initial t_i and final t_f times it is simply a scalar field on Ω. Its value at a point x represents how stretched nearby initial conditions will get when following the flow (1.8) from time t_i to t_f. Put another way, if a blob of fluid is centered at x, it will eventually evolve into an ellipsoidal entity, and the FTLE value represents the exponential stretching rate that this ellipsoid has experienced in the direction in which it has stretched most. Now, exponential stretching is often associated with *chaotic transport*, but then usually in the sense of sensitive dependence on initial conditions. Nearby intial conditions lose sight of one another exponentially quickly in chaotic regimes [9]. On the other hand, one might have a purely nonchaotic flow such as (1.9) with no nonlinear terms, in which exponential stretching occurs, and hence directly thinking of a Lyapunov exponent as measuring chaos may be disingenuous. What is certainly true is that the FTLE measures exponential stretching, and such exponential stretching may be thought of as promoting mixing in a fluid. Thus, some researchers (see the references in Section 1.4 on FTLEs) use FTLEs as a proxy for the mixing in a fluid; for example, the scalar field could be averaged over Ω to give *one* measure if that is what is required.

The above description provides some details on global mixing measures which are used in the literature. The interested reader is referred to the review articles [400, 144] for a more extensive collection of global mixing measures.

At this point, it is worth discussing the difference between "mixing" and "transport" that is often implicit in the literature. The word "mixing" is used by many to mean diffusive mixing, in contrast to "transport," which might be a purely advective mechanism from an equation such as (1.8). In the presence of large exponential stretching (due to advection from a velocity field such as (1.8), say), fluid stretches rapidly and forms filaments. Eventually the width of the filaments is sufficiently small to be affected by diffusion, which "breaks up" the filament and causes its constituent fluid to "mix" into the adjacent fluid. Therefore, the usage of the FTLE, which is based on a purely advective equation, to quantify fluid mixing is not unreasonable. It should be pointed out that the concentration variance associated with an advection-diffusion process (1.18) already *does* explicitly incorporate diffusion, and thus that quantity can without prejudice be termed a mixing measure and not merely a transport measure. In fluids in which the diffusion is *small*, which is often legitimate in oceanic flows away from boundaries, and in microfluidic devices, it is probably correct to say that mixing is achieved by an *advection-driven diffusion mechanism* as described above: advection promotes fluid filamentation and transport into anomalous areas, with diffusion becoming effective when the filaments have sufficiently small scales. Thus, in such flows, understanding mixing cannot be done without understanding transport, i.e., the advective process. Therefore, subsequent chapters are devoted to quantifying, analyzing, and optimizing transport from the advective mechanism.

Now let us turn from global transport to transport in a much more specific sense: *transport across flow barriers.* "Transport barriers" or "flow barriers" are a common theme in unsteady flows that are in our consciousness: the edge of Jupiter's Great Red Spot, the boundary of the Antarctic Circumpolar Vortex, the edge of an oceanic eddy, the outer extent of spread of an oil spill, the interface between two miscible fluids in a microfluidic device, the outer boundary of a plankton bloom, etc. While these *barriers* are in reality not absolutely well-defined, they are often associated with a fairly abrupt transition of something (vorticity, temperature, oil concentration, volume fraction of fluid #1, plankton density). It is often of physical or biological interest to understand how this quantity varies across the "barrier" and how much and how quickly the quantity of interest "leaks across" the "barrier." This would, for example, quantify the transport of heat or energy between an atmospheric or oceanic vortex and its surroundings, thereby impacting weather, or the intermingling of two fluids across their mutual interface in a microfluidic device in which the mixing between the fluids is the goal. Put another way, the *transport flux* across the "barrier" often has relevance to scientists. The global mixing measures described previously do not offer any obvious help in this endeavor. On the other hand, even if achieving good global mixing is the goal, one would first need to achieve transport across these flow barriers to begin the process. Thus, transport *across flow barriers* is crucial.

Given the ambiguity in defining flow barriers in unsteady flows, how can one therefore define a *transport* across such an entity? Indeed, if there actually is transport across it, the term "barrier" would seem inappropriate! Unsteady flows which have dominant coherent structures will then not have boundaries as clearly defined as those in Figure 1.1. Instead of the codimension-1 red curves/surfaces displayed in this figure, the boundaries or barriers will be some fuzzy version of these, which moreover move around in time because the velocity is nonautonomous. Despite the obvious relevance, trying to quantify transport across these would apparently be an ill-defined exercise. It would seem that that there is little insight into this issue in many of the techniques suggested in the literature for quantifying mixing (these tend to be global) or in diagnostics for identifying flow barriers (which tend not to assess transport across them). Hence, this monograph. A rigorous way of defining these flow barriers as time-varying stable and unstable manifolds will be essential; Chapter 2 delves into this in great detail. Care is taken to ensure that time-periodicity is not a requirement in the theory, which means that standard techniques such as Poincaré maps are inapplicable. Given that these flow barriers will intersect each other in many ways while also moving around with time, Chapter 3 approaches the issue of how one can quantify transport across flow barriers in a meaningful way that respects the time-variation. Chapter 4 asks questions—motivated by microfluidic applications—of how one might optimize transport across flow barriers and obtains theoretical results with practical consequences. Again with microfluidics in mind, Chapter 5 addresses whether flow barriers can be controlled in the sense of having them evolve with time in a way that we specify.

1.6 ▪ Hyperbolic trajectories and exponential dichotomies

The importance of stable and unstable manifolds as time-varying flow barriers has been highlighted in the previous sections. In defining such entities, it is important to know *what* they are stable and unstable manifolds of. For example, to what entities are trajectories on the stable manifold attracted as time goes to infinity? Knowing what these are is clearly important in any intuitive understanding of a stable manifold. For autonomous flows, in the simplest situation one thinks of stable and unstable manifolds as being the attracting and repelling sets to *fixed points* (or stagnation points, or critical points). However, stable and unstable manifolds to more general entities, such as a periodic orbits, can of course be defined. In nonautonomous flows, much more general entities can also have attached to them the attracting and repelling sets that one thinks of as stable and unstable manifolds. A type of object to which stable and unstable manifolds are attached in such general situations, encompassing many of the examples mentioned, is called a *hyperbolic trajectory*. This includes, for example,

1. A fixed point in a two-dimensional steady flow with one positive and one negative eigenvalue;
2. A periodic orbit in a three-dimensional steady flow to which is attached a two-dimensional stable manifold and a two-dimensional unstable manifold;
3. A trajectory $x(t)$ in a general nonautonomous flow which is not periodic but to which is attached both a stable and an unstable manifold.

The above intuitive statements are somewhat unsatisfactory, and hence a more formal statement of a hyperbolic trajectory is necessary. To state this, consider a general differential equation

$$\dot{x} = F(x,t) \tag{1.20}$$

for $x \in \mathbb{R}^n$. Then, the linearized dynamics around $x(t)$ is governed by the equation of variations, which in this general case is given by [183]

$$\dot{y} = DF\,(x(t),t)\,y\,. \tag{1.21}$$

(This can be obtained by the formal process of putting in a solution $x(t)+y(t)$ into (1.20) and expanding to order $|y(t)|$.) A trajectory $x(t)$ of (1.20) is said to be *hyperbolic* if the variational equation (1.21) admits an *exponential dichotomy* [115, 305, 61], i.e., if there exists positive constants A_u and A_s, constants $\alpha_u > 0$ and $\alpha_s < 0$, and a projection matrix P such that

$$\begin{aligned} \left\|Y(t)\,P\,Y^{-1}(s)\right\| \quad & \leq A_u e^{\alpha_u(t-s)} \quad \text{for } t \leq s \text{ and} \\ \left\|Y(t)\,(I-P)\,Y^{-1}(s)\right\| \quad & \leq A_s e^{\alpha_s(t-s)} \quad \text{for } t \geq s\,, \end{aligned} \tag{1.22}$$

where I is the identity and Y is a fundamental matrix solution to the nonautonomous linear system (1.21), satisfying $Y(0)=I$. A projection operator P is characterized by the condition $P^2=P$. The particular projection in (1.22) relates to a projection, at the time 0, onto the unstable manifold of the hyperbolic trajectory $x(t)$. The "remaining" projection $I-P$ is a projection onto the stable manifold. The reason for this can be explained in terms of the decay of solutions. It is easy to verify that the solution to (1.21) will be given by

$$y(t) = Y(t)Y^{-1}(s)y_s$$

where $Y(t)$ is a fundamental matrix solution to (1.21) with $y(s)=y_s$. Now let w be an arbitrary vector in $\mathbb{R}^n$, and suppose $y_s = Y(s)Pw$. Then,

$$\begin{aligned} \|y(t)\| &= \|Y(t)Pw\| = \|Y(t)PPw\| = \left\|Y(t)PY^{-1}(s)Y(s)Pw\right\| \\ &\leq \left\|Y(t)PY^{-1}(s)\right\| \|Y(s)Pw\| \leq A_u e^{\alpha_u(t-s)} \|y_s\|, \end{aligned}$$

where $P^2 = P$ and the first exponential dichotomy condition of (1.22) has been used for $t \leq s$. Note in particular that if $s = 0$, then $y_0 = Y(0)Pw = Pw$, and that the corresponding solution $y(t)$ decays according to

$$\|y(t)\| \leq A_u e^{\alpha_u t} \|y_0\| \text{ for } t \leq 0, \quad y_0 = Pw.$$

Initial conditions in the range of P therefore decay exponentially in backward time. The evolution of these solutions is given by

$$y(t) = Y(t)Y^{-1}(0)y_0 = Y(t)Pw,$$

and thus in a time-slice t, conditions chosen in the range of $Y(t)P$ decay exponentially in backward time. Thus, $\mathscr{R}\{Y(t)P\}$ represents the t-variation of the direction of the unstable manifold, that is, the tangent vector drawn to the unstable manifold at the hyperbolic trajectory $x(t)$. This is indeed the t-variation of the (unstable) Oseledets space associated with the trajectory $x(t)$. Similarly examining the second exponential dichotomy condition gives the understanding that $I-P$ is the projection onto the stable subspace at time 0. In a general time-slice t, conditions chosen in $\mathscr{R}\{Y(t)(I-P)\}$ represent the stable manifold, in which there is exponential decay in backwards time. For more details on these intuitions, see also Appendix A in [33].

It is possible that P is either the identity or the zero matrix, both of which are trivial projections. If this is so, one or the other of the exponential dichotomy conditions (1.22) become vacuous, while the other states that the full space is the stable (or unstable) manifold. In other words, the hyperbolic trajectory will possess either a full co-dimensional stable manifold or a full co-dimensional unstable manifold. While this is considered in Section 5.2, much of the subsequent development will relate to saddle-like behavior; hyperbolic trajectories will usually possess both a stable and an unstable manifold.

One simple situation in which exponential dichotomies are easily verifiable is in autonomous flows in which $x(t)$ is a fixed point. Then, $Df(x(t),t)$ is a constant matrix, and its eigenvalues govern motion of (1.20) near to the fixed point. If the eigenvalues have a real part which is separated by the imaginary axis, then $x(t)$ will satisfy exponential dichotomy conditions. In using (1.22), one can use α_u to be the real part of the eigenvalue which lies in the right-half plane which is closest to the imaginary axis, and α_s the real part of the eigenvalue in the left-hand plane closest to the imaginary axis. This works since these values govern the smallest exponential expansion/repulsion rates associated with the unstable and stable eigenspaces.

However, in any *realistic* (observational, experimental, numerically computed, etc.) system with nontrivial time-dependence, the exponential dichotomy conditions are usually impossible to use in identifying hyperbolic trajectories. One reason is that these trajectories are defined only implicitly through the definition. The utility of exponential dichotomies arises from theoretical considerations which, among other things, enable quantification of the persistence of hyperbolic trajectories under perturbations. Thus, they have usage in determining whether some observed object actually *is* a stable or an unstable manifold. Since exponential dichotomies as given in

(1.22) additionally apply only to idealized flows in which $t \in \mathbb{R}$ (that is, the exponential decay rates occur as $t \to \pm\infty$), there is considerable interest in extending the concept to more realistic situations. In particular, since any data set is only available for a finite time, can a restricted definition for exponential dichotomies be provided for finite times? Simply restricting t in (1.22) does not work, since *any* function can be bounded by an exponential with sufficiently large prefactor and/or exponent on a finite interval. Thus, finite-time hyperbolicity has become an emerging theoretical area of research [332, 169, 219, 127, 58, 66], in the sense of "restricting" the exponential dichotomy conditions (1.22) to finite times.

It has already been argued in the discussion associated with equation (1.7) that hyperbolic trajectories—the time-varying analogue of saddle fixed points—cannot be obtained by examining instantaneous fixed points in a flow. Nevertheless, in situations in which the time-variation is *small*, some methods [265, 212] do attempt to utilize instantaneous stagnation points as proxies, or initial guesses, for determining hyperbolic trajectories. Here is one situation, though, in which the hyperbolicity of a trajectory can be verified exactly using exponential dichotomies:

Example 1.4. Let c be a constant, and consider the system

$$\frac{d}{dt}\begin{pmatrix} x_1 \\ x_2 \end{pmatrix} = \begin{pmatrix} x_1 + ct - c + 2(x_1+ct)^2 - 3x_2^2 \\ -2x_2 - 5(x_1+ct)^3 + x_2^4 \end{pmatrix}. \tag{1.23}$$

It will be claimed that $(x_1, x_2) = (-ct, 0)$ is a hyperbolic trajectory of (1.23). With the right-hand side of (1.23) being considered the F in (1.20),

$$DF|_{(x_1,x_2,t)} = DF|_{(-ct,0,t)} = \left.\begin{pmatrix} 1+4(x_1+ct) & -6x_2 \\ -15(x_1+ct)^2 & -2+4x_2^3 \end{pmatrix}\right|_{(-ct,0,t)} = \begin{pmatrix} 1 & 0 \\ 0 & -2 \end{pmatrix},$$

and hence the variational equation (1.21) in this case becomes

$$\frac{d}{dt}\begin{pmatrix} y_1 \\ y_2 \end{pmatrix} = \begin{pmatrix} 1 & 0 \\ 0 & -2 \end{pmatrix}\begin{pmatrix} y_1 \\ y_2 \end{pmatrix}.$$

The variational equation is therefore autonomous. The alert reader would have realized that this example is contrived in the sense that the affine terms in (1.23) are exactly those of (1.7), which is a simple translation of the saddle point flow which indeed appears as the variational equation above. Except for situations such as this, the exponential dichotomy conditions are usually impossible to work with. Here, however, the $(1,0)^\top$ direction is clearly the unstable one, associated with the positive eigenvalue, and $(0,1)^\top$ the stable direction. The fundamental matrix solution and the projection P to the unstable manifold can be written immediately as

$$Y(t) = \begin{pmatrix} e^t & 0 \\ 0 & e^{-2t} \end{pmatrix} \quad \text{and} \quad P = \begin{pmatrix} 1 & 0 \\ 0 & 0 \end{pmatrix}.$$

Now, in a general time-slice t, the range of $Y(t)P$ represents the t-variation of the unstable fiber. In this case,

$$Y(t)Pw = \begin{pmatrix} e^t & 0 \\ 0 & 0 \end{pmatrix}\begin{pmatrix} w_1 \\ w_2 \end{pmatrix} = w_1 e^t \begin{pmatrix} 1 \\ 0 \end{pmatrix},$$

indicating that $\mathcal{R}\{Y(t)P\}$ does not change with t and is always in the x_2 direction. Thus, this characterization is consistent with directly reading off from the variational equation, which could be done in this case. Moreover, note that

$$Y(s) = \begin{pmatrix} e^s & 0 \\ 0 & e^{-2s} \end{pmatrix} \quad \text{and} \quad Y^{-1}(s) = \begin{pmatrix} e^{-s} & 0 \\ 0 & e^{2s} \end{pmatrix}.$$

The first exponential dichotomy condition in (1.22) is now analyzed for $t \leq s$:

$$\begin{aligned} \left\| Y(t)PY^{-1}(s) \right\| &= \left\| \begin{pmatrix} e^{t-s} & 0 \\ 0 & 0 \end{pmatrix} \right\| = \sup_{w_1^2+w_2^2=1} \left\| \begin{pmatrix} e^{t-s} & 0 \\ 0 & 0 \end{pmatrix} \begin{pmatrix} w_1 \\ w_2 \end{pmatrix} \right\| \\ &= \sup_{w_1^2+w_2^2=1} \left\| \begin{pmatrix} w_1 e^{t-s} \\ 0 \end{pmatrix} \right\| = \sup_{w_1^2+w_2^2=1} |w_1| e^{t-s} = e^{t-s}, \end{aligned}$$

where the matrix norm induced by Euclidean distance is the norm used for matrices in the definition for exponential dichotomies [33, Appendix A]. Thus, the first exponential dichotomy condition (1.22) is satisfied with the choice $A_u = 1$ and $\alpha_u = 1$. Verification of the second exponential dichotomy condition is left as an exercise; it should not be a surprise that in this case $A_s = 1$ and $\alpha_s = -2$. ∎

Chapter 2

Melnikov theory for stable and unstable manifolds

I've always been more comfortable sinking while clutching a good theory, than swimming with an ugly fact.
—David Mamet

2.1 ▪ Classical Melnikov results

Stable and unstable manifolds have a celebrated role in establishing chaos in a dynamical system via the Smale-Birkhoff Theorem [183]. This is stated for a *discrete* dynamical system (or a *difference equation*, or a *map*) of the form

$$x_{n+1} = P(x_n), \tag{2.1}$$

where $x \in \Omega$ which is two-dimensional, and P is sufficiently smooth and subject to several other technical assumptions [183]. Suppose a is a fixed point of (2.1), i.e., $a = P(a)$. Moreover, suppose that the linearization identifies a as a saddle fixed point. To express this in the simplest form, suppose $DP(a)$, the Jacobian derivative matrix evaluated at a, has eigenvalues which are positive, with one being smaller than 1 and the other being greater than 1. This indicates that a has a stable manifold in the local direction given by the eigenvector associated with the first eigenvalue and an unstable manifold in the direction of the eigenvector of the second. Unlike in the continuous-time instances addressed so far, the manifolds here are not trajectories themselves. A trajectory is in fact a discrete sequence of points, and if an initial condition is chosen on the stable manifold, then its iterates will converge towards a by hopping along the stable manifold. Now comes the key assumption: suppose the stable and unstable manifolds of a intersect transversally (not tangentially) at a point. Then the Smale-Birkhoff Theorem guarantees that the system (2.1) has chaotic trajectories [183]. Exactly what is meant by *chaotic trajectories* is elucidated in texts such as [9]; briefly, the idea is that highly complicated dynamics exhibiting properties, such as (i) infinitely many periodic trajectories, (ii) uncountably many aperiodic trajectories, (iii) a dense trajectory, and (iv) high sensitivity to initial conditions, occur in a region of Ω. While the proof of the Smale-Birkhoff Theorem is beyond the scope of this monograph, a quick insight is that one intersection point p must imply infinitely many, since $P^n(p)$ for $n \in \mathbb{Z}$ must lie on both the stable and unstable manifolds. This results in a *homoclinic tangle* with the

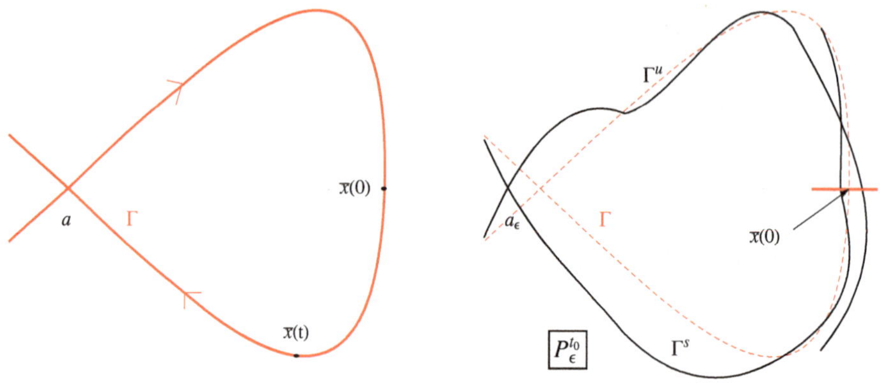

(a) Homoclinic (b) Stable/Unstable manifolds

Figure 2.1. *Classical Melnikov method: (a) homoclinic manifold for Poincaré map P^{t_0} for (2.3), and (b) stable and unstable manifolds for $P_\varepsilon^{t_0}$ for (2.2).*

stable and unstable manifolds intersecting each other infinitely many times, resulting in stretching and folding of the areas in between these manifolds, i.e., complicated motion. What is exciting about this result is that it provides a method for *proving chaos*; an intersection between the stable and unstable manifold needs to be established.

Melnikov theory provides a mechanism for providing exactly such an intersection [183], thereby enabling a proof that a given system is chaotic. One obvious deficiency of the above discussion on (2.1) is that the dynamical system is discrete, whereas it should be *continuous* (i.e., a differential equation rather than a difference equation) to describe fluid motion. Standard Melnikov theory is able to bridge the gap by imagining a differential equation whose velocity field is time-periodic with some period, say T. Then, by sampling the trajectories of this equation at times which are multiples of T, one generates discrete trajectories in phase space. The process of sampling generates a map (as in (2.1)) which is called a *Poincaré map*. Thus, by applying the discrete dynamical systems ideas above to the Poincaré map, one can obtain a proof that the continuous, time-periodic dynamical system has chaotic motion.

The most fundamental Melnikov theory is now described and is due to Melnikov [271]. Descriptions are also available in standard texts such as [183, 17, 421, 354]. Consider the system

$$\dot{x} = f(x) + \varepsilon g(x,t), \tag{2.2}$$

where $x \in \Omega \subset \mathbb{R}^2$, f and g are as smooth as we like, and $|\varepsilon|$ is small. Moreover, $g(x,t)$ is assumed to be periodic in t; that is, there exists $T > 0$ such that $g(x,t+T) = g(x,t)$ for all $(x,t) \in \Omega \times \mathbb{R}$. Before addressing (2.2), consider its $\varepsilon = 0$ version, i.e., the equation

$$\dot{x} = f(x). \tag{2.3}$$

Since Ω is two-dimensional, chaotic motion is prohibited in the system (2.3) by the Poincaré-Bendixson Theorem [9, 183]. The question is: can chaos be *proven* for (2.2) for nonzero values of ε, when the system is now nonautonomous?

Some assumptions on (2.3) will be necessary. First, it is required that (2.3) be area-preserving, in that regions of Ω advected by (2.3) conserve their area. This requirement turns out to translate to requiring that $\mathrm{Tr}\, Df = 0$; that is, the trace of the Jacobian derivative matrix of f must be zero. Another way of stating this is that $\nabla \cdot f = 0$; that is, f needs to be divergence-free. From a fluid mechanical perspective, if f is thought of as a velocity, this is the statement that the fluid is incompressible. Thus, it has wide validity in applications. Second, suppose (2.3) has a fixed point a such that $Df(a)$ has a positive and a negative eigenvalue. This guarantees that a has an unstable and a stable manifold, each of which is one-dimensional. The third condition is that a branch of the stable manifold of a must coincide with a branch of the unstable manifold of a. This is therefore a *homoclinic manifold*, Γ, of (2.3), where the "homo-" relates to the fact that the manifold connects a fixed point to *itself*. The geometry associated with this is shown in Figure 2.1(a), and it should be borne in mind that (2.3) is purely autonomous. (Note that Figure 1.1(b), which is a toy model for a steady eddy in the ocean, has the relevant geometry.) Now, if one picks any point on Γ as an initial condition, the corresponding trajectory both forwards- and backwards-asymptotes to a since it lies on the stable, and simultaneously the unstable, manifold. This trajectory of (2.3) can be represented as $\bar{x}(t)$, thereby enabling Γ to be parametrically represented by $\{\bar{x}(t) : t \in \mathbb{R}\}$. Note that if a different point on Γ is chosen as an initial condition, the relevant $\bar{x}(t)$ would simply be a shift in t; this issue is not highly relevant in the classical Melnikov approach. Whatever choice is made, $\lim_{t\to\pm\infty} \bar{x}(t) = a$, and the decay is exponential with rate given by the eigenvalues of $Df(a)$.

While the fixed point and its stable/unstable manifold structure elucidated so far is with respect to a continuous flow, they can be recast in terms of a discrete flow. Define the Poincaré map P^{t_0}, which takes points in Ω at time t_0, and then follows the flow for a time T, the period of g. Repeated applications of P^{t_0} are then described by (2.1), for this specific choice of P^{t_0}. Now, since a was a fixed point of (2.3), it would be a fixed point for (2.1) as well. Moreover, since any point x_0 on Γ asymptotes towards a in forward time under the flow (2.3), a sequence $\{x_0, x_1, x_2, \cdots\}$ generated by (2.1) must also approach a. Therefore, Γ will be a branch of the stable manifold of a under the map P^{t_0}. Similarly, it will be a branch of the unstable manifold as well, and here the understanding is that one considers backwards iterations of P^{t_0}, that is, the map $(P^{t_0})^{-1}$, associated with flowing (2.3) backwards in time by time units T. Thus, the heteroclinic manifold structure shown in Figure 2.1(a) is appropriate even for the discrete viewpoint, where the Poincaré map P^{t_0} is considered. Of course, the arrows on the heteroclinic manifold need a different interpretation in this case, since the iterates associated with P do not continuously flow along the manifold, but rather jump along it. Now, the eigensystem determination of stability from $Df(a)$ for the continuous system (2.3) can be replaced by the eigensystem $DP^{t_0}(a)$ for the map (2.1). The fact that $Df(a)$ has a negative eigenvalue, with corresponding eigenvector giving the local direction of the stable manifold, translates to $DP^{t_0}(a)$ having an eigenvalue between 0 and 1, with corresponding eigenvector being identical to that of $Df(a)$. The positive eigenvalue of $Df(a)$ is therefore associated with an eigenvalue greater than 1 of $DP^{t_0}(a)$, with the corresponding eigenvectors of these matrices coinciding.

The reason for choosing a Poincaré map with this particular period is to enable the extension to small but nonzero $|\varepsilon|$. Let $P_\varepsilon^{t_0}$ be the Poincaré map which now considers points in Ω and flows them forward from a time t_0 to a time T beyond that, but now the flow is (2.2) rather than (2.3). Since $P_\varepsilon^{t_0}$ is a small perturbation on P^{t_0}, a fixed point a_ε which is $\mathcal{O}(\varepsilon)$-close to a will exist. This can be proven as a simple consequence of the Implicit Function Theorem, say. Moreover, since $DP_\varepsilon^{t_0}(a_\varepsilon)$ would also be $\mathcal{O}(\varepsilon)$-close to $DP^{t_0}(a)$, its eigenvalues will also change only by $\mathcal{O}(\varepsilon)$. Consequently, there will continue to be one eigenvalue less than 1 and the other eigenvalue greater than

1. Therefore, there will continue to be a stable, Γ^s, and an unstable, Γ^u, manifold. These are continuous curves emanating from a_ε, and near a_ε are $\mathcal{O}(\varepsilon)$-close to the original manifolds of a as shown in Figure 2.1. Thus, the manifolds (with respect to the discrete map $P_\varepsilon^{t_0}$) will closely follow the manifolds of P^{t_0}. However, there is something quite different when $\varepsilon \neq 0$: there is no longer any guarantee that the manifolds will coincide to form the heteroclinic manifold as in Figure 2.1(a). Generically, they will *not* coincide, and a typical picture is shown in Figure 2.1(b). If the manifolds can be shown to intersect at just *one* point in a nondegenerate manner, then the Smale-Birkhoff Theorem would apply, and it could be concluded that the system is chaotic. Melnikov theory provides a method—the Melnikov function—for proving this one intersection point. A version of this is provided below.

Theorem 2.1 (Classical Melnikov theorem [271, 183, 17, 421]). *Suppose* $\mathrm{Tr}\, Df = 0$, *g is periodic with period $T > 0$, and f and g are smooth in (2.2), where $x \in \Omega \subset \mathbb{R}^2$. Moreover, suppose there exists a one-dimensional homoclinic manifold of (2.3) which can be represented as a trajectory $\bar{x}(t)$ of (2.3). Define the* Melnikov function

$$M(t_0) := \int_{-\infty}^{\infty} f(\bar{x}(\tau - t_0)) \wedge g(\bar{x}(\tau - t_0), \tau)\, d\tau, \tag{2.4}$$

where the wedge product is defined componentwise for vectors $y = (y_1, y_2)$ and $z = (z_1, z_2)$ in $\mathbb{R}^2$ by

$$y \wedge z := y_1 z_2 - y_2 z_1 . \tag{2.5}$$

If there exists a value t_0 such that $M(t_0) = 0$ and $M'(t_0) \neq 0$, then a transverse (nontangential) intersection between the stable, Γ^s, and unstable, Γ^u, manifolds of (2.2) associated with the T-periodic Poincaré map $P_\varepsilon^{t_0}$ occurs for small enough $|\varepsilon|$.

Proof. This version of the Melnikov method is closest to that given in [183], to which the reader is referred to for the proof. □

The Melnikov function $M(t_0)$ in (2.4) has a geometric interpretation: it is a scaled, leading-order in ε, signed distance function between the stable and the unstable manifolds associated with $P_\varepsilon^{t_0}$. This signed distance is measured along a normal vector drawn to Γ (thick red line) at the point $\bar{x}(0)$. The variation with t_0 is associated with choosing different initial times t_0 for implementation of the Poincaré map. Thus, the presence of a value t_0 satisfying $M(t_0) = 0$ and $M'(t_0) \neq 0$ would imply that the stable and unstable manifolds intersect near the point $\bar{x}(0)$ when considering the Poincaré map as operating on the time-slice t_0. A considerably expanded geometric description of the Melnikov function, in a more general setting, will follow.

In "typical" applications, one would try to explicitly evaluate the Melnikov function for given f and g (and in many such cases g is taken to be a separable function of the form $h(x)\cos\omega t$). This usually requires computing the improper integral (2.4) using a contour integration process. Once evaluated, the goal would be to show that $M(t_0)$ has a simple zero. Many such applications are available in the literature [183, 178, 225, 200, 421, 341, 202, 99, 354, e.g.].

So how can the Melnikov approach of Theorem 2.1 be extended? One minor point is that, while Theorem 2.1 is stated for a *homoclinic* situation in which the stable and unstable manifolds are associated with the *same* fixed point a, it turns out that the theorem as stated is valid even in a *heteroclinic* case in which Γ is simultaneously a branch of the unstable manifold of a and a branch of the stable manifold of a *different* saddle fixed point b. The proof actually requires no modification whatsoever, yet this possibility is often not stated in discussions on the Melnikov method.[5] Another slight issue is that the standard proof available in most textbooks [183, 421] is formal in nature; it relies on expansions in ε in which higher-order terms are simply neglected. Would it be possible to keep these and obtain the results in a rigorous way? In seeking rigor, a functional-analytic approach in terms of the Fredholm alternative has been developed, from which one can obtain that a transverse intersection exists if the perturbing vector field g is in the null-space of a bounded operator [106, 305, 61, 355]. This effectively obtains the fact that (2.4) needs to have a zero. These developments—unlike the intuitive ones available in [183, 17, 421]—however, lose the geometric interpretation of the Melnikov function, which as shall be seen is important in understanding its role of transport across flow barriers. Having said that, the functional analytic approach can be advantageous in situations in which a geometric interpretation is impossible. There are many extensions of Melnikov theory: to higher-dimensions [180, 50], to higher-order [133], for implicitly defined differential equations [60], for stochastic perturbations [366], singular perturbations [181, 292], nonhyperbolic situations [415, 436], fixed points at infinity [415], and discontinuous vector fields [133, 95, 59, 230].

The methods to be described will enable extensions to the classical theory in several other fronts, in addition to simply strengthening the mathematics behind a formal ε-expansion. First, the requirement of f being area-preserving will be relaxed. Area-preservation has been a widely used restriction since it enables the dynamics to be couched in terms of Hamiltonian mechanics [183, 421]. Indeed, a common version of Theorem 2.1 requires additionally that g be area-preserving and writes the integrand of (2.4) in terms of a Poisson bracket between an unperturbed and a perturbed Hamiltonian function [421, 264, 225, 417, 202]. While area-preservation seems natural for incompressible flows, there are situations in geophysical flows in which area-preservation does *not* occur. Examples include the atmosphere (air is compressible), and two-dimensional ocean models in which each two-dimensional sheet has an exchange of fluid with adjacent sheets to maintain incompressibility (in this situation, it is the fully three-dimensional fluid system which must be volume-preserving). Hence relaxing area-preservation is important for fluids applications.

A second relaxation is in relation to the velocity field being time-periodic. This was necessary to define a Poincaré map. General fluid flows are of course not periodic and will have arbitrary variation with time. Can the Melnikov results be obtained in this case? It will be shown that this is indeed possible, but the viewpoint requires a change. Rather than as a fixed point of a Poincaré map, one needs to think of a_ε as a hyperbolic trajectory in the sense of exponential dichotomies, as has been defined in Section 1.6. Attached to it will be stable and unstable manifolds which themselves need to be thought of as genuinely time-varying entities, rather than invariant curves associated with a map.

A third modification requires another change of attitude. Here, the concern is with *flow barriers*, that is, the stable and unstable manifolds. As discussed above, these need to be thought of as evolving with time. However, the Melnikov theory of Theorem 2.1 gives us leading-order information not on the locations of the stable and unstable manifolds but on the distances (subject to some scaling which has not been elucidated yet) *between them*. Can the proof of the Melnikov method be adapted to instead quantify how the stable and the unstable manifolds evolve? This is achieved in Sections 2.3 and

[5]While Theorem 2.1 applies for heteroclinic manifolds, the Smale-Birkhoff Theorem does not, and hence a simple zero of the Melnikov function would not automatically guarantee chaos; this is perhaps the reason for the popularity of the homoclinic statement of Melnikov theory.

2.4. Since the classical Melnikov approach is associated with displacements in relation to a *normal to* Γ (i.e., the thick red line in Figure 2.1(b)), it is required to not only separately characterize the normal displacements of the stable and unstable manifolds but also determine the modifications necessary in the *tangential direction* to Γ. Chapter 2 explores many aspects of this issue, dealing, for example, also with flows which are discontinuous in time in Sections 2.6 and 2.7.

A fourth—and particularly important from the fluid transport perspective—viewpoint requires connecting fluid flux to the Melnikov function. The work by Rom-Kedar and collaborators [341, 421] has provided an elegant transport connection between the integral of the Melnikov function and *lobe dynamics* transport that occurs because of complicated intersection patterns between stable and unstable manifolds. This depends on the flow being time-periodic, and indeed the time-periodicity needs to be much more restrictive to actually be able to use this. Can one obtain a Melnikov characterization of the fluid flux in general time-aperiodic situations, given also that the time-varying stable and unstable manifolds may intersect once, a finite number of times, an infinite number of times, or not at all? Additionally, is it possible to represent the Melnikov function in a way which is easier to evaluate? These issues are the focus of Chapter 3.

The possibility of relating Melnikov methods in a direct way to fluid transport leads to the fifth group of extensions: can one obtain insights from the Melnikov development to either *optimize* (Chapter 4) or *control* (Chapter 5) fluid flows? These chapters adapt and apply the Melnikov approaches of the previous chapters to answer questions on, for example, the best velocity agitation to enhance transport across the flow interface in Figure 1.1(c) or how to decide on the velocity perturbation to make any of the stable manifolds shown in Figure 1.1 display a (nearby) time-varying evolution that is *specified.*

2.2 ▪ Nonautonomously perturbed 2D flows

The basic setting throughout much of this chapter is the nonautonomously perturbed flow

$$\dot{x} = f(x) + \varepsilon g(x,t), \tag{2.6}$$

in which $x \in \Omega$, a two-dimensional open connected set, and $|\varepsilon| \ll 1$. This system will play a significant role throughout this monograph. In this section, the basic framework for its hyperbolic trajectories will be established, with the stable and unstable manifold determination performed in the next sections.

Hypothesis 2.2 (smoothness of f). *The vector field $f \in C^2(\Omega)$, with second derivative tensor $D^2 f$ being uniformly bounded on Ω.*

Notice that it is not assumed that f is area-preserving (i.e., that $\operatorname{Tr} Df = 0$) as in the classical approaches [271, 183, 17, 421]. The perturbing nonautonomous field *is not required to be time-periodic.*

Hypothesis 2.3 (smoothness and boundedness of g). *The perturbing vector field g satisfies the following:*

(a) $g(\cdot,t) \in C^2(\Omega)$ *for any $t \in \mathbb{R}$, and $g(x,\cdot)$ is continuous and piecewise smooth in $\mathbb{R}$ for each $x \in \Omega$, and*

(b) g *and* Dg *are both bounded in* $\Omega \times \mathbb{R}$.

Remark 2.4 (Sobolev norms). The boundedness conditions in Hypotheses 2.2 and 2.3 could alternatively be stated in terms of Sobolev norms $||\bullet||_{m,\infty}$ defined for nonnegative integers m by

$$||f||_{m,\infty} := \sum_{|i|\leq m} \left\| D^i f \right\|_\infty \tag{2.7}$$

for functions f on Ω, with the L^∞ norms being defined by

$$||h||_\infty = \text{ess sup}_{x\in\Omega} |h(x)| \,. \tag{2.8}$$

Under these definitions, Hypothesis 2.2 could be restated as there existing a C_f such that $||f||_{2,\infty} \leq C_f$, whereas Hypothesis 2.3(b) would promise the presence of a C_g such that $\sup_{t\in\mathbb{R}} ||g(\bullet,t)||_{1,\infty} \leq C_g$.

Certain properties will be assumed for the unperturbed system, obtained by setting $\varepsilon = 0$ in (2.6), which yields the autonomous system

$$\dot{x} = f(x). \tag{2.9}$$

Hypothesis 2.5 (saddle point). *There exists a point $a \in \Omega$ such that $f(a) = 0$, and $Df(a)$ possesses eigenvalues $\lambda_s < 0$ and $\lambda_u > 0$, with corresponding normalised eigenvectors v_s and v_u.*

The presence of a saddle point a in (2.9) implies that a possesses both a one-dimensional stable manifold Γ^s (locally tangential to v_s) and a one-dimensional unstable manifold Γ^u (locally tangential to v_u). The eigenvectors $v_{s,u}$ shall be assumed to have been normalized (i.e., $||v_{s,u}|| = 1$). In this section, the time-varying modification to the saddle point a will be characterized. It is to this entity that is attached stable and unstable manifolds, whose time-varying locations will be determined in the next sections.

Now, using (1.21), the equation of variations for the special trajectory $x(t) = a$ of (2.9) is the autonomous equation

$$\dot{y} = Df(a)y, \tag{2.10}$$

which by virtue of Hypothesis 2.5 can be immediately said to possess an exponential dichotomy (1.22) with the choice $\alpha_\sigma = \lambda_\sigma$ for $\sigma = u, s$. The projection P corresponds to projection to the unstable manifold when $t = 0$, and its complement $I - P$ is the projection to the stable manifold. When $\varepsilon \neq 0$ but is small, persistence of integral manifolds [433, 434, 186] implies that the manifold (a, t) persists as an ε-close entity $(a_\varepsilon(t), t)$ and moreover retains exponential dichotomy conditions. This means that its equation of variations

$$\dot{y} = [Df(a_\varepsilon(t)) + \varepsilon Dg(a_\varepsilon(t), t)]y \tag{2.11}$$

will also satisfy the exponential dichotomy conditions, but for different constants. While this argument guarantees the *existence* of such an entity, a method for *finding* it is not clear yet since it is defined implicitly.

Since the range of $Y(t)P$ (with Y being the fundamental matrix solution to (2.11), and P the projection guaranteed by the exponential dichotomy conditions (1.22)) represents the projection to the unstable manifold direction at each time t, this will in general change with t. Moreover, the range of $Y(t)(I - P)$, representing the stable manifold direction, will also change. The following has therefore been established.

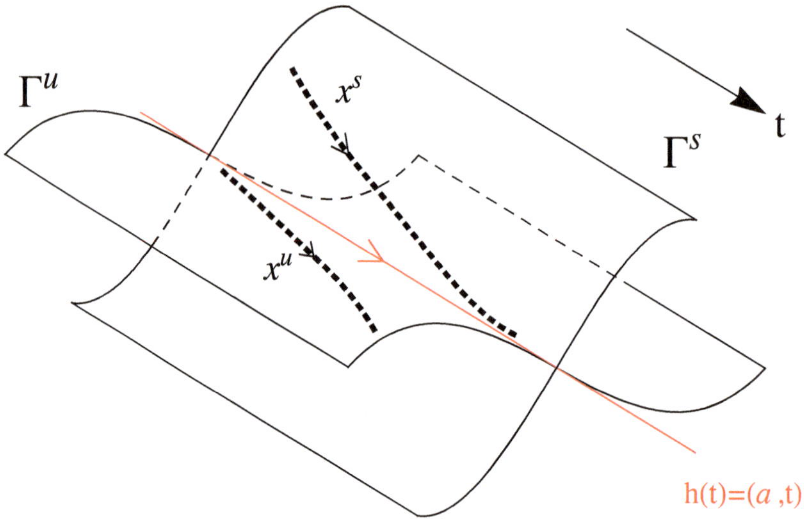

Figure 2.2. *The hyperbolic trajectory [in red] $h(t) = (a, t)$ of (2.12) when $\varepsilon = 0$. The stable and unstable manifolds are foliated with trajectories—which are simple shifts of one another upon each manifold—of which one on each of the manifolds is shown [dotted lines].*

Theorem 2.6 (persistence of hyperbolic trajectory). *When $|\varepsilon|$ is sufficiently small, the saddle point a of (2.9) perturbs to a hyperbolic trajectory $a_\varepsilon(t)$ of (2.6), which continues to possess both a stable and an unstable manifold.*

The above result was obtained by using fairly weak conditions on g. For alternative persistence results which require trajectories to remain in a compact set, or enforce this via periodicity, the reader is referred to [141, 199, 305]. Once the presence of $a_\varepsilon(t)$ is established, the next question is determining an expression for it. In doing so, it is best to view the nonautonomous system (2.6) as an augmented (or appended) system

$$\left.\begin{aligned} \dot{x} &= f(x) + \varepsilon g(x,t) \\ \dot{t} &= 1 \end{aligned}\right\}, \tag{2.12}$$

which renders (2.6) autonomous at the cost of adding a dimension to the phase space, which is now $\Omega \times \mathbb{R}$. As a consequence of Hypothesis 2.5, the system (2.12) when $\varepsilon = 0$ possesses a trajectory $h(t) := (a, t)$ which is a straight line in the augmented phase space, as shown in Figure 2.2. This has *two*-dimensional stable and unstable manifolds, as shown locally. These manifolds are foliated with trajectories; a trajectory on each of the stable and unstable manifolds, denoted, respectively, by x^s and x^u, are also shown. Given the autonomous nature of (2.12) when $\varepsilon = 0$, the trajectory foliations on each manifold are obtained by simple shifts in t of the pictured trajectories. Now, the consequence of Theorem 2.6 is that for small enough $|\varepsilon|$, there will be a trajectory $h_\varepsilon(t) := (a_\varepsilon(t), t)$ of (2.12) such that the corresponding picture will be a "wobbled" version of Figure 2.2. In other words, the straight line (a, t) will wobble around somewhat, and its attached manifolds—which are regular in t in Figure 2.2—will acquire some t-variation. This statement is only true *locally* near $h_\varepsilon(t)$; nothing can be said about the manifolds beyond a tubular neighborhood around $h_\varepsilon(t)$.

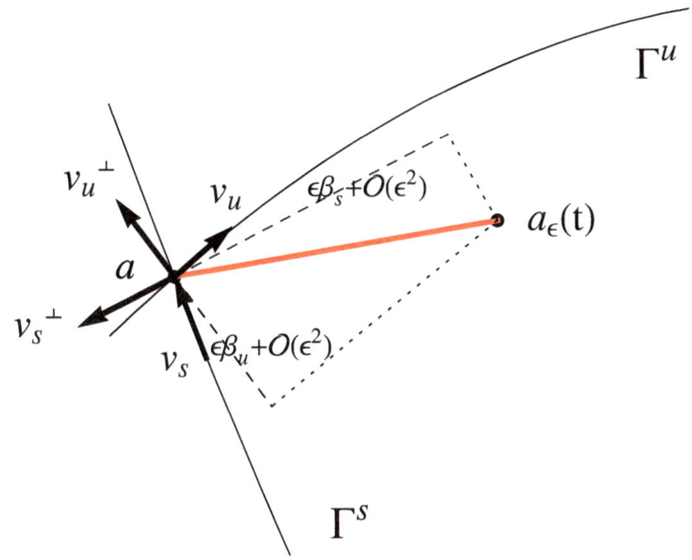

Figure 2.3. *The quantities of Theorem 2.7 in a time-slice t, needed in determining the perturbed hyperbolic trajectory $a_\varepsilon(t)$.*

It will be convenient to introduce the following notation for the *perpendicular vector* for vectors $G=(G_1,G_2)\in\mathbb{R}^2$:

$$G^\perp := \begin{pmatrix} 0 & -1 \\ 1 & 0 \end{pmatrix} G = \begin{pmatrix} -G_2 \\ G_1 \end{pmatrix}. \tag{2.13}$$

The vector $G^\perp$ is obtained by rotating and vector G by $+\pi/2$. Now consider the time-slice t of Figure 2.2, as shown in Figure 2.3. The perturbed hyperbolic trajectory $a_\varepsilon(t)$ will be $\mathcal{O}(\varepsilon)$-close to a, and its location can be characterized in terms of an orthogonal coordinate system comprising $(v_u^\perp, v_u)$, as shown in Figure 2.3, by the following theorem.

Theorem 2.7 (perturbed hyperbolic trajectory [32]). *The perturbed hyperbolic trajectory $a_\varepsilon(t)$ of (2.12) can be represented by*

$$a_\varepsilon(t) = a + \varepsilon\left[\beta_u(t)v_u^\perp + \frac{\beta_u(t)(v_u\cdot v_s)-\beta_s(t)}{v_u^\perp\cdot v_s}v_u\right] + \mathcal{O}(\varepsilon^2), \tag{2.14}$$

where

$$\begin{aligned} \beta_s(t) &:= -\int_0^\infty g(a,t+\tau)\cdot v_s^\perp e^{-\lambda_u\tau}\,\mathrm{d}\tau \quad \text{and} \\ \beta_u(t) &:= \int_0^\infty g(a,t-\tau)\cdot v_u^\perp e^{\lambda_s\tau}\,\mathrm{d}\tau \end{aligned} \tag{2.15}$$

are, respectively, the leading-order projections of $a_\varepsilon(t)-a$ in the $v_s^\perp$ and $v_u^\perp$ directions.

Proof. The proof will be postponed to Section 2.4.1, since it will use theorems on the stable and unstable manifolds which shall be developed in subsequent sections. □

As illustrated in Figure 2.3, the quantities $\beta_{s,u}(t)$ comprise the $\mathcal{O}(\varepsilon)$ components of the projection of the vector $a_\varepsilon(t)-a$ in the directions $v_{s,u}^\perp$, which are locally tangential to the stable/unstable manifolds $\Gamma^{s,u}$. That is,

$$\beta_i(t)=\lim_{\varepsilon\to 0}\frac{a_\varepsilon(t)-a}{\varepsilon}\cdot v_i^\perp \quad i=s,u. \tag{2.16}$$

Elementary geometry then enables the representation (2.14). As is reasonable for a leading-order theory, only *local* information on f is needed, as represented by $\lambda_{s,u}$ and $v_{s,u}$. However, the entire time-variation of g has an impact on determining the hyperbolic trajectory at each instance in time.

Remark 2.8. A representation for $a_\varepsilon(t)$ in terms of an alternative orthogonal coordinate system $\left(v_s^\perp, v_s\right)$ can be obtained by simply interchanging the "s" and the "u" throughout in all the subscripts in (2.14).

Remark 2.9. An "elementary" method of obtaining an expression for the hyperbolic trajectory might be suggested by using the following idea, which was successfully pursued in a different context in [52, 53]. Since $h_\varepsilon(t)$ continues to possess stable and unstable manifolds, it is the only trajectory in a tubular neighborhood around $h(t)$ which continues to remain within such a tubular neighborhood for all t. Other trajectories will eventually be influenced by the stable manifold (in backward time) or the unstable manifold (in forward time) and be pulled out of such a tubular neighborhood. Thus, it is the only trajectory which remains $\mathcal{O}(\varepsilon)$-close to $h(t)$ for all $t\in\mathbb{R}$. Therefore, a formal expansion $h_\varepsilon(t)=h(t)+\varepsilon h^1(t)+\mathcal{O}\left(\varepsilon^2\right)$ may be substituted into (2.12) or, equivalently, an expansion $a_\varepsilon(t)=a+\varepsilon a^1(t)+\mathcal{O}\left(\varepsilon^2\right)$ into (2.6). The leading-order term then gives the linear differential equation

$$\dot{a}^1(t)=Df(a)a^1(t)+g(a,t)$$

for the unknown leading-order term $a^1(t)$. While a solution to this can be written down easily, the tricky issue is knowing the correct initial condition to apply. It turns out that the $\mathcal{O}(\varepsilon)$-term in (2.14) is consistent with this formal approach.

Example 2.10 (linear saddle). If $g=(g_1,g_2)$ satisfies Hypothesis 2.3 and $0<\varepsilon\ll 1$, consider

$$\frac{d}{dt}\begin{pmatrix} x_1 \\ x_2 \end{pmatrix}=\begin{pmatrix} x_1 \\ -x_2 \end{pmatrix}+\varepsilon\begin{pmatrix} g_1(x_1,x_2,t) \\ g_2(x_1,x_2,t) \end{pmatrix}. \tag{2.17}$$

When $\varepsilon=0$, the flow is of a linear saddle point at the origin, with

$$\lambda_s=-1,\, v_s=\begin{pmatrix} 0 \\ -1 \end{pmatrix},\, \lambda_u=1,\, v_u=\begin{pmatrix} 1 \\ 0 \end{pmatrix},$$

where the branches chosen of the unstable and stable manifolds are those lying along the boundaries of the first quadrant. Now, applying Theorem 2.7, the perturbed hyperbolic trajectory when $\varepsilon\neq 0$ is given by

$$a_\varepsilon(t)=\varepsilon\begin{pmatrix} -\int_0^\infty g_1(0,0,t+\tau)e^{-\tau}\,d\tau \\ \int_0^\infty g_2(0,0,t-\tau)e^{-\tau}\,d\tau \end{pmatrix}+\mathcal{O}\left(\varepsilon^2\right),$$

for which g's information is only required at the origin. For example, if taking $g_1 = \cos\omega t$ and $g_2 = \sin\omega t$ at $(0,0)$ for some frequency $\omega > 0$, the hyperbolic trajectory follows the temporal variation

$$a_\varepsilon(t) = \frac{\varepsilon}{1+\omega^2}\begin{pmatrix} \omega\sin(\omega t) - \cos(\omega t) \\ \sin(\omega t) - \omega\cos(\omega t) \end{pmatrix} + \mathcal{O}\left(\varepsilon^2\right)$$
$$= \frac{\varepsilon}{\sqrt{1+\omega^2}}\begin{pmatrix} \sin(\omega t - \phi) \\ -\cos(\omega t + \phi) \end{pmatrix} + \mathcal{O}\left(\varepsilon^2\right),$$

where $\phi = \tan^{-1}\omega^{-1}$. ∎

Example 2.11 (Duffing oscillator). Consider the Duffing oscillator in the form [203, 421, 265, 215, 416, 253]

$$\ddot{\theta} + \gamma\dot{\theta} - \theta + \theta^3 = \varepsilon\phi(t), \tag{2.18}$$

where $|\varepsilon| \ll 1$, $\phi(t)$ represents a general bounded forcing function and $\gamma \geq 0$ is a damping coefficient. Setting $\theta = x_1$ and $\dot{\theta} = x_2$, this can be written as

$$\frac{d}{dt}\begin{pmatrix} x_1 \\ x_2 \end{pmatrix} = \begin{pmatrix} x_2 \\ x_1 - x_1^3 - \gamma x_2 + \varepsilon\phi(t) \end{pmatrix}. \tag{2.19}$$

When $\gamma \equiv 0$ and $\varepsilon \equiv 0$, the origin is a saddle fixed point, whose eigensystem is given by

$$\lambda_s = -1,\ v_s = \frac{1}{\sqrt{2}}\begin{pmatrix} -1 \\ 1 \end{pmatrix},\ \lambda_u = 1,\ v_u = \frac{1}{\sqrt{2}}\begin{pmatrix} 1 \\ 1 \end{pmatrix}.$$

When γ continues to be zero but when $\varepsilon \neq 0$, the hyperbolic trajectory moves from the zero location to

$$\begin{pmatrix} x_1(t) \\ x_2(t) \end{pmatrix} = \frac{\varepsilon}{2}\begin{pmatrix} -\int_0^\infty [\phi(t-\tau) + \phi(t+\tau)]e^{-\tau}\,d\tau \\ \int_0^\infty [\phi(t-\tau) - \phi(t+\tau)]e^{-\tau}\,d\tau \end{pmatrix} + \mathcal{O}\left(\varepsilon^2\right) \tag{2.20}$$

by applying Theorem 2.7. So, for example, if $\phi(t) = \cos t$, then the above simplifies to give

$$x_1(t) = -\frac{\varepsilon}{2}\cos t + \mathcal{O}\left(\varepsilon^2\right),$$

indicating that in this instance of no damping but small forcing, the hyperbolic trajectory to leading-order is exactly in phase with the forcing. ∎

2.3 ▪ Unstable manifolds for non-area-preserving 2D flows

In Section 2.2, the location of the hyperbolic trajectory under the influence of the nonautonomous perturbation was stated. In this section, the modification to the unstable manifold Γ^u will be derived, with the analogous result for the stable manifold presented in the next section.

The presence of the term $\varepsilon g(x,t)$ in (2.6) causes the hyperbolic trajectory to move to a location $a_\varepsilon(t)$ from a, as quantified in Theorem 2.7. Now, given the conditions on g as stated in Hypothesis 2.3 and Theorem 2.6, the preservation of hyperbolicity of

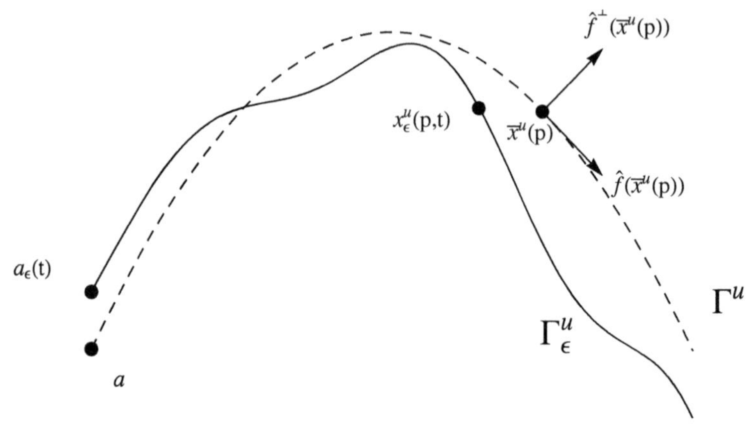

Figure 2.4. *The unstable manifold associated with (2.12) in a time-slice t.*

$a_\varepsilon(t)$ implies the persistence of stable and unstable manifolds. In this nonautonomous context, this is best thought of in terms of the augmented phase space of (2.12). The previously existing hyperbolic trajectory (a, t) as shown in Figure 2.2 perturbs to the hyperbolic trajectory $(a_\varepsilon(t), t)$, which is $\mathcal{O}(\varepsilon)$-close to (a, t) for $t \in \mathbb{R}$, which merely causes Figure 2.2 to "wobble." The stable and unstable manifolds, which are uniform in t in the unperturbed picture of Figure 2.2, acquire a t-dependence. The issue now is to characterize this dependence explicitly. Since the stable and unstable manifolds are two-dimensional entities in the augmented phase space, it helps to think of taking a t-slice in which the manifolds are to be characterized. It is such a t-slice which is shown in Figure 2.3, but this is a local picture near $a_\varepsilon(t)$. A picture which shows the local perturbation of one branch of the the unstable manifold in this general time-slice t is shown in Figure 2.4. The dashed curve Γ^u is the unperturbed ($\varepsilon = 0$) unstable manifold, which has now moved to a perturbed unstable manifold given by the solid curve Γ^s_ε. In determining the location of this in each time-slice t, it proves convenient to first introduce some notation related to the unperturbed unstable manifold.

When examining the unperturbed unstable manifold, its dynamical system (2.9) is autonomous. The fixed point a satisfies $f(a) = 0$, and its unstable manifold is the set of points which gets attracted towards a as $t \to -\infty$. Suppose a point $\bar{x}^u(0)$ were picked on the stable manifold as an initial condition to (2.9). This generates a trajectory—which shall be called $\bar{x}^u(t)$—such that $\bar{x}^u(t) \to a$ as $t \to -\infty$ (indeed, this decay is exponential with rate given by the eigenvalue λ_u). In this trajectory, if t is replaced by a parameter p, the quantity $\bar{x}^u(p)$ therefore gives a parametric representation of this branch of the unstable manifold. Thus, $p \in (-\infty, 0]$ gives the backward trajectory emanating from $\bar{x}^u(0)$, and this can be extended to $p \in [-\infty, 0]$ with the understanding that $\bar{x}^u(-\infty) = a$ since $\bar{x}(p) \to a$ as $p \to -\infty$. Moreover, there is no reason not to extend this such that $p \in [-\infty, P]$ for $P > 0$ since $p \in [0, P]$ gives the forwards trajectory—which also continues to remain on the stable manifold—through $\bar{x}^u(0)$. Now, P will be taken to be as large as needed but will *not* be taken to be ∞. The reason for avoiding this limit is to allow for the possibility of the unstable manifold having many different behaviors in this limit. For example, it may connect up with the stable manifold of another fixed point to form a heteroclinic trajectory, or indeed with a's own stable manifold to form a homoclinic trajectory. Alternatively, the unstable manifold may spiral towards a limit cycle, or escape to infinity. By keeping P finite, $\bar{x}^u(p)$ for $p \in [-\infty, P]$ will parametrize the unstable manifold curve up to some point, without having to worry about where the other end of the unstable manifold goes. Thus, this discussion enables the parametrization of the unperturbed unstable manifold according to

$$\Gamma^u := \{\bar{x}^u(p) \, : \, p \in [-\infty, P]\} \tag{2.21}$$

with P chosen to be as large as needed, but not ∞. This shall also be referred to as *clipping* the manifold; (2.21) is a *clipped unstable manifold.*

This parametric representation of a branch of the unstable manifold in the regular phase space can be of help in Figure 2.4, in which a time-slice t of the augmented phase space is shown. The unstable manifold shown here as the dashed curve is the same in every t-slice since (2.9) is autonomous. Thus, $\bar{x}^u(p)$ can be used to parametrize this dashed curve, with parameter $p \in [-\infty, P]$. A point $\bar{x}^u(p)$ (i.e., a particular choice of the parameter p) has been shown in Figure 2.4, and an orthogonal coordinate system is constructed using the unperturbed vector field f. Since the flow is in the direction of f in the unperturbed flow (2.9), the vector $f(\bar{x}^u(p))$ drawn at the point $\bar{x}^u(p)$ must be in the direction of flow along the unstable manifold. Thus, f must be pointing towards the right in the unperturbed unstable manifold Γ^u shown in Figure 2.4. The perpendicular vector $f^\perp$, corresponding to a rotation of $+\pi/2$, must therefore point upwards. The "hat" notation shall be used to denote a unit vector in the relevant direction, i.e.,

$$\hat{f} := \frac{f}{|f|} \quad \text{and} \quad \hat{f}^\perp := \frac{f^\perp}{|f|}, \tag{2.22}$$

enabling the unit vector representation as given in Figure 2.4. Another necessary definition is that of the *projected rate-of-strain*, which will be defined by

$$R^u(\xi) := \left(\hat{f}^\perp\right)^\top (\bar{x}^u(\xi)) \left[(Df)^\top (\bar{x}^u(\xi)) + (Df)(\bar{x}^u(\xi)) \right] \hat{f}(\bar{x}^u(\xi)), \tag{2.23}$$

in which the superscript $\top$ is the transpose.

The orthonormal system $(\hat{f}, \hat{f}^\perp)$ shall be used to represent the *perturbed* unstable manifold's location. This manifold, shown by the solid curve in Figure 2.4, culminates at the location $a_\varepsilon(t)$ in this time-slice. While in the unperturbed (autonomous) situation, the vector field f was tangential to the dashed curve, in the perturbed (nonautonomous) situation, the corresponding vector field $f + \varepsilon g$ need *not* be tangential to the solid curve. This is since the solid curve is itself changing with t; for example, the whole picture may be moving in a direction normal to the curve. What can be said is that if a trajectory $(y(t), t)$ intersects the solid curve in the time-slice t, then $|y(t) - a_\varepsilon(t)| \to 0$ as $t \to -\infty$, since the trajectory, lying on the unstable manifold, must approach the corresponding hyperbolic entity in backward time. This is true for all trajectories intersecting the solid curve in the time-slice t. One such point, $x_\varepsilon^u(p, t)$, has been shown in Figure 2.4. One characterization of this point is that it is $\mathcal{O}(\varepsilon)$-close to $\bar{x}^u(p)$ in this t time-slice. This by itself does not uniquely identify $x_\varepsilon^u(p, t)$; the subtleties of choosing this point are elucidated in the proofs of the statements to follow. As long as it is done consistently, $x_\varepsilon^u(p, t)$ can be used as a parametric representation of

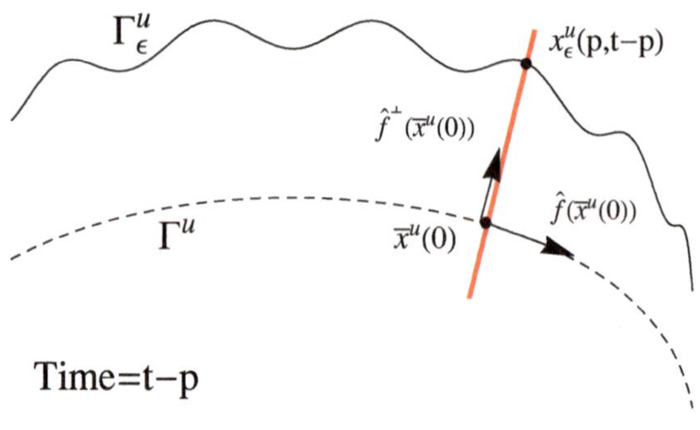

Figure 2.5. *Choice of parametrization for unstable manifold in time-slice $t-p$, reflecting the parametrization condition (2.24).*

the perturbed unstable manifold, in terms of the parameters (p,t). The unique choice of $x_\varepsilon^u(p,t)$ satisfies the conditions:

(a) The first argument of x_ε^u will identify a trajectory on the unstable manifold, and the second argument its temporal variation.

(b) The temporal argument $t \in (-\infty, T_u]$ for some chosen $T_u > 0$.

(c) For each time choice $t \in (-\infty, T_u]$, the p-parametrization for $x_\varepsilon^u(p,t)$ will be chosen such that

$$\hat{f}(\bar{x}^u(0)) \cdot [x_\varepsilon^u(p,t-p) - \bar{x}(0)] = 0 \tag{2.24}$$

for all $p \in (-\infty, P]$.

This final condition is merely a choice of parametrization of the trajectories on the unstable manifold and guarantees uniqueness. To understand why, suppose a $t \in (-\infty, T_u]$ is chosen. Then, for each p, consider a time-slice at $t-p$, as shown in Figure 2.5, and consider the point $\bar{x}^u(0)$ lying on the unperturbed unstable manifold Γ^u. Now, knowing that the perturbed unstable manifold Γ_ε^u is $\mathcal{O}(\varepsilon)$-close to Γ^u, consider drawing a normal vector (in the direction $\hat{f}^\perp(\bar{x}^u(0))$) from $\bar{x}^u(0)$. This will meet Γ_ε^u at one point within the $\mathcal{O}(\varepsilon)$-close neighborhood, and this intersection point corresponds to a trajectory of (2.12) which lies on the unstable manifold. Label this point $x_\varepsilon^u(p,t-p)$, indicating from the second argument that this is in the time-slice $t-p$ and from the first argument that this is the trajectory labeled by p. For example, the trajectory labeled by $p=0$ is that which is given by $x_\varepsilon^u(0,t)$; that is, in the time-slice t it lies in the normal direction from the point $\bar{x}^u(0)$. The trajectory $p=1$ is that given by $x_\varepsilon^u(1,t-1)$, which is characterized by lying along the normal at $\bar{x}^u(0)$ in the time-slice $t-1$. Since every trajectory on the unstable manifold will eventually cross the normal vector at $\bar{x}^u(0)$ in *some* time-slice, this procedure can be used to label all of the trajectories on Γ_ε^u. This parametrization may seen somewhat unusual but enables a nice characterization of Γ_ε^u as follows.

Theorem 2.12 (unstable manifold location [32]). *Consider (2.6) under Hypotheses 2.2, 2.3, and 2.5. The perturbed unstable manifold of the hyperbolic trajectory $a_\varepsilon(t)$*

of (2.6) has a parametric representation with parameters $(p,t) \in (-\infty, P] \times (-\infty, T_u]$ for arbitrarily large but fixed P and T_u, given by

$$x_\varepsilon^u(p,t) = \bar{x}^u(p) + \varepsilon \left[\frac{M^u(p,t)}{|f(\bar{x}^u(p))|} \hat{f}^\perp(\bar{x}^u(p)) + \frac{B^u(p,t)}{|f(\bar{x}^u(p))|} \hat{f}(\bar{x}^u(p)) \right] + \mathcal{O}(\varepsilon^2), \quad (2.25)$$

in which the unstable Melnikov function is

$$M^u(p,t) := \int_{-\infty}^{p} \exp\left[\int_\tau^p \mathrm{Tr}[Df(\bar{x}^u(\xi))]\, \mathrm{d}\xi \right] f^\perp(\bar{x}^u(\tau)) \cdot g(\bar{x}^u(\tau), \tau + t - p)\, \mathrm{d}\tau \quad (2.26)$$

and

$$B^u(p,t) := |f(\bar{x}^u(p))|^2 \int_0^p \frac{R^u(\tau) M^u(p, \tau+t-p) + f(\bar{x}^u(\tau)) \cdot g(\bar{x}^u(\tau), \tau+t-p)}{|f(\bar{x}^u(\tau))|^2} \mathrm{d}\tau. \quad (2.27)$$

Proof. See Section 2.3.1. □

Remark 2.13. Here, the $M^u(p,t)$ is a *Melnikov function* which usually [183, 17, 421] characterizes distances between stable and unstable manifolds. Unlike in that situation in which the integration is over $(-\infty, \infty)$, locating just the unstable manifold requires an integration over $(-\infty, p)$ in (2.26). As is clear from the proof in Section 2.3.1, the unstable Melnikov function is obtained by using the definition

$$M^u(p,\tau) := \lim_{\varepsilon \to 0} f(\bar{x}^u(\tau - t + p))^\perp \cdot \left[\frac{x_\varepsilon^u(p,\tau) - \bar{x}^u(p)}{\varepsilon} \right], \quad (2.28)$$

at fixed t, where the issue of (2.28) being well-defined is tackled in the proof. Thus, $M^u(p,t)$ will by definition give a projection in the $f^\perp$ direction at the point $\bar{x}^u(p)$. The point of Theorem 2.12 is that the quantity (2.28) can indeed be written in terms of *unperturbed* quantities in the form (2.26).

Remark 2.14. Theorem 2.12 characterizes the movement of the unstable manifold in the *normal* direction (quantified by M^u), as well as in the *tangential* direction (quantified by B^u). The necessity for the tangential correction was only recently recognized in [32, Theorem 2.2]. Theorem 2.12 is a combination of Theorems 2.1 and 2.2 from [32].

Remark 2.15 (connection to Theorem 2.1). Note that the quantity $f^\perp \cdot g$ in the integrand of the Melnikov function is the same as $f \wedge g$ in terms of wedge products [183, 421] as also given in Theorem 2.1. The change of integration variable $\tau + t - p \to \tau$ in (2.26) gives the alternative representation

$$M^u(p,t) = \int_{-\infty}^{t} \exp\left[\int_\tau^t \mathrm{Tr}[Df(\bar{x}^u(\xi - t + p))]\, \mathrm{d}\xi \right] f^\perp(\bar{x}^u(\tau - t + p)) \cdot g(\bar{x}^u(\tau - t + p), \tau)\, \mathrm{d}\tau; \quad (2.29)$$

using "equivalent under change of variable" forms of Melnikov functions such as (2.26) and (2.29) will turn out to be useful in other sections of this monograph as well. If

$p = 0$, i.e., a particular choice of location is made for measuring the signed distance, note that the argument of $f^{\perp} \cdot g$ is now expressed in exactly the same way as that of $f \wedge g$ in Theorem 2.1. The integration domain in (2.29) is $(-\infty, t)$ as opposed to $(-\infty, \infty)$ in (2.4); this is since the *location* of the unstable manifold is characterized rather than its displacement relative to the stable manifold. The extra exponential term in the integrand is discussed in the next remark.

Remark 2.16 (incompressibility). If f is area-preserving (i.e., divergence-free, or incompressible), then $\mathrm{Tr} Df = 0$ in the Melnikov integrand (2.26). This converts the integrand of the Melnikov function into the standard forms that one observes [183, 17, 421]. Under this condition, in terms of rectangular coordinates (x_1, x_2) in the plane, one can express

$$f = \left(\frac{\partial \psi_0}{\partial x_2}, -\frac{\partial \psi_0}{\partial x_1} \right) \tag{2.30}$$

for a streamfunction (or Hamiltonian) $\psi_0(x_1, x_2)$. If the perturbation g is itself incompressible, it too is associated with a streamfunction $\psi_1(x_1, x_2, t)$ such that

$$g = \left(\frac{\partial \psi_1}{\partial x_2}, -\frac{\partial \psi_1}{\partial x_1} \right). \tag{2.31}$$

(Some conventions choose the streamfunctions such that *negative* the above expressions are used for both f and g as will indeed be addressed in Section 3.8; the following then would need correction by a negative sign.) Then, the integrand $f^{\perp} \cdot g$ of the Melnikov integral (2.26) can also be written as

$$\{\psi_0, \psi_1\} := \frac{\partial \psi_0}{\partial x_1} \frac{\partial \psi_1}{\partial x_2} - \frac{\partial \psi_0}{\partial x_2} \frac{\partial \psi_1}{\partial x_1} =: J(\psi_0, \psi_1)$$

as is commonly seen in the literature [421, 264, 225, 417]; the two representations given above are the Poisson bracket (left) and Jacobian (right) notations.

Remark 2.17 (primary unstable manifold). The expression (2.25) does not give a parametric representation of the full, global unstable manifold of $a_\varepsilon(t)$. Instead, it characterizes what shall be called the *primary unstable manifold*; this is the segment of the unstable manifold which remains $\mathcal{O}(\varepsilon)$-close to the unperturbed steady unstable manifold Γ^u. Once the manifold's spatial extend goes beyond the clipped unstable manifold Γ^u as represented in (2.21)—that is, it has extended beyond $\bar{x}^u(P)$—Theorem 2.12 is no longer able to assess its location.

Remark 2.18. The arclength of the unstable manifold that can be determined from (2.25) diminishes as t decreases from T_u. This is because the particular unperturbed trajectory near the location $\bar{x}^u(P)$ in the time-slice T_u would have traveled to near the location $\bar{x}^u(P - T_u + t)$ in a time-slice $t < T_u$ since it "shadows" the unperturbed trajectory. But this point is closer to a (along the unperturbed unstable manifold) than the point $\bar{x}^u(P)$; the relevant arclength of the unstable manifold representible therefore shrinks as $t \to -\infty$.

Remark 2.19. It can be shown that $B^u(p, t) \to 0$ as $p \to -\infty$ [32], indicating that the tangential correction becomes less significant as the hyperbolic trajectory is approached within any time-slice.

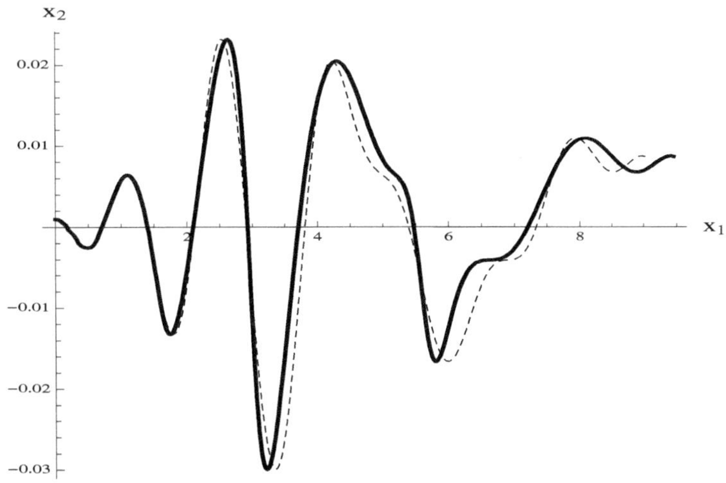

Figure 2.6. *The unstable manifold for the linear saddle as described in Example 2.20 in the time-slice $t = 0$ with $\varepsilon = 0.1$. (The dashed curve is that obtained by neglecting the tangential term B^u.)*

Example 2.20 (linear saddle cont.). In the linear saddle (2.17) with $\varepsilon = 0$, consider the unstable manifold branch emanating in the positive x_1 direction. This is given parametrically by $\bar{x}^u(p) = (e^p, 0)$, and for this situation many simplifications occur:

$$f(\bar{x}^u(p)) = \begin{pmatrix} e^p \\ 0 \end{pmatrix}, \hat{f} = \begin{pmatrix} 1 \\ 0 \end{pmatrix}, \hat{f}^\perp = \begin{pmatrix} 0 \\ 1 \end{pmatrix}, \operatorname{Tr} Df = 0, R^u = 0.$$

Using (2.26) and (2.27), the expressions

$$M^u(p,t) = \int_{-\infty}^{p} g_2(e^\tau, 0, \tau + t - p) e^\tau \, d\tau,$$

$$B^u(p,t) = e^{2p} \int_0^p g_1(e^\tau, 0, \tau + t - p) e^{-\tau} \, d\tau,$$

are obtainable. Thus, by (2.25) the perturbed unstable manifold is parametrized by

$$x_\varepsilon^u(p,t) = \begin{pmatrix} e^p \\ 0 \end{pmatrix} + \varepsilon \begin{pmatrix} e^p \int_0^p g_1(e^\tau, 0, \tau + t - p) e^{-\tau} \, d\tau \\ e^{-p} \int_{-\infty}^p g_2(e^\tau, 0, \tau + t - p) e^\tau \, d\tau \end{pmatrix} + \mathcal{O}(\varepsilon^2).$$

In Figure 2.6, the unstable manifold computed from the above is shown for the choice $g = (5\cos(5x_2)\sin(10t)\tanh t, \cos(5x_1)\cos(10t))$ and $\varepsilon = 0.1$, in the time-slice $t = 0$. The tangential component becomes more important for larger p. ■

Example 2.21 (Duffing oscillator cont.). Return to the Duffing system (2.19) and consider as the unperturbed system that with no forcing ($\phi(t) = 0$) or damping ($\gamma = 0$). This possesses a phase portrait as given in Figure 2.7, which is easily obtainable by sketching level curves of the conserved quantity $H(x_1, x_1) = x_2^2/2 - x_1^2/2 + x_1^4/4$,

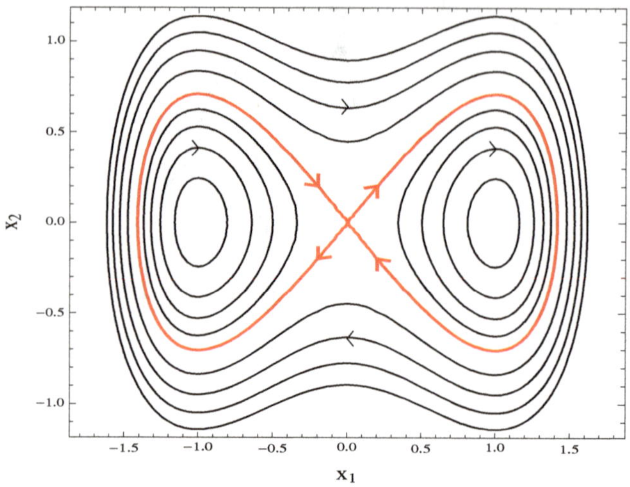

Figure 2.7. *Phase portrait of the Duffing oscillator (2.19) with no damping or forcing.*

which serves as a Hamiltonian for the unperturbed vector field $f = \left(x_2, x_1 - x_1^3\right)$. The red curves consist of when the stable and unstable manifolds of the origin coincide. Indeed, these can be numerically determined by using FTLEs, and in fact approximations to the stable and unstable manifold appear as ridges in Figures 1.5(e) and (f), respectively (these pictures are for a *moving* version of the Duffing equation, but at $t = 0$ its manifolds are located exactly as shown in Figure 2.7). Now, the unstable manifold emanating from the origin to the first quadrant can be written as a solution to the unperturbed version of (2.19) as

$$\bar{x}^u(t) = \begin{pmatrix} \bar{x}_1(t) \\ \bar{x}_2(t) \end{pmatrix} = \begin{pmatrix} \sqrt{2}\,\mathrm{sech}\, t \\ -\sqrt{2}\,\mathrm{sech}\, t \tanh t \end{pmatrix} \tag{2.32}$$

for $t \in [-\infty, T_u]$ for T_u as large as needed. This corresponds to choosing the point on the unstable manifold which is on the positive x_1-axis as the initial condition; that is, $\bar{x}^u(0) = \left(\sqrt{2}, 0\right)$. A different initial condition would result only in a shift in the argument in (2.32), which has no impact on the final answer, as it is associated with a parametrization which is a simple shift of the one that is considered here. In this particular case, this unstable manifold is also the stable manifold of the same fixed point at the origin, forming a homoclinic trajectory. The goal is to determine a parametric representation of the perturbed unstable manifold under the inclusion of both forcing and damping.

First, suppose that $\phi(t)$ continues to be zero but that $0 < \gamma \ll 1$. Thus, the damping constant γ plays the role of the small parameter ε, and moreover $g = (0, -x_2)$. Here, the unperturbed system is Hamiltonian, $\mathrm{Tr}\, Df = 0$, and here $f^\perp \cdot g = -x_2^2$, and so

$$M^u(p,t) = -\int_{-\infty}^{p} \left[\bar{x}_2(\tau)\right]^2 \mathrm{d}\tau = -2\int_{-\infty}^{p} \mathrm{sech}^2\tau \tanh^2\tau\, \mathrm{d}\tau = -\frac{2}{3}\left(1 + \tanh^3 p\right).$$

The lack of t-dependence in M^u (and also in B^u) is unsurprising since the perturbed system is itself autonomous. The negativity of M^u for all p implies that the unstable

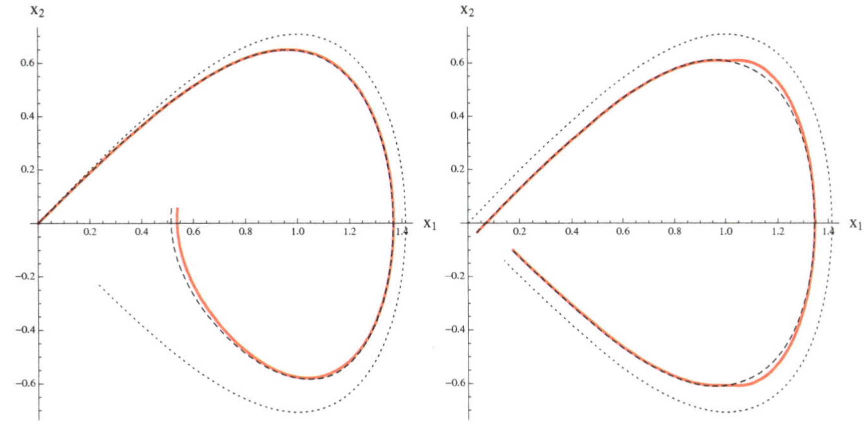

(a) $\gamma = 0.1$, $\phi(t) \equiv 0$ (b) $\gamma = 0$, $\phi(t) = \tanh t$, $\varepsilon = 0.07$

Figure 2.8. *Unstable manifold [solid red] of the Duffing system computed using Theorem 2.12 in the time-slice $t = -3$, for two different choices of parameters. The dotted curve is the unperturbed manifold, and the dashed curve is computed by ignoring the tangential component B^u in (2.25).*

manifold moves in the direction of $-\hat{f}^{\perp}$, i.e., inwards in Figure 2.7. This is consistent with the standard Lyapunov function analysis (i.e., decay of energy) of the damped Duffing oscillator. To find the tangential movement using (2.27), additional required quantities are $f \cdot g$, $|f|$ and R^u, which turn out to give messy expressions such as (2.27) requiring numerical evaluation [32]. In Figure 2.8(a), the unstable manifold computed using the leading-order expression (2.25) is shown by the solid curve, with the dotted curve being the unperturbed ($\gamma = 0$) unstable manifold. Also shown by the dashed curve—very close to the solid curve—is the unstable manifold computed using (2.25) but with the tangential component ignored.

Next, suppose there is a forcing but no damping. The standard situation of sinusoidal ϕ is examined in detail in [32], but here a more general time-dependence will be addressed: $\phi(t) = \tanh t$. Note that this is a legitimate choice by Hypothesis 2.3, since the perturbation as expressed in (2.19) remains bounded. In this case, there is genuine t-dependence in both M^u and B^u, and thus the unstable manifold needs to be shown in a specific t-slice. Numerically, evaluation is necessary in this case, and the resulting picture in the time-slice $t = -3$ is shown in Figure 2.8(b) for $\varepsilon = 0.07$. In both situations in Figure 2.8, ignoring the tangential component leads to a small error in the observed unstable manifold. Much more dramatic errors will occur in neglecting B^u in situations in which $f \cdot g$ is large but $f^{\perp} \cdot g$ is small [41, e.g.]; such choices could not be made for the Duffing system (2.19), in which the perturbations appear in a prescribed manner due to their physical meaning. ∎

Remark 2.22 (obtaining wave profiles). Information from both the tangential and normal components of the perturbed stable and unstable manifolds turns out to have an interesting application in the determination of wave profiles in traveling/stationary waves which have a "soliton-like" structure. For an example in the usage of Theorem 2.12 in determining wave profiles in in the Korteweg-de Vries equation subject to spatially dependent forcing, the reader is referred to [41, 71]. The method relies also on determining intersections of the stable and unstable manifolds of a perturbed heteroclinic connection (to be addressed in Section 3.3), which provides an extra condition for the *existence* of such traveling/stationary waves in perturbed diffusive partial differential equations, and which has been addressed in combustion [46, 55] and ecology [45, 30]. Applying Theorem 2.12 would be the next step in determining the wave profiles.

2.3.1 ▪ Proof of Theorem 2.12 (unstable manifold location)

The proof here mirrors that given in [32], with some slight modifications. It possesses elements of the Melnikov function derivation by Guckenheimer and Holmes [183], and particularly by Holmes [200] in which the divergence-free nature of f is relaxed. However, rather than doing a formal ε-expansion, an approach which rigorously establishes that the error terms are $\mathcal{O}(\varepsilon^2)$ in t is followed. The presence of a nonzero $\operatorname{Tr} Df$ term leads also to a subtlety in discarding a boundary term (cf. the argument in going from (20) to (21) in Holmes [200]), since this term may diverge.

The idea is to think of p as fixed, and consider $x_\varepsilon^u(p,\cdot)$ as being a solution to the system (2.6) which lies on the unstable manifold. Define x_1^ε through

$$x_\varepsilon^u(p,\tau)=\bar{x}^u(\tau-t+p)+\varepsilon\, x_1^u(p,\tau,\varepsilon), \tag{2.33}$$

in which τ is used as the temporal variable since the time-slice t is fixed. In particular, when $\tau=t$, this implies

$$x_\varepsilon^u(p,t)=\bar{x}^u(p)+\varepsilon\, x_1^u(p,t,\varepsilon),$$

i.e., $x_\varepsilon^u(p,t)$ is a point which is "close" to $\bar{x}^u(p)$. Now, since $\bar{x}^u(\tau-t+p)$ approaches a, and similarly $x_\varepsilon^u(p,\tau)$ approaches $a_\varepsilon(\tau)$ as $\tau\to-\infty$ by virtue of being on the unstable manifold of the relevant hyperbolic trajectory, $x_\varepsilon^u(p,\tau)$ and $\bar{x}^u(\tau-t+p)$ remain $\mathcal{O}(\varepsilon)$-close for $\tau\in(-\infty,T_u]$, and thus

$$|x_1^u(p,\tau,\varepsilon)|+\left|\frac{\partial x_1^u(p,\tau,\varepsilon)}{\partial\tau}\right|\le K \quad \text{for } \tau\in[-\infty,T_u], \tag{2.34}$$

for a K independent of both τ and $\varepsilon\in[0,\varepsilon_0]$ for some ε_0. The τ-derivative bound included above is automatic by taking the derivative of (2.33), in view of the fact that the velocity fields f and $f+\varepsilon g$ are uniformly $\mathcal{O}(\varepsilon)$-close. Now defining the preliminary unstable Melnikov function by

$$M_\varepsilon^u(p,\tau):=f\left(\bar{x}^u(\tau-t+p)\right)^\perp\cdot x_1^u(p,\tau,\varepsilon) \tag{2.35}$$

leads to the observation that

$$\hat{f}^\perp(\bar{x}^u(p))\cdot\left[x_\varepsilon^u(p,t)-\bar{x}^u(p)\right]=\varepsilon\,\frac{f\left(\bar{x}^u(p)\right)^\perp}{\left|f\left(\bar{x}^u(p)\right)\right|}\cdot x_1^u(p,t,\varepsilon)=\varepsilon\,\frac{M_\varepsilon^u(p,t)}{\left|f\left(\bar{x}^u(p)\right)\right|}, \tag{2.36}$$

and hence $M_\varepsilon^u(p,t)$ carries information on the movement of the manifold in the direction perpendicular to $f(\bar{x}^u(p))$ in the time-slice t. Taking the derivative of $M_\varepsilon^u(p,\tau)$

in (2.35) with respect to τ (at fixed t and p) yields

$$\begin{aligned}\frac{\partial M_\varepsilon^u}{\partial \tau} &= \left[Df\left(\bar{x}^u(\tau-t+p)\right) \frac{\partial \bar{x}^u(\tau-t+p)}{\partial \tau}\right]^\perp \cdot x_1^u(p,\tau,\varepsilon) \\ &\quad + f\left(\bar{x}^u(\tau-t+p)\right)^\perp \cdot \frac{\partial x_1^u(p,\tau,\varepsilon)}{\partial \tau} \\ &= \left[Df\left(\bar{x}^u(\tau-t+p)\right) f\left(\bar{x}^u(\tau-t+p)\right)\right]^\perp \cdot x_1^u(p,\tau,\varepsilon) \\ &\quad + f\left(\bar{x}^u(\tau-t+p)\right)^\perp \cdot \frac{\left[f\left(x_\varepsilon^u(p,\tau)\right) + \varepsilon\, g\left(x_\varepsilon^u(p,\tau),\tau\right) - f\left(\bar{x}^u(\tau-t+p)\right)\right]}{\varepsilon}\end{aligned} \tag{2.37}$$

since $\bar{x}^u(\tau-t+p)$ solves the unperturbed flow (2.9) and $x_\varepsilon^u(p,\tau)$ solves the perturbed flow (2.6). Applying Taylor's theorem remainder terms to expansions of both f and g around $\bar{x}^u(\tau-t+p)$, this can be rewritten as

$$\begin{aligned}\frac{\partial M_\varepsilon^u}{\partial \tau} &= \left[Df\left(\bar{x}^u(\tau-t+p)\right) f\left(\bar{x}^u(\tau-t+p)\right)\right]^\perp \cdot x_1^u(p,\tau,\varepsilon) \\ &+ f\left(\bar{x}^u(\tau-t+p)\right)^\perp \cdot \left[Df\left(\bar{x}^u(\tau-t+p)\right) x_1^u(p,\tau,\varepsilon)\right] \\ &+ f\left(\bar{x}^u(\tau-t+p)\right)^\perp \cdot g\left(\bar{x}^u(\tau-t+p),\tau\right) \\ &+ \varepsilon\, f\left(\bar{x}^u(\tau-t+p)\right)^\perp \cdot \left\{ \left[\frac{1}{2} x_1^u(p,\tau,\varepsilon)^\top D^2 f(y_1) + Dg(y_2,\tau)\right] x_1^u(p,\tau,\varepsilon)\right\},\end{aligned} \tag{2.38}$$

where y_1 and y_2 are points located within εK of $\bar{x}^u(\tau-t+p)$. Now, if A is a 2×2 matrix and b and c are 2×1 vectors, it is easy to verify that

$$(A\,b)^\perp \cdot c + b^\perp \cdot (A\,c) = (\operatorname{Tr} A)\left(b^\perp \cdot c\right). \tag{2.39}$$

Choosing $A = Df\left(\bar{x}^u(\tau-t+p)\right)$, $b = f\left(\bar{x}^u(\tau-t+p)\right)$, and $c = x_1^u(p,\tau,\varepsilon)$ enables (2.38) to be written as

$$\begin{aligned}\frac{\partial M_\varepsilon^u}{\partial \tau} &= \operatorname{Tr} Df\left(\bar{x}^u(\tau-t+p)\right) M_\varepsilon^u + f\left(\bar{x}^u(\tau-t+p)\right)^\perp \cdot g\left(\bar{x}^u(\tau-t+p),\tau\right) \\ &+ \varepsilon\, f\left(\bar{x}^u(\tau-t+p)\right)^\perp \cdot \left\{ \left[\frac{1}{2} x_1^u(p,\tau,\varepsilon)^\top D^2 f(y_1) + Dg(y_2,\tau)\right] x_1^u(p,\tau,\varepsilon)\right\}.\end{aligned} \tag{2.40}$$

Now suppose that $M^u(p,\tau)$ is given by

$$M^u(p,\tau) := \lim_{\varepsilon\to 0} M_\varepsilon^u(p,\tau),$$

which is equivalent to the definition given in (2.28). This limit is well-defined because of (2.34). One can take the τ-derivative of this by interchanging the derivative with the ε-limit, again using (2.34) to justify the process. If doing so, the equation that $M^u(p,\tau)$ satisfies would be obtained by simply taking the $\varepsilon \to 0$ limit of (2.40), which is then

$$\frac{\partial M^u}{\partial \tau} = \operatorname{Tr} Df\left(\bar{x}^u(\tau-t+p)\right) M^u + f\left(\bar{x}^u(\tau-t+p)\right)^\perp \cdot g\left(\bar{x}^u(\tau-t+p),\tau\right). \tag{2.41}$$

The solution of this linear equation for M^u involves using the integrating factor

$$\mu(\tau) := \exp\left[-\int_0^\tau \operatorname{Tr} Df\left(\bar{x}^u(\xi-t+p)\right) \mathrm{d}\xi \right],$$

after which one obtains

$$\frac{\partial}{\partial \tau}\left[\mu(\tau) M^u(p,\tau)\right] = \mu(\tau) f\left(\bar{x}^u(\tau-t+p)\right)^{\perp} \cdot g\left(\bar{x}^u(\tau-t+p),\tau\right). \tag{2.42}$$

Integration of (2.42) from L to t leads to

$$\mu(t)M^u(p,t)-\mu(L)M^u(p,L)=\int_L^t \mu(\tau) f\left(\bar{x}^u(\tau-t+p)\right)^{\perp} \cdot g\left(\bar{x}^u(\tau-t+p),\tau\right) \mathrm{d}\tau. \tag{2.43}$$

Taking the limit $L \to -\infty$ needs some care, since the integrating factor μ is generically unbounded. However, note using (2.28) that

$$\begin{aligned} &\left|\mu(L)M^u(p,L)\right| \\ &= \left| f\left(\bar{x}^u(L-t+p)\right)^{\perp} \cdot \lim_{\varepsilon\to 0} x_1^u(p,L,\varepsilon)\right| \exp\left[\int_L^0 \operatorname{Tr} Df\left(\bar{x}^u(\xi-t+p)\right)\mathrm{d}\xi\right]. \end{aligned}$$

As $L \to -\infty$, $f\left(\bar{x}^u(L-t+p)\right)$ converges to zero with exponential rate λ_u, since this is an approach to a along the unstable manifold. The boundedness of x_1^u expressed in (2.34) enables the first term above to be bounded by $K_1 e^{\lambda_u L}$, where K_1 is a constant. On the other hand, the integrand of μ approaches $\lambda_u + \lambda_s$ as $L \to -\infty$, since this is the sum of the eigenvalues at the limiting point a. Thus, μ is bounded by $K_2 e^{-(\lambda_u+\lambda_s)L}$. Hence

$$\lim_{L\to-\infty} \left|\mu(L)M^u(p,L)\right| \le \lim_{L\to-\infty} K_1K_2 e^{-\lambda_s L} = 0$$

since $\lambda_s < 0$. Thus, the term involving L on the left-hand side of (2.43) vanishes in this limit. On the other hand, L appears as a limit of the integral on the right-hand side of (2.43), and the examining the convergence of the associated improper integral is necessary. The term $\left|f^{\perp}\cdot g\right|$ is bounded by $K_3 e^{\lambda_u \tau}$ since f has exponential decay to zero and g is bounded, and μ is bounded by $K_2 e^{-(\lambda_u+\lambda_s)\tau}$. Thus, the integrand is bounded by $K_3K_2 e^{-\lambda_s \tau}$, which is integrable over $(-\infty, 0)$. Taking the limit $L \to -\infty$ in (2.43) yields

$$M^u(p,t)=\int_{-\infty}^t \exp\left[\int_\tau^t \operatorname{Tr} Df\left(\bar{x}^u(\xi-t+p)\right)\mathrm{d}\xi\right] f\left(\bar{x}^u(\tau-t+p)\right)^{\perp}\cdot g\left(\bar{x}^u(\tau-t+p),\tau\right)\mathrm{d}\tau, \tag{2.44}$$

in which the improper integral converges. By utilizing the change of variable $\tau - t + p \to \tau$ above, (2.26) is derived. (It must be mentioned that the derivation in Holmes [200] for the distance between stable and unstable manifolds in non-divergence-free flows closely anticipates the above formula, but the details needed to neglect the boundary term were first developed in [32].)

The next task is to show that $M_\varepsilon^u(p,t)$ is within $\mathcal{O}(\varepsilon)$ of $M^u(p,t)$, which is not immediate since integration of (2.40) and (2.41) over the noncompact domain $(-\infty, t)$ is needed for their determination. Subtracting (2.41) from (2.40) gives

$$\frac{\partial}{\partial \tau}\left[M_\varepsilon^u - M^u\right] = \varepsilon f^{\perp} \cdot \left\{\left[\frac{1}{2} x_1^u(p,\tau,\varepsilon)^{\top} D^2 f(y_1) + Dg(y_2,\tau)\right] x_1^u\right\}, \tag{2.45}$$

where (when omitted) the spatial argument is $\bar{x}^u(\tau-t+p)$ and the temporal argument is τ. When considering $\tau \in (-\infty, t)$, $x_1^u(p,\tau,\varepsilon)$ remains bounded by K. The tensor

$D^2 f(y_1)$ also remains bounded by Hypothesis 2.2. A similar argument based on Hypothesis 2.3 works for the first argument of $Dg(y_2,\tau)$, and the second argument offers no difficulty since it is assumed in Hypothesis 2.3 that Dg is bounded in time as well. The prefactor term $f(\bar{x}^u(\tau-t+p))$ has a stronger bound because of the exponential decay with rate λ_u. Thus, there exists a constant K_4 such that the right-hand side of (2.45) is bounded by a term $\varepsilon K_4 e^{\lambda_u \tau}$. Hence

$$-\varepsilon K_4 e^{\lambda_u \tau} \leq \frac{\partial}{\partial \tau}\left[M_\varepsilon^u(p,\tau,\varepsilon) - M^u(p,\tau)\right] \leq \varepsilon K_4 e^{\lambda_u \tau}.$$

Now, (2.28) indicates that $M^u \to 0$ as $\tau \to -\infty$—a condition which, from (2.35), is also shared by M_ε^u—and the integration of the above from $-\infty$ to t enables the bound

$$|M_\varepsilon^u(p,t) - M^u(p,t)| \leq \varepsilon \frac{K_4}{\lambda_u} e^{\lambda_u t}, \tag{2.46}$$

and thus $M_\varepsilon^u(p,t) = M^u(p,t) + \mathcal{O}(\varepsilon)$. From (2.35) this means that

$$\begin{aligned} M^u(p,t) &= f(\bar{x}^u(p))^\perp \cdot x_1^u(p,t,\varepsilon) + \mathcal{O}(\varepsilon) \\ &= \frac{1}{\varepsilon} f(\bar{x}^u(p))^\perp \cdot \left[x_\varepsilon^u(p,t) - \bar{x}^u(p)\right] + \mathcal{O}(\varepsilon), \end{aligned}$$

and therefore,

$$\hat{f}^\perp(\bar{x}^u(p)) \cdot \left[x_\varepsilon^u(p,t) - \bar{x}^u(p)\right] = \varepsilon \frac{M^u(p,t)}{|f(\bar{x}^u(p))|} + \mathcal{O}\left(\varepsilon^2\right), \tag{2.47}$$

justifying the *normal* component in the unstable manifold expression (2.25).

To obtain the *tangential* component in (2.25), consider the analogous version of (2.35) defined by

$$B_\varepsilon^u(p,\tau) := f(\bar{x}^u(\tau-t+p)) \cdot x_1^u(p,\tau,\varepsilon) = f^\top(\bar{x}^u(\tau-t+p))\, x_1^u(p,\tau,\varepsilon). \tag{2.48}$$

Putting $\tau = t - p$ into the above and comparing with the parametrization condition (2.24) yields the condition

$$B_\varepsilon^u(p,t-p) = 0 \tag{2.49}$$

for all $p \in (-\infty, P]$, that is, for any trajectory on the unstable manifold. Also note that

$$\hat{f}(\bar{x}^u(p)) \cdot \left[x_\varepsilon^u(p,t) - \bar{x}^u(p)\right] = \varepsilon \frac{B_\varepsilon^u(p,t)}{|f(\bar{x}^u(p))|},$$

implying that B^u characterizes the tangential displacement of the unstable manifold.

To determine B_ε^u, the τ-derivative of $B_\varepsilon^u(p,\tau)$ in (2.48) is taken at fixed t and p. Utilizing the expression for the derivative of x_1^u as used in the derivation of (2.38), one obtains

$$\begin{aligned} \frac{\partial B_\varepsilon^u}{\partial \tau} &= f^\top(\bar{x}^u(\tau-t+p))\left[Df(\bar{x}^u(\tau-t+p))\, x_1^u(p,\tau,\varepsilon) + g(\bar{x}^u(\tau-t+p),\tau)\right] \\ &\quad + \frac{\partial f^\top(\bar{x}^u(\tau-t+p))}{\partial \tau} x_1^u(p,\tau,\varepsilon) \\ &\quad + \varepsilon f(\bar{x}^u(\tau-t+p))^\top \left[\frac{1}{2} x_1^u(p,\tau,\varepsilon)^\top D^2 f(y_1) + Dg(y_2,\tau)\right] x_1^u(p,\tau,\varepsilon), \end{aligned} \tag{2.50}$$

where $y_{1,2}$ are as described in (2.38). Now note that

$$\frac{\partial f\left(\bar{x}^u(\tau-t+p)\right)}{\partial \tau} = Df\left(\bar{x}^u(\tau-t+p)\right)\frac{\partial \bar{x}^u(\tau-t+p)}{\partial \tau}$$
$$= Df\left(\bar{x}^u(\tau-t+p)\right) f\left(\bar{x}^u(\tau-t+p)\right);$$

i.e., f satisfies the equation of variations associated with the trajectory $\bar{x}^u(\tau-t+p)$ of (2.9). Therefore,

$$\frac{\partial}{\partial \tau} f = (Df)\, f \quad \text{and} \quad \frac{\partial}{\partial \tau} f^\top = f^\top (Df)^\top \tag{2.51}$$

with each quantity being evaluated at $\bar{x}^u(\tau-t+p)$. Substituting these into (2.50),

$$\frac{\partial}{\partial \tau}\left[f^\top x_1^u\right] = f^\top\left[(Df)^\top + Df\right]x_1^u + f^\top g + \varepsilon f^\top\left[\frac{1}{2}(x_1^u)^\top D^2 f(y_1) + Dg(y_2,\tau)\right]x_1^u, \tag{2.52}$$

where (when omitted) the spatial arguments are $\bar{x}^u(\tau-t+p)$ and the temporal argument is τ. Even if the ε term were discarded, (2.52) is not a closed equation for $B^u = f^\top x_1^u$, and the Melnikov strategy adopted for solving for M_ε^u cannot be used. Instead, x_1^u can be written in terms of its components in the two perpendicular directions $\hat{f}$ and $\hat{f}^\perp$ by

$$x_1^u = \frac{f \cdot x_1^u}{|f|}\hat{f} + \frac{f^\perp \cdot x_1^u}{|f|}\hat{f}^\perp = \frac{B_\varepsilon^u}{|f|^2} f + \frac{M_\varepsilon^u}{|f|^2} f^\perp .$$

With the definition of R^u in (2.23), (2.52) can be written as

$$\frac{\partial B_\varepsilon^u}{\partial \tau} = \frac{f^\top\left[(Df)^\top + Df\right]f}{|f|^2} B_\varepsilon^u + R^u(\tau-t+p) M_\varepsilon^u(p,\tau) + f^\top g$$
$$+\varepsilon f^\top\left[\frac{1}{2}(x_1^u)^\top D^2 f(y_1) + Dg(y_2,\tau)\right]x_1^u . \tag{2.53}$$

The linear coefficient in (2.53) simplifies to

$$\frac{f^\top\left[(Df)^\top + Df\right]f}{|f|^2} = \frac{\frac{\partial f^\top}{\partial \tau} f + f^\top \frac{\partial f}{\partial \tau}}{|f|^2} = \frac{\frac{\partial}{\partial \tau}\left[f^\top f\right]}{|f|^2} = \frac{1}{|f|^2}\frac{\partial}{\partial \tau}|f|^2 = \frac{\partial}{\partial \tau}\left(\ln|f|^2\right)$$

through the usage of the variational equations (2.51). Moreover,

$$R^u(\tau-t+p)M_\varepsilon^u(p,\tau) = R^u(\tau-t+p)M^u(p,\tau) + R^u(\tau-t+p)\left[M_\varepsilon^u(p,\tau) - M^u(p,\tau)\right],$$

enabling (2.53) to become

$$\frac{\partial B_\varepsilon^u}{\partial \tau} = \frac{\partial}{\partial \tau}\left(\ln|f|^2\right) B_\varepsilon^u + R^u(\tau-t+p)M^u(p,\tau) + f^\top g$$
$$+\varepsilon f^\top\left[\frac{1}{2}(x_1^u)^\top D^2 f(y_1) + Dg(y_2,\tau)\right]x_1^u$$
$$+R^u(\tau-t+p)\left[M_\varepsilon^u(p,\tau) - M^u(p,\tau)\right]. \tag{2.54}$$

Now, the last two lines of (2.54) are $\mathcal{O}(\varepsilon)$ terms—a point which will be elaborated on later. Define

$$B^u(p,\tau) := \lim_{\varepsilon \to 0} B_\varepsilon^u(p,\tau),$$

and then, by the same argument used for M^u, the equation of evolution for B^u can be obtained by ignoring the higher-order terms, i.e.,

$$\frac{\partial B^u}{\partial \tau} = \frac{\partial}{\partial \tau}\left(\ln |f|^2\right) B^u + \Omega^u(\tau - t + p) M^u(p,\tau) + f^{\top} g\,, \tag{2.55}$$

which moreover satisfies the same "initial" condition as B^u_ε in (2.49), namely, $B^u(p, T_u) = 0$ for all $p \in (-\infty, P]$. Using the integrating factor

$$\mu(\tau) := \exp\left[-\int_t^\tau \frac{\partial}{\partial u} \ln |f(\bar{x}^u(u - t + p))|^2 \,\mathrm{d}u\right] = \frac{|f(\bar{x}^u(p))|^2}{|f(\bar{x}^u(\tau - t + p))|^2}\,,$$

(2.55) can be represented by

$$\frac{\partial}{\partial \tau}\left[\frac{B^u(p,\tau)}{|f(\bar{x}^u(\tau - t + p))|^2}\right] = \frac{R^u(\tau - t + p) M^u(p,\tau) + f^{\top} g}{|f(\bar{x}^u(\tau - t + p))|^2}\,. \tag{2.56}$$

While it is clear from (2.48) that $B^u \to 0$ as $\tau \to -\infty$, this condition cannot be applied to (2.56) because the denominator $|f|^2$ also goes to zero (and indeed does so faster than the numerator). Hence, (2.56) cannot be integrated from $-\infty$ as was done for the normal component. This highlights the necessity of another condition at a finite time, which in this case is at $t - p$, as given by (2.24). Integrating (2.56) from a time $t - p$ to t and utilizing (2.49) to get rid of the lower boundary term yields

$$\frac{B^u(p,t)}{|f(\bar{x}^u(p))|^2} = \int_{t-p}^{t} \frac{R^u(\tau - t + p) M^u(p,\tau) + f^{\top}(\bar{x}^u(\tau - t + p))\, g(\bar{x}^u(\tau - t + p), \tau)}{|f(\bar{x}^u(\tau - t + p))|^2} \mathrm{d}\tau\,.$$

By changing the variable of integration in the form $\tau - t + p \to \tau$, (2.27) is obtained.

It remains to be argued that the terms neglected in (2.54) yield higher-order corrections to B^u. The terms explicitly with an ε in front in (2.54) have a specific bound, exactly as argued for the function M^u, and since the integration here is over a finite interval, their contribution remains $\mathcal{O}(\varepsilon)$. The final term in (2.54) can be bounded by

$$|R^u(\tau - t + p)[M^u_\varepsilon(p,\tau) - M^u(p,\tau)]| \le \varepsilon\, |R^u(\tau - t + p)|\, \frac{K_4}{\lambda_u} e^{\lambda_u \tau} \le \varepsilon K_5 e^{\lambda_u \tau}\,.$$

This is through the usage of (2.46) at a general point τ, and then by realizing from (2.23) that R^u remains bounded since it approaches a constant in the limit as $\tau \to -\infty$. Indeed, Df converges to the local linearization at a, and $\hat{f} \to v_u$. Therefore, the contribution from this term also remains $\mathcal{O}(\varepsilon)$, implying that $B^u_\varepsilon(p,t)$ and $B^u(p,t)$ differ by at most $\mathcal{O}(\varepsilon)$. Thus,

$$\hat{f}(\bar{x}^u(p)) \cdot [x^u_\varepsilon(p,t) - \bar{x}^u(p)] = \varepsilon\, \frac{B^u(p,t)}{|f(\bar{x}^u(p))|} + \mathcal{O}(\varepsilon^2)\,, \tag{2.57}$$

verifying that $B^u(p,t)$ does in fact give the leading-order tangential movement of the unstable manifold as given in (2.25). Thus, both the normal and the tangential components in (2.25) have been established.

2.4 ▪ Stable manifolds for non-area-preserving 2D flows

Here, the modification to the stable manifold under the influence of the perturbation in the form (2.6) will be presented. The results are closely analogous to the unstable manifold discussion in the previous section and thus will be addressed more briefly. The clipped unperturbed stable manifold can be represented parametrically in the autonomous phase space by

$$\Gamma^s := \{\bar{x}^s(p) \,:\, p \in [-P, \infty]\} \tag{2.58}$$

with P being as large as needed, but not ∞.

Suppose $\bar{x}^s(t)$ is a solution on Γ^s such that $x^s(t) \to a$ as $t \to \infty$. The idea—exactly as in the previous section—is to represent the perturbed stable manifold via $x^s_\varepsilon(p,t)$ where p is the location (i.e., a point close to $\bar{x}^s(p)$) and t is the time-slice. The temporal argument will be specified by $t \in [T_s, \infty)$ for some $T_s < 0$, and for each $t \in ([T_s, \infty)$, the p-parametrization for $x^u_\varepsilon(p,t)$ will be chosen such that

$$\hat{f}(\bar{x}^s(0)) \cdot [x^s_\varepsilon(p, t-p) - \bar{x}(0)] = 0 \tag{2.59}$$

for all $p \in [-P, \infty)$. The relevant projected rate-of-strain is now

$$R^s(\xi) := \left(\hat{f}^\perp\right)^\top (\bar{x}^s(\xi)) \left[(Df)^\top (\bar{x}^s(\xi)) + (Df)(\bar{x}^s(\xi)) \right] \hat{f}(\bar{x}^s(\xi)) \,. \tag{2.60}$$

Theorem 2.23 (stable manifold location [32]). *Consider (2.6) under Hypotheses 2.2, 2.3, and 2.5. The perturbed stable manifold of the hyperbolic trajectory $a_\varepsilon(t)$ of (2.6) has a parametric representation with parameters $(p,t) \in [-P, \infty) \times [T_u, \infty)$ for arbitrarily large but fixed P and $-T_u$, given by*

$$x^s_\varepsilon(p,t) = \bar{x}^s(p) + \varepsilon \left[\frac{M^s(p,t)}{|f(\bar{x}^s(p))|} \hat{f}^\perp(\bar{x}^s(p)) + \frac{B^s(p,t)}{|f(\bar{x}^s(p))|} \hat{f}(\bar{x}^s(p)) \right] + \mathcal{O}\left(\varepsilon^2\right), \tag{2.61}$$

in which the stable Melnikov function is

$$M^s(p,t) := -\int_p^\infty \exp\left[\int_\tau^p \mathrm{Tr}[Df(\bar{x}^s(\xi))]\,\mathrm{d}\xi \right] f^\perp(\bar{x}^s(\tau)) \cdot g(\bar{x}^s(\tau), t+\tau-p)\,\mathrm{d}\tau \tag{2.62}$$

and

$$B^s(p,t) := |f(\bar{x}^s(p))|^2 \int_0^p \frac{R^s(\tau) M^s(p, \tau+t-p) + f(\bar{x}^s(\tau)) \cdot g(\bar{x}^s(\tau), \tau+t-p)}{|f(\bar{x}^s(\tau))|^2}\,\mathrm{d}\tau \,. \tag{2.63}$$

Proof. This proof shall be omitted since it is substantially equivalent to that of Theorem 2.12, with minor differences in integration limits. □

Remark 2.24 (primary stable manifold). As for the unstable manifold, Theorem 2.23 is only able to characterize the *primary* segment of the perturbed stable manifold. There is no information about the stable manifold once it has extended spatially beyond $\bar{x}^s(-P)$.

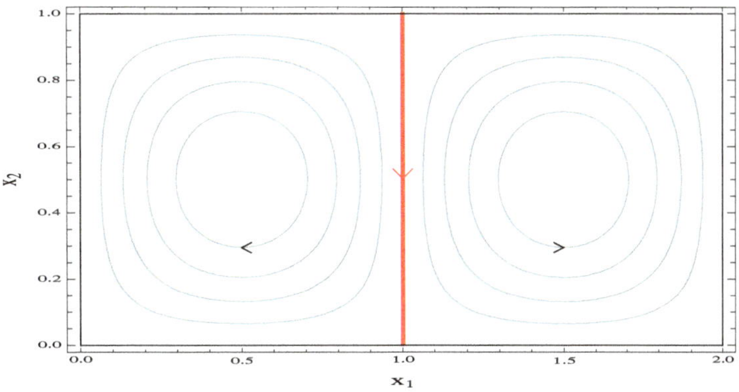

Figure 2.9. *Phase portrait of the double-gyre (2.64) with $\varepsilon = 0$, displaying [in red] the heteroclinic manifold separating the left and the right gyres.*

Example 2.25 (double-gyre). A highly used [156, 150, 391, e.g.] test bed for numerical methods in computing stable and unstable manifolds in nonautonomous systems is the double-gyre originally proposed in [364], which is given by

$$\left.\begin{aligned} \dot{x}_1 &= -\pi A \sin[\pi\phi(x_1,t)]\cos[\pi x_2] \\ \dot{x}_2 &= \pi A \cos[\pi\phi(x_1,t)]\sin[\pi x_2]\frac{\partial \phi}{\partial x_1}(x_1,t) \end{aligned}\right\}, \tag{2.64}$$

in which $A > 0$ and $0 < \varepsilon \ll 1$,

$$\phi(x_1,t) := \varepsilon q(t) x_1^2 + (1 - 2\varepsilon q(t)) x_1,$$

and the standard choice is $q(t) = \sin(\omega t)$, which leads to time-periodic flow [364]. The advantage of time-periodicity is the presence of a Poincaré map which strobes the flow (2.64) at time intervals which are $2\pi/\omega$ apart, essentially enabling the nonautonomous flow (2.64) to be viewed as an autonomous map instead [341, 421, 183, 17, cf.]. Of course, imposing time-periodicity is a substantial limitation. This will be relaxed by permitting $q(t)$ to be arbitrary but bounded. The system (2.64) is usually considered in the spatial domain $[0,2]\times[0,1]$, when $\varepsilon = 0$ the phase portrait consists of two counter-rotating gyres (vortices) in the squares $[0,1]\times[0,1]$ (the "left gyre") and $[1,2]\times[0,1]$ (the "right gyre"). The boundary between these gyres, the line $x_1 = 1$, $0 \le x_2 \le 1$, is a branch of the stable manifold of the saddle fixed point $b = (1,0)$ and simultaneously a branch of the unstable manifold of the saddle fixed point $a = (1,1)$. The $\varepsilon = 0$ phase portrait of the double-gyre is shown in Figure 2.9. When ε is turned on, the stable and unstable manifolds evolve with time t in some nontrivial fashion, and in particular no longer need to coincide. The relative positioning of these manifolds, and their intersections with each other, have a profound influence on the movement of fluid between the left and the right gyre.

To compute these manifolds, the system (2.64) needs to be first written in the perturbed form (2.6), with $\varepsilon \ll 1$. Setting $\varepsilon = 0$ in (2.64) leads to the identification of the autonomous unperturbed velocity

$$f = \begin{pmatrix} -\pi A \sin(\pi x_1)\cos(\pi x_2) \\ \pi A \cos(\pi x_1)\sin(\pi x_2) \end{pmatrix}, \tag{2.65}$$

based on which Figure 2.9 is drawn. The corresponding flow is equivalent to the classical Taylor-Green cellular flow [330, 100, 25, 26, 42, 54]. Straightforward Taylor expansions lead to the $\mathcal{O}(\varepsilon)$-component

$$g = \begin{pmatrix} -\pi^2 A\left(x_1^2 - 2x_1\right)\cos(\pi x_2)\cos(\pi x_1)q(t) \\ \pi A \sin(\pi x_2)\left[2\cos(\pi x_1)(x_1 - 1) - \pi\left(x_1^2 - 2x_1\right)\sin(\pi x_1)\right]q(t) \end{pmatrix}. \tag{2.66}$$

Several quantities need to be computed to be able to use Theorems 2.12 and 2.23. On the heteroclinic orbit $x_1 = 1$, $f^\perp = (\pi A \sin(\pi x_2), 0)$ and

$$f^\perp \cdot g\Big|_{x_1=1} = -\frac{\pi^3 A^2}{2}\sin(2\pi x_2)q(t). \tag{2.67}$$

Moreover, $\operatorname{Tr} Df = 0$, and it will be left as an exercise to show that both $R^{s,u}$ and $f \cdot g$ are zero on the heteroclinic. Thus, the tangential Melnikov functions $B^{s,u} = 0$. Now, by solving the $\dot{x}_2$ equation in (2.64) when $\varepsilon = 0$ with the initial condition $x_2 = 1/2$ at time zero, the unperturbed heteroclinic trajectory can be represented by

$$\bar{x}^{s,u}(t) = \begin{pmatrix} 1 \\ \frac{2}{\pi}\cot^{-1} e^{\pi^2 A t} \end{pmatrix}.$$

Thus, $\bar{x}^{s,u}(p)$ provides a symmetric parametrization of the heteroclinic, and the inverse of the second component above can be written as

$$p = \frac{1}{\pi^2 A}\ln\left(\cot\frac{\pi x_2}{2}\right), \tag{2.68}$$

which establishes the monotonic relationship between the p-parametrization and the x_2-coordinate.

Now, by using Theorem 2.12, the x_1-coordinate of the perturbed unstable manifold of $a = (1, 1)$, in each time-slice t, can be expressed to leading-order by

$$x_1^u(p, t) = 1 - \varepsilon\frac{\pi^2 A}{\operatorname{sech}\left(\pi^2 A p\right)}\int_{-\infty}^{p}\tanh\left(\pi^2 A\tau\right)\operatorname{sech}\left(\pi^2 A\tau\right)q(\tau + t - p)\,\mathrm{d}\tau. \tag{2.69}$$

In the above (and in the subsequent computations in this example), the fact that there should actually be a $\mathcal{O}\left(\varepsilon^2\right)$ error in the expressions will be neglected. Exploiting the relationship between p and x_2 and shifting the integration variable leads to the equivalent representation

$$x_1^u(x_2, t) = 1 - \varepsilon\frac{\pi^2 A}{2\sin(\pi x_2)}\int_{-\infty}^{t}\sin\left[4\cot^{-1}\left(\cot\frac{\pi x_2}{2}\exp\left[\pi^2 A(\tau - t)\right]\right)\right]q(\tau)\,\mathrm{d}\tau, \tag{2.70}$$

for $x_2 \in (0, 1]$, where (with an abuse of notation) x_2 is now included as the first argument. This is more natural in that it gives the unstable manifold as a graph from x_2 to x_1, at each time t. Either by applying the limit $x_2 \to 1$ using L'Hôpital's rule to the above, or by directly applying Theorem 2.7, the location of the hyperbolic trajectory $a_\varepsilon(t) = \left(x_1^u(1, t), 1\right)$ to which this unstable manifold is attached can be determined as

$$x_1^u(1, t) = 1 + \varepsilon\pi^2 A\exp\left(-\pi^2 A t\right)\int_{-\infty}^{t}\exp\left(\pi^2 A\tau\right)q(\tau)\,\mathrm{d}\tau. \tag{2.71}$$

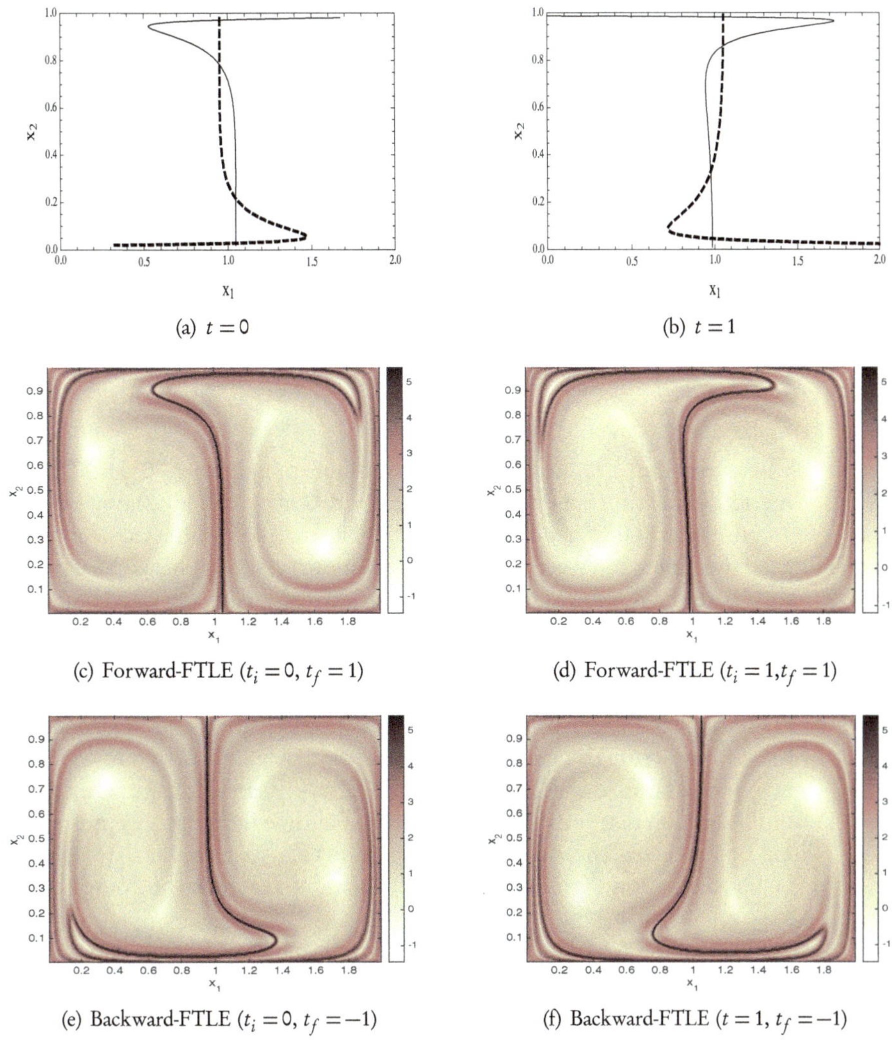

(a) $t = 0$ (b) $t = 1$

(c) Forward-FTLE ($t_i = 0$, $t_f = 1$) (d) Forward-FTLE ($t_i = 1$, $t_f = 1$)

(e) Backward-FTLE ($t_i = 0$, $t_f = -1$) (f) Backward-FTLE ($t = 1$, $t_f = -1$)

Figure 2.10. *(a) and (b): Stable manifold [solid] and unstable manifold [dashed] of the double-gyre for $q(t) = \sin(\omega t)$, $A = 1$, $\omega = 15$, and $\varepsilon = 0.1$, using (2.73) and (2.70). (c) and (d): Forward-time FTLE fields. (e) and (f): Backward-time FTLE fields.*

(The structure of the system (2.64) ensures that the x_2-coordinate of $a_\varepsilon(t)$ is always 1.) A similar application of Theorem 2.23 leads to

$$x_1^s(p,t) = 1 + \varepsilon \frac{\pi^2 A}{\operatorname{sech}(\pi^2 A p)} \int_p^\infty \tanh\left(\pi^2 A \tau\right) \operatorname{sech}\left(\pi^2 A \tau\right) q(\tau + t - p)\,\mathrm{d}\tau, \quad (2.72)$$

as the perturbed stable manifold of $b = (1,0)$, in a time-slice t. Its graphical representation is

$$x_1^s(x_2,t) = 1 + \varepsilon \frac{\pi^2 A}{2\sin(\pi x_2)} \int_t^\infty \sin\left[4\cot^{-1}\left(\cot\frac{\pi x_2}{2}\exp\left[\pi^2 A(\tau - t)\right]\right)\right] q(\tau)\,\mathrm{d}\tau \quad (2.73)$$

for $x_2 \in [0,1)$, and this is attached to the hyperbolic trajectory $b_\varepsilon(t) = \left(x_1^s(t,0),0\right)$ in which

$$x_1^s(0,t) = 1 + \varepsilon \pi^2 A \exp\left(\pi^2 A t\right) \int_t^\infty \exp\left(-\pi^2 A \tau\right) q(\tau) \mathrm{d}\tau . \tag{2.74}$$

While the expressions above hold for general $q(t)$, in the remainder of this example the standard choice [364] $q(t) = \sin(\omega t)$ will be followed; subsequent examples (see Examples 2.56, 2.33, and 2.48) will address more general q. In this case, the perturbed upper and lower hyperbolic trajectories to leading-order can be computed from (2.71) and (2.74) as

$$a_\varepsilon(t) = \begin{pmatrix} 1 + \varepsilon \cos\theta \sin(\omega t - \theta) \\ 1 \end{pmatrix}, \quad b_\varepsilon(t) = \begin{pmatrix} 1 + \varepsilon \cos\theta \sin(\omega t + \theta) \\ 0 \end{pmatrix}, \tag{2.75}$$

where $\theta = \tan^{-1}[\omega/(A\pi^2)]$. The manifold expressions (2.70) and (2.73) require numerical evaluation; snapshots of these at $t = 0$ and $t = 1$ are shown in Figures 2.10(a) and (b) for $\varepsilon = 0.1$, $A = 1$, and $\omega = 15$. The manifolds intersect transversely, and indeed they must intersect infinitely often in each t-slice. This is since the flow (2.64) is time-periodic in this case, and thus Figure 2.10(a) is equivalent to the phase portrait of a Poincaré map P which samples the flow (2.64) at times $0, \pm 2\pi/\omega, \pm 4\pi/\omega$, etc. The manifolds pictured are also the manifolds associated with the fixed points $a_\varepsilon(0)$ and $b_\varepsilon(0)$ of this Poincaré map, and one intersection of these manifolds implies infinitely many since each intersection must get mapped to another intersection of the manifolds. The picture in Figure 2.10(b) can be similarly interpreted in terms of a Poincaré map which samples the flow at times $1, 1 \pm 2\pi/\omega, 1 \pm 4\pi/\omega$, etc. Using Poincaré maps in this fashion disguises the fact that an initial time (i.e., a phase) must be chosen; having t parametrize the manifolds incorporates this issue seamlessly, while of course also being able to deal with time-aperiodic situations in which the Poincaré map approach fails.

Figures 2.10(c) and (d) show the forward-time FTLE field, computed at the same times as in (a) and (b), with the computations performed over a time-period of 1 unit, that is, $|t_f - t_i| = 1$ in (1.14). The ridge of the forward-time FTLE is expected to line up with the stable manifold. Comparison between the solid curve in (a) and the ridge in (c) shows that this is indeed the case, and a similar comparison between (b) and (d) is possible. The point of emanation of the stable manifold, slightly bigger than $x_1 = 1$ in (a) and (c), and slightly smaller than 1 in (b) and (d), is consistently captured. The *details* of the ridges in Figures 2.10(c) and (d) are different from the perturbative expansions in (a) and (b). The backward-time FTLE fields at these same times is shown in Figures 2.10(e) and (f), which displays good alignment with the (dashed) unstable manifold of Figures 2.10(a) and (b).

One important observation from Figure 2.10 is that the FTLE ridges indicate that the stable and unstable manifolds extend beyond those obtainable from the perturbative theory. They wrap around the double-gyre in various ways. The theory, using Theorems 2.23 and 2.12, is only able to capture *primary* segments of the manifolds, that is, segments which can be represented as graphs over the unperturbed (steady) manifolds. Indeed, the actual stable and unstable manifolds for the double-gyre and complex, winding around in a nontrivial fashion and causing chaotic transport. It is possible to capture longer versions of the manifolds approximated in Figures 2.10(c)–(f) by simply running the FTLEs over longer time periods. For illustrative purposes, the forward-time FTLE fields (whose ridges represent the stable manifold) run over longer

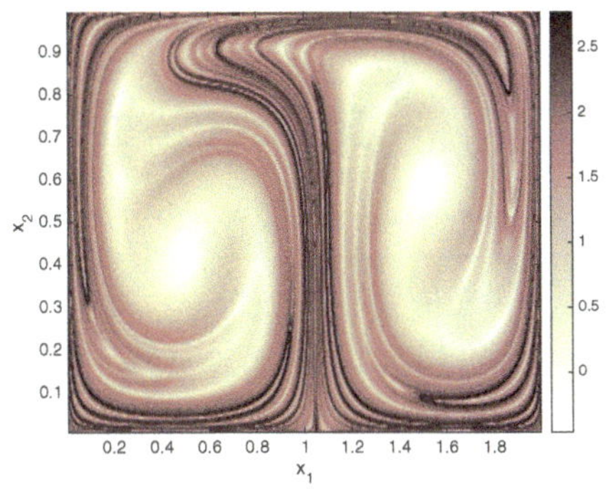

(a) $t_i = 0, t_f = 2$

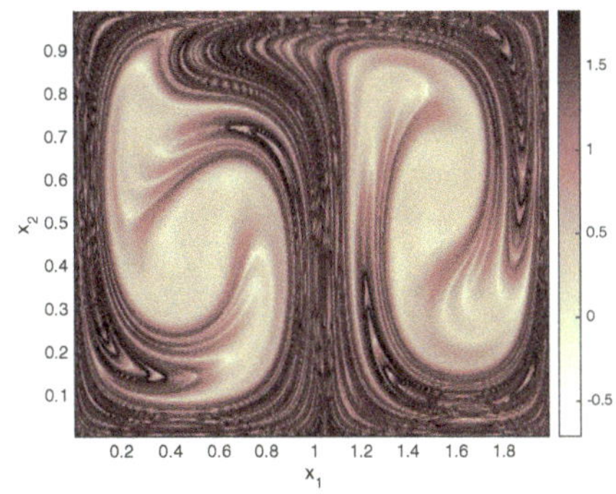

(b) $t = 0, t_f = 3$

Figure 2.11. *Forward-time FTLE fields for the double-gyre with the same parameter values as in Figure 2.10, with two different choices of integration interval.*

periods are shown in Figure 2.11. The calculations are performed at $t_i = 0$, exactly as in Figure 2.10(c), but with the time-interval over which the integration is performed increased from $[0,1]$ to (a) $[0,2]$ and (b) $[0,3]$. As the time-of-integration is increased, the fate of extended parts of the stable manifold is revealed. It wraps back and forth around the boundary of the double-gyre and keeps traversing regions infinitesimally close to itself. To obtain the behavior fully, one would need to perform the integration over $[0,\infty)$, which would reveal additional striations of the stable manifold. Of course, increased spatial resolution is required to see these. For a detailed investigation of the complicated motion of these manifolds in the standard double-gyre, and the resulting chaotic transport, the reader is referred to [325]. ∎

At this point, expressions for both the unstable and stable manifolds in a time-slice t are available from Theorems 2.12 and 2.23. By applying $p \to -\infty$ to (2.25) and $p \to \infty$ to (2.61), one may approach the "foot" of the relevant manifold, that is, the hyperbolic trajectory. This strategy can be used to determine the location of the hyperbolic trajectory $a_\varepsilon(t)$ in each time-slice t, as given in Theorem 2.7. This proof is given in Section 2.4.1.

2.4.1 ▪ Proof of Theorem 2.7 (perturbed hyperbolic trajectory)

From the expression (2.36) obtained in the proof of the unstable manifold theorem, it can be said that

$$\hat{f}^{\perp}(\bar{x}^u(p))\cdot[x_\varepsilon^u(p,t)-\bar{x}^u(p)] = \varepsilon\frac{M^u(p,t)}{|f(\bar{x}^u(p))|}+\mathcal{O}(\varepsilon^2),$$

for $(p,t)\in(-\infty,P]\times(-\infty,T_u]$. Now, going back along the unstable manifold towards the hyperbolic trajectory in each time-slice t is equivalent to considering the limit $p\to-\infty$ in the above. Since $\hat{f}\to v_u$ in this limit, $\hat{f}^{\perp}(\bar{x}^u(p))\to v_u^{\perp}$. Thus, the left-hand side above converges to $v_u^{\perp}\cdot[a_\varepsilon(t)-a]$. In analyzing the limit on the $\mathcal{O}(\varepsilon)$ term on the right, it is convenient to represent M^u by employing the change of integration variable $\tau\to\tau-t+p$ as given in (2.44) in the proof of Theorem 2.12.

The limit of this leading-order term is then

$$\begin{aligned}\Xi_u &= \lim_{p\to-\infty}\int_{-\infty}^{t}\exp\left[\int_{\tau}^{t}\operatorname{Tr}Df\left(\bar{x}^u(\xi-t+p)\right)d\xi\right]\frac{f^{\perp}(\bar{x}^u(\tau-t+p))\cdot g\left(\bar{x}^u(\tau-t+p),\tau\right)}{|f\left(\bar{x}^u(p)\right)|}d\tau\\ &=\lim_{p\to-\infty}\int_{-\infty}^{t}\exp\left[\int_{\tau}^{t}\operatorname{Tr}Df\left(\bar{x}^u(\xi-t+p)\right)d\xi\right]\frac{|f\left(\bar{x}^u(\tau-t+p)\right)|}{|f\left(\bar{x}^u(p)\right)|}g_u^n\left(\bar{x}^u(\tau-t+p),\tau\right)d\tau,\end{aligned} \quad (2.76)$$

in which g_u^n is the component of the nonautonomous perturbation g in the direction of $\hat{f}^{\perp}$. Now, for suitably negative p, it is claimed that the integrand above is bounded by

$$H(\tau) := C\,e^{\lambda_s(t-\tau)}\,g\,(a,\tau)\cdot v_u^{\perp}, \quad (2.77)$$

for some constant C. To see why, observe first that $g\left(\bar{x}^u(\tau-t+p,\tau\right)\to g(a,\tau)$ in this limit. Moreover, the quantity

$$\frac{|f\left(\bar{x}^u(\tau-t+p)\right)|}{|f\left(\bar{x}^u(p)\right)|}\to\frac{A\,e^{\lambda_u(\tau-t+p)}}{A\,e^{\lambda_u p}}=e^{\lambda_u(\tau-t)}$$

and thus can be bounded by a constant times the function on the right. Finally, $\operatorname{Tr}Df\to\operatorname{Tr}Df\,(a)=\lambda_u+\lambda_s$, and hence there exists K_4 such that

$$\exp\left[\int_{\tau}^{t}\operatorname{Tr}Df\left(\bar{x}^u(\xi-t+p)\right)d\xi\right]\leq\exp\left[\int_{\tau}^{t}K_4(\lambda_u+\lambda_s)\,d\xi\right]=e^{K_4}e^{(\lambda_u+\lambda_s)(t-\tau)}.$$

Putting these bounds together yields a bounding function of the form (2.77). Now, (2.77) is integrable over $(-\infty,t)$, and by the Lebesgue dominated convergence theorem, the limit $p\to-\infty$ can be moved inside the integral in (2.76). Applying a term-by-term limit analogously to what has been described above,

$$\Xi_u=\int_{-\infty}^{t}\lim_{p\to-\infty}e^{(\lambda_u+\lambda_s)(t-\tau)}\frac{A\,e^{\lambda_u(\tau-t+p)}}{A\,e^{\lambda_u p}}\,g_u^n\,(a,\tau)\,d\tau=\int_{-\infty}^{t}e^{\lambda_s(t-\tau)}g_u^n\,(a,\tau)\,d\tau,$$

after which a straightforward shift in the integration variable yields the second equation in (2.15). Thus, β_u represents the $\mathcal{O}(\varepsilon)$-component of the vector $a_\varepsilon(t)-a$ in the direction of $v_u^{\perp}$.

The first equation in (2.15) is obtained by applying the limit $p\to\infty$ to

$$\hat{f}^{\perp}(\bar{x}^s(p))\cdot\left[x_\varepsilon^s(p,t)-\bar{x}^s(p)\right]=\varepsilon\frac{M^s(p,t)}{|f\left(\bar{x}^s(p)\right)|}+\mathcal{O}(\varepsilon^2),$$

which arises from the analogous equation to (2.36) for the stable manifold. The details are similar to the above and will be omitted. Thus, β_s is the $\mathcal{O}(\varepsilon)$-component of $a_\varepsilon(t)-a$ in the direction of $v_s^{\perp}$.

Obtaining the expression (2.14) relies on applying straightforward geometry, since the projections of $a_\varepsilon(t)-a$ in the potentially nonorthogonal directions $v_u^{\perp}$ and $v_s^{\perp}$ are known via (2.15). The linear independence of v_u and v_s ensures that $v_u^{\perp}\cdot v_s\neq 0$. Once again, the details will be omitted.

Remark 2.26. The above calculations used the normal displacements as given in Theorems 2.12 and 2.23. It is *not* possible to use the tangential displacement expressions in Theorems 2.12 and 2.23 to arrive at the displacement of the saddle point. This is caused by the boundary term arising from integrating (2.56), $B^u(p,t-p)/|f\left(\bar{x}^u(p)\right)|$, being indeterminate in this limit, and thus the expression (2.27) is valid strictly for finite p.

2.5 ▪ Tangent vectors to invariant manifolds

In the setting of the nonautonomously perturbed system (2.6) which obeys Hypotheses 2.2, 2.3, and 2.5, the location of the perturbed hyperbolic trajectory and primary segments of its stable and unstable manifolds have now been characterized. In this section, the local tangent vectors to the stable and unstable manifolds at the hyperbolic trajectory location are addressed. This is important from the perspective of the classical theory of stable and unstable manifolds for autonomous systems, which are usually defined *locally*; the *global* manifolds are then defined by a temporal evolution of the local manifolds [183, Theorem 1.3.2, e.g.]. These local stable and unstable manifolds are well-approximated by the tangent vectors at the hyperbolic point. In such an autonomous setting, the tangent vectors have another, simpler characterization: they are the eigenvectors associated with the linearized flow about the hyperbolic fixed point. For example, in (2.6) with $\varepsilon = 0$, the eigenvectors v_u and v_s are, respectively, the tangent vectors to the unstable and stable manifolds at the hyperbolic fixed point a.

In a nonautonomous setting, if a hyperbolic trajectory exists, it possesses stable and unstable manifolds, and thus it is possible in each fixed time-slice t to define their tangent vectors. However, it is important to note that these tangent vectors are *not* given by eigenvectors of the frozen-time Jacobian matrix of the nonautonomous vector field. The stable and unstable manifolds are Lagrangian entities which flow with the nonautonomous vector field, and thus the stable and unstable manifolds are not static objects; the time-history of trajectories on them contributes to their evolution. Thus, the *analogous* entity to eigenvectors in autonomous flows are tangent vectors to the stable/unstable manifolds in nonautonomous flows. Since these depend on the time-slice t, they can be thought of as vectors $v_s(t)$ and $v_u(t)$, representing the local tangent vectors to the stable and unstable manifold, respectively. An alternative characterization in terms of the projection operator in the exponential dichotomy relations (1.22) is possible. If P is the projection to the unstable manifold at time 0 in these conditions, then $\mathcal{R}\{Y(t)P\}$, in which $Y(t)$ is the fundamental matrix solution to the variational equation at the relevant hyperbolic trajectory, defines the direction of $v_u(t)$. Moreover, $\mathcal{R}\{Y(t)(I-P)\}$ defines the direction of $v_s(t)$ (see Section 1.6 for these results). These conditions are, however, very difficult to use, since they require full knowledge of the hyperbolic trajectory, its variational equations, its solution, and the exponential dichotomy projection associated with it. Another alternative is to use the t-parametrized linear spaces arising from the ergodic-theoretical idea of Oseledets splitting [299, 150, 153, 240, 170], which again examines the linearized motion (the equation of variations) around trajectories. While computationally more attractive (particularly in the finite-time setting—see Section 2.8) and intuitively connected to the idea of exponential dichotomies, this connection is only gradually coming into the coherent structure literature. When thought of as linearized on a hyperbolic trajectory, the projections arising from exponential dichotomies, the Oseledets spaces and the local tangent vectors to the stable and unstable manifolds, are different ways of examining the *same* thing. Following the *geometric* spirit of this monograph, the focus in this section shall be in thinking of these entities as tangent vectors in each time-slice t. The results of this section are mainly associated with [39], to which the reader is referred for additional discussion.

Expressions for the stable and unstable manifolds have already been derived for (2.6) in the form of Theorems 2.12 and 2.23. In each fixed time-slice, these are curves described parametrically in terms of a parameter p. Therefore, the relevant tangent vector can be obtained by taking the p-derivative of these parametric forms and then

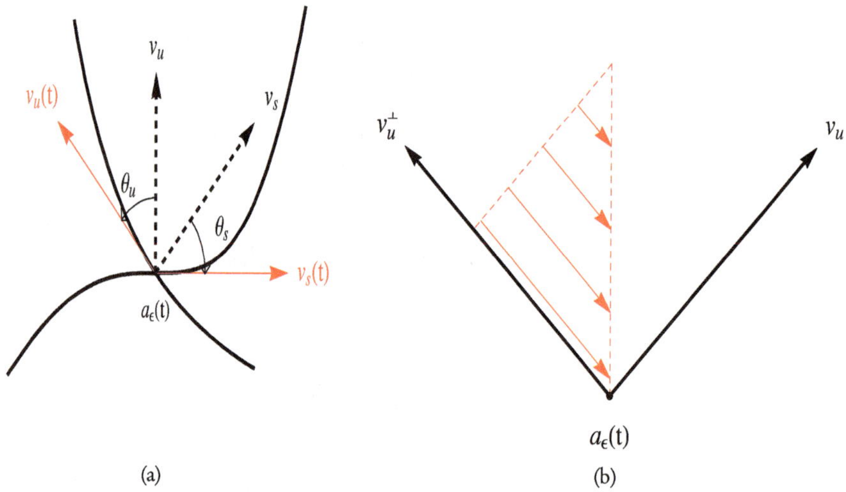

Figure 2.12. *(a) Tangent vectors (red) to stable and unstable manifolds in a time-slice t, as expressed by (2.78) and (2.79). The dashed arrows are the $\varepsilon = 0$ eigenvectors. (b) Illustration of a shear velocity profile tangential to the $v_u^\perp$ direction which might intuitively be thought to rotate v_u in the counterclockwise direction. [39] Reproduced with permission from Springer Science+Business Media.*

taking the limit as p approaches the hyperbolic trajectory location $a_\varepsilon(t)$. The way in which the $\varepsilon = 0$ eigenvectors rotate to their nonautonomous analogues $v_s(t)$ and $v_u(t)$ when $\varepsilon \neq 0$ can be represented by angles $\theta_s(t)$ and $\theta_u(t)$, as shown in Figure 2.12(a). These are to be thought of as the rotations in the counterclockwise direction, and thus at the time pictured in Figure 2.12(a), $\theta_s(t) < 0$ and $\theta_u(t) > 0$.

Theorem 2.27 (tangent vectors [39]). *Consider the flow (2.6) under Hypotheses 2.2, 2.3, and 2.5. The local tangential direction $v_s(t)$ to the stable manifold Γ_ε^s at the point $a_\varepsilon(t)$ for $t \in [T_s, \infty)$ (for finite T_s) can be obtained by a counterclockwise rotation of v_s by an angle*

$$\theta_s(t) = -\varepsilon\, e^{(\lambda_u - \lambda_s)t} \int_t^\infty e^{(\lambda_s - \lambda_u)\tau} v_s \cdot \nabla\left[g \cdot v_s^\perp\right](a, \tau)\, \mathrm{d}\tau + \mathcal{O}\left(\varepsilon^2\right). \tag{2.78}$$

Similarly, the local tangential direction $v_u(t)$ to the unstable manifold Γ_ε^u at $a_\varepsilon(t)$ for $t \in (-\infty, T_u]$ for any finite T_u is obtained by rotating v_u by a counterclockwise angle

$$\theta_u(t) = \varepsilon e^{(\lambda_s - \lambda_u)t} \int_{-\infty}^t e^{(\lambda_u - \lambda_s)\tau} v_u \cdot \nabla\left[g \cdot v_u^\perp\right](a, \tau)\, \mathrm{d}\tau + \mathcal{O}\left(\varepsilon^2\right). \tag{2.79}$$

Proof. See Section 2.5.1. $\square$

The quantity which is being integrated over all relevant time in Theorem 2.27 (either backwards or forwards from time t, depending on whether the stable or unstable manifold's tangent vector is being considered) is the *velocity shear* defined by

$$\sigma_i(t) := v_i \cdot \nabla\left[g \cdot v_i^\perp\right](a, \tau), \quad i = u, s. \tag{2.80}$$

The reason this is a shear is that it is the directional derivative of $g \cdot v_i^\perp$ in the direction perpendicular to $v_i^\perp$. For intuition as to why the shear affects the rotation of the tangent vectors, see the right panel of Figure 2.12(b), in which a velocity profile for g is shown in relation to the v_u and $v_u^\perp$ vectors. In this picture, g is purely in the $-v_u^\perp$ direction, and it is increasing in the coordinate along the v_u direction. Thus, σ_u is positive. However, the velocity situation in Figure 2.12(b) intuitively will push the vector v_u down more near $a_\varepsilon(t)$ than further along its length, and thus the vector v_u would be expected to rotate in the counterclockwise direction. This is the positive direction of rotation; positive σ_u corresponds to positive θ_u, as is clear from (2.79). A similar intuition can be used to describe the influence of the shear on the tangent vector v_s.

Example 2.28 (double-gyre cont.). Return to the double-gyre as expressed in Example 2.25, for general aperiodic functions q. The fixed points $(1,0)$ and $(1,1)$ have been shown to perturb to hyperbolic trajectories given to leading-order by (2.74) and (2.71), respectively. The leading-order tangent vector movement at time t for the stable and unstable manifolds emanating from these hyperbolic trajectories is now determined. For the fixed point $(1,0)$, $\lambda_s = -\pi^2 A$, $v_s = \begin{pmatrix} 0 & -1 \end{pmatrix}^\top$, $\lambda_u = \pi^2 A$, and $v_u = \begin{pmatrix} 1 & 0 \end{pmatrix}^\top$, and thus $v_s^\perp = \begin{pmatrix} 1 & 0 \end{pmatrix}^\top$, and using (2.66), the shear takes the form

$$
\begin{aligned}
\sigma_s(t) &= \begin{pmatrix} 0 \\ -1 \end{pmatrix} \cdot \left\{ \nabla \left[-\pi^2 A (x_1^2 - 2x_1) \cos(\pi x_2) \cos(\pi x_1) q(t) \right] \right\}_{(x_1,x_2)=(1,0)} \\
&= -\pi^2 A q(t) \begin{pmatrix} 0 \\ -1 \end{pmatrix} \cdot \left. \begin{pmatrix} \cos(\pi x_2)\left[(2x_1 - 2)\cos(\pi x_1) - \pi(x_1^2 - 2x_1)\sin(\pi x_1)\right] \\ -\pi(x_1^2 - 2x_1)\cos(\pi x_1)\sin(\pi x_2) \end{pmatrix} \right|_{(1,0)} \\
&= 0.
\end{aligned}
$$

Therefore, there is *no* shear in the velocity field to leading-order in the double-gyre flow (2.64), for any choice of $q(t)$. Thus, from (2.78), the stable manifold emanating from the perturbed version of $(1,0)$ has *no* rotation to leading-order in ε, for any bounded choice of function $q(t)$. This is certainly consistent with Figure 2.10, where the emanating stable manifolds for $q(t) = \sin(\omega t)$ at both times pictured appear to be vertical. Now, a similar analysis to that above can be performed for the unstable manifold emanating from the perturbed version of $(1,1)$, and in this case one obtains that $\sigma_u(t) = 0$, which indicates that the unstable manifold also undergoes no rotation. This is also consistent with Figure 2.10 and with all the double-gyre simulations available in the literature for $q(t) = \sin(\omega t)$. ■

Remark 2.29. The (unscaled) version of the stable tangent vector can itself be represented in the form

$$
v_s(t) = v_s(1 + \gamma_s(t)) + v_s^\perp \theta_s(t) + \mathcal{O}(\varepsilon^2) \tag{2.81}
$$

for some $\mathcal{O}(\varepsilon)$-function $\gamma_s(t)$ which (perusing the proof in Section 2.5.1) is associated with the p-derivatives of the tangential displacement B^s specified in Theorem 2.23. However, the expression for $\gamma_s(t)$ is cumbersome and, as can be seen from the proof, is irrelevant in determining the eigenvector rotation to leading-order. Therefore, it has not been explicitly stated. The unstable tangent vector's characterization is similarly (2.81) with the subscript s replaced by u.

2.5.1 ▪ Proof of Theorem 2.27 (tangent vectors)

The proof shall be first done for the tangent vector to the stable manifold in a time-slice t. The stable manifold is given by the expression for $x_\varepsilon^s(p,t)$ in (2.61) in Theorem 2.23, and $\lim_{p\to\infty} x_\varepsilon^s(p,t) = a_\varepsilon(t)$. Taking the p-derivative leads to

$$x_p^s = \bar{x}_p^s + \varepsilon\left[\frac{M^s f_p^\perp}{|f|^2} + \left(\frac{M_p^s}{|f|^2} - \frac{2M^s f\cdot f_p}{|f|^4}\right) f^\perp\right] + \varepsilon\left[\frac{B^s f_p + B_p^s f}{|f|^2} - \frac{2B^s\left(f\cdot f_p\right)}{|f|^4} f\right] + \mathcal{O}\left(\varepsilon^2\right),$$

where the p-subscript represents the partial derivative and the arguments (p,t) for M^s and B^s and the argument $\bar{x}^s(p)$ for f have been suppressed. Since to approach $a_\varepsilon(t)$, the limit $p\to\infty$ is required, the large p approximation

$$\bar{x}^s(p) \sim a + c v_s e^{\lambda_s p} \tag{2.82}$$

for a constant $c \neq 0$ is valid. The c represents a scaling on the linearized tangent vector of the $\varepsilon = 0$ flow, and this representation arises since the variational equation for the trajectory a of (2.6) when $\varepsilon = 0$ is $\dot{y} = Df(a)y$, and $Df(a)$ has an eigenvalue λ_s associated with the eigendirection v_s. Now, $f(\bar{x}^s(p)) = \bar{x}_p^s(p)$ since $\bar{x}^s(p)$ is a solution to (2.6) when $\varepsilon = 0$, and thus

$$f(\bar{x}^s(p)) \sim c\lambda_s v_s e^{\lambda_s p},\; f^\perp(\bar{x}^s(p)) \sim c\lambda_s v_s^\perp e^{\lambda_s p} \text{ and } f_p(\bar{x}^s(p)) \sim c\lambda_s^2 v_s e^{\lambda_s p}. \tag{2.83}$$

These large p value approximations substituted into the expression for x_p^s yields, after some algebra,

$$x_p^s = c\lambda_s e^{\lambda_s p} v_s + \frac{\varepsilon}{c\lambda_s e^{\lambda_s p}}\left[\left(M_p^s - \lambda_s M^s\right) v_s^\perp + \left(B_p^s - \lambda_s B^s\right) v_s\right] + \mathcal{O}\left(\varepsilon^2\right).$$

The rotational angle θ_s from v_s towards $v_s^\perp$ is $\mathcal{O}(\varepsilon)$ and thus equal to $\tan\theta_s$ to leading-order. This is essentially the slope of the above tangent line in an axis system $(v_s, v_s^\perp)$ as illustrated in Figure 2.12. Thus

$$\theta_s = \frac{\varepsilon\left(M_p^s - \lambda_s M^s\right) + \mathcal{O}\left(\varepsilon^2\right)}{c\lambda_s e^{\lambda_s p}\left[c\lambda_s e^{\lambda_s p} + \frac{\varepsilon(B_p^s - \lambda_s B^s)}{c\lambda_s e^{\lambda_s p}} + \mathcal{O}\left(\varepsilon^2\right)\right]} = \varepsilon\frac{M_p^s - \lambda_s M^s}{c^2\lambda_s^2 e^{2\lambda_s p}} + \mathcal{O}\left(\varepsilon^2\right) \tag{2.84}$$

as $p\to\infty$. Fortunately, the contribution of B^s towards θ_s is higher-order and can be ignored. The p-derivative of the Melnikov function (2.62) is now required in the limit $p\to\infty$. In this limit, as $\bar{x}^s$ approaches a, $\nabla\cdot f(\bar{x}^s(p)) \to \lambda_s + \lambda_u$ since the trace of Df at a is the sum of its eigenvalues. Putting this along with the other large p estimates in (2.83) into (2.62) gives the large p estimate

$$M^s(p,t) = -c\lambda_s e^{\lambda_u t} e^{\lambda_s p}\int_t^\infty e^{\lambda_u \tau} g_s^n\left(a + c v_s e^{\lambda_s(\tau - t + p)}, \tau\right) \mathrm{d}\tau,$$

where $g_s^n = g\cdot v_s^\perp$. When computing $M_p^s - \lambda_s M^s$, the fact that M^s is a product of $e^{\lambda_s p}$ with another function of p leads to cancellations and results in

$$\begin{aligned} M_p^s - \lambda_s M^s &= -c\lambda_s e^{\lambda_u t} e^{\lambda_s p}\int_t^\infty e^{-\lambda_u \tau}\nabla g_s^n\left(a + c v_s e^{\lambda_s(\tau - t + p)}, \tau\right)\cdot\left(c\lambda_s v_s e^{\lambda_s(\tau - t + p)}\right) \mathrm{d}\tau \\ &= -c^2\lambda_s^2 e^{(\lambda_u - \lambda_s)t} e^{2\lambda_s p}\int_t^\infty e^{(\lambda_s - \lambda_u)\tau} v_s\cdot\nabla g_s^n\left(a + c v_s e^{\lambda_s(\tau - t + p)}, \tau\right) \mathrm{d}\tau. \end{aligned}$$

Substituting this into (2.84) leads to (2.78), as required. The parameter c disappears at this point, as it must; the rotation angle should be the same for all local points on the stable manifold. Now, (2.79) can be similarly derived, by using the unstable manifold formulation of Theorem 2.12, and will be left as an exercise.

2.6 ▪ Jump discontinuities

In everything so far, Hypothesis 2.3 was assumed for the nonautonomous part of the perturbation. Of particular note was the fact that g was assumed *continuous* in the temporal variable. This ensured that the computed stable and unstable manifolds were smooth in t, as is the expectation when the term "manifold" is used.

Here, following the development in [28], a situation in which the smoothness of g is slightly compromised is considered. Suppose g has a finite number of jump discontinuities in t. Now, this does not prevent solutions from existing, since the integral representation of solutions to (2.6) is given by

$$x(t) = x(0) + \int_0^t \left[f\left(x(\tau)\right) + \varepsilon\, g\left(x(\tau), \tau\right) \right] \mathrm{d}\tau\,. \tag{2.85}$$

A jump discontinuity in the t-variable in g still ensures continuous solutions $x(t)$ are formed in phase space. However, such a jump can mean that the smoothness of the solutions $x(t)$ in t is reduced at the t-values at which the jump occurs; for example, $x(t)$ may not be in $\mathrm{C}^1(\mathbb{R})$. From the perspective of solutions of (2.6) traversing the (x,t) appended phase space, this is not a problem. In other words, the flow remains well-defined, despite the fact that $g(x,t)$ may be ill-defined at a value $t = t_0$ because of a jump discontinuity. This section will outline how stable and unstable manifolds can still be rationalized in a situation of jump discontinuities, though their smoothness is lessened.

For this section, Hypothesis 2.3 will be replaced by the following.

Hypothesis 2.30 (jump discontinuities in g). *There exists $\mathcal{J} = \{t_1, t_2, \ldots, t_n\}$ such that $t_1 < t_2 < \cdots < t_n$, and*

(a) $g(\cdot, t) \in \mathrm{C}^2(\Omega)$ *for any* $t \in \mathbb{R} \setminus \mathcal{J}$;

(b) $g(x, \cdot) \in \mathrm{C}^0(\mathbb{R} \setminus \mathcal{J})$ *for all* $x \in \Omega$, *and moreover*

$$\lim_{t \to t_i^-} g(x,t) \quad \textit{and} \quad \lim_{t \to t_i^+} g(x,t)$$

both exist for any $x \in \Omega$ *and* $i \in \{1, 2, \ldots, n\}$; *and*

(c) g *and* Dg *are both bounded in* $\Omega \times (\mathbb{R} \setminus \mathcal{J})$.

Under the above conditions, the unstable manifold of the perturbed version of a shall exist in the classical sense for $t < t_1$. As t passes through t_1, the hyperbolic trajectory $a_\varepsilon(t)$ evolves continuously, though not necessarily in a C^1-fashion in time. Consider this one-dimensional curve with initial conditions at $t = t_1^-$; this evolves into a one-dimensional curve at $t = t_1^+$ in a continuous fashion. This curve, evolving according to (2.6), shall be *defined* to be the unstable manifold of $a_\varepsilon(t)$ for $t \in (t_1, t_2)$. As this curve, and the hyperberlic trajectory, cross t_2, once again they are well-defined as solutions to (2.6). This process serves to define both the hyperbolic trajectory and the unstable manifold for times as large as needed. In a similar vein, the stable manifold of $a_\varepsilon(t)$ also is well-defined for as negative a time as needed.

When carefully examining the proofs of Theorems 2.12 and 2.23 to determine the locations of the unstable and stable manifolds, it is clear that the taking of t-derivatives is not legitimate at the times t_i. However, in computing the quantities $M^{u,s}$ an integral of these terms is necessary; essentially, this is akin to considering the integral representation (2.85) rather than the differential formulation. The continuing solutions to the differential equations are obtained by integrating through the jump discontinuities, which can be performed legitimately. Thus, this process will give solutions which are, by definition, the stable and unstable manifolds as explained above. Hence, the expressions obtained in Theorems 2.12 and 2.23 can be used with no additional modifications, even though Hypothesis 2.3 has been relaxed to Hypothesis 2.30.

Remark 2.31 (nonsmoothness of stable/unstable manifolds). The "manifolds" that one defines through this process are continuous across the jump values but are *not differentiable* in general. Thus, using the term "manifolds" might be questionable; perhaps *pseudomanifolds* might be more appropriate. A formal definition will not be attempted here since these will be defined under far more serious discontinuities (impulsive ones) in the next section.

Example 2.32 (Duffing oscillator cont.). The Duffing oscillator examined in Examples 2.11 and 2.21 is now considered once again, but with no damping and with the forcing being a jump given by $\phi(t) = \mathbb{1}_{(0,\infty)}(t)$, where $\mathbb{1}$ is the indicator function. Essentially, all derived formulae for continuous perturbations are valid in this case. From (2.20), the hyperbolic trajectory location will be

$$x_1(t) = \begin{cases} -\frac{\varepsilon}{2} e^t & \text{if } t \leq 0, \\ -\frac{\varepsilon}{2}(2 - e^{-t}) & \text{if } t > 0, \end{cases}$$

with $x_2(t) = \dot{x}_1(t)$. It is clear that $(x_1(t), x_2(t))$ is continuous in t at the jump value $t = 0$, but, as expected, the time-derivative (specifically $\dot{x}_2(t)$) has a discontinuity. The unstable Melnikov function is, from (2.26),

$$\begin{aligned} M^u(p,t) &= -\int_{-\infty}^{p} \sqrt{2} \operatorname{sech} \tau \tanh \tau \, \mathbb{1}_{(p-t,\infty)}(\tau) d\tau \\ &= \sqrt{2} [\operatorname{sech}(p) - \operatorname{sech}(p-t)] \, \mathbb{1}_{(0,\infty)}(t), \end{aligned}$$

in which M^u's continuity at $t = 0$ is clear. The next step is computing B^u, but in this case it turns out that neither of the integrals resulting from (2.27) are explicitly computable (unlike for M^u). However, B^u will also be continuous in t, and thus the unstable manifold location $x_\varepsilon^u(p,t)$ computed from M^u and B^u using (2.25) shall retain this continuity. As the final numerical computations are not particularly inspiring, they shall be skipped and left as an exercise for the reader. ∎

Example 2.33 (double-gyre cont.). Previously addressed in Examples 2.25, 2.28, and 2.56, the double-gyre is now considered with $q(t) = \mathbb{1}_{[0,T]}(t)$ for some $T > 0$. This

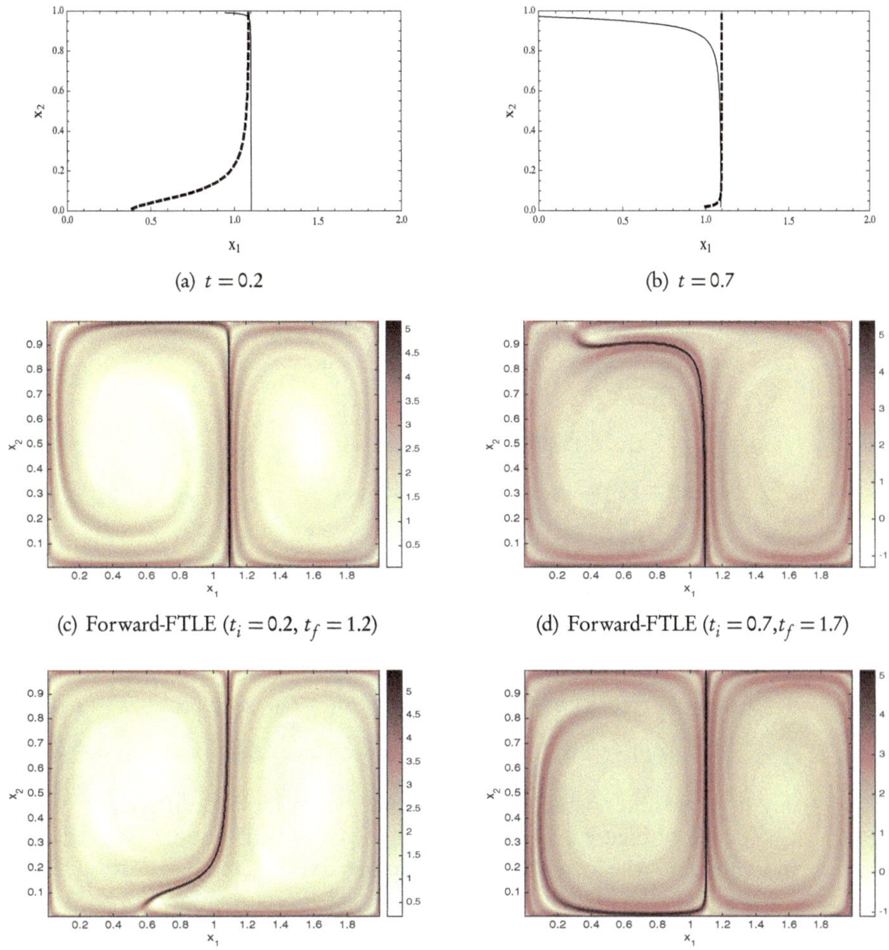

(a) $t = 0.2$

(b) $t = 0.7$

(c) Forward-FTLE ($t_i = 0.2$, $t_f = 1.2$)

(d) Forward-FTLE ($t_i = 0.7, t_f = 1.7$)

(e) Backward-FTLE ($t_i = 0.2$, $t_f = -0.8$)

(f) Backward-FTLE ($t_i = 0.7$, $t_f = -0.3$)

Figure 2.13. *(a) and (b): Stable manifold [solid] and unstable manifold [dashed] of the double-gyre for $q(t) = \mathbb{1}_{[0,T]}(t)$, $A = 1$, $\varepsilon = 0.1$, and $T = 1$, using (2.87) and (2.86). (c) and (d): Forward-time FTLE fields. (e) and (f): Backward-time FTLE fields.*

corresponds to a transitory perturbation [281, cf.] which is zero outside a finite interval in time. Consider first the unstable manifold equation (2.70). If $t > T$, we get from (2.69)

$$\begin{aligned} x_1^u(p,t) &= 1 - \varepsilon \frac{\pi^2 A}{\operatorname{sech}\,(\pi^2 A p)} \int_{-\infty}^{p} \tanh\left(\pi^2 A \tau\right) \operatorname{sech}\left(\pi^2 A \tau\right) \mathbb{1}_{[0,T]}(\tau + t - p)\,\mathrm{d}\tau \\ &= 1 - \varepsilon \frac{\pi^2 A}{\operatorname{sech}\,(\pi^2 A p)} \int_{p-t}^{T+p-t} \tanh\left(\pi^2 A \tau\right) \operatorname{sech}\left(\pi^2 A \tau\right) \mathrm{d}\tau \\ &= 1 + \varepsilon \cosh\left(\pi^2 A p\right)\left(\operatorname{sech}\left[\pi^2 A (T - t + p)\right] - \operatorname{sech}\left[\pi^2 A (p - t)\right]\right) + \mathcal{O}\left(\varepsilon^2\right). \end{aligned}$$

For $0 < t \leq T$, the upper integration limit above needs adjustment, leading to

$$x_1^u(p,t) = 1 + \varepsilon\left(1 - \operatorname{sech}\left[\pi^2 A (p-t)\right] \cosh\left[\pi^2 A p\right]\right) + \mathcal{O}\left(\varepsilon^2\right),$$

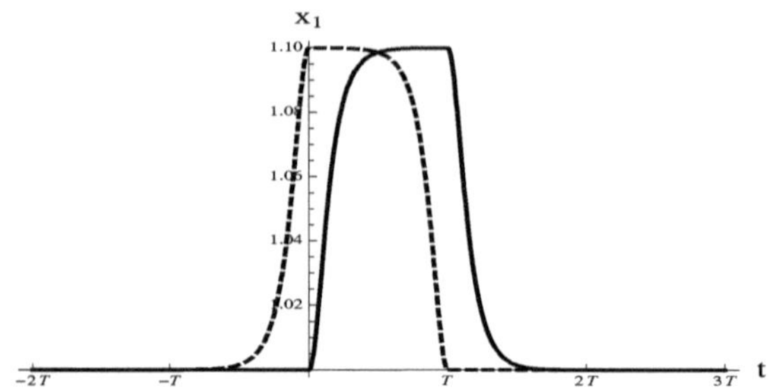

Figure 2.14. *Variation of x_1-coordinate of hyperbolic trajectory a_ε (solid) and b_ε (dashed) for the double-gyre with $q(t) = \mathbb{1}_{[0,T]}(t)$, $T = 1$ and $\varepsilon = 0.1$.*

while if $t \leq 0$, the $\mathcal{O}(\varepsilon)$ term vanishes. In summary,

$$x_1^u(p,t) = 1 + \varepsilon\left(1 - \operatorname{sech}\left[\pi^2 A(p-t)\right]\cosh\left[\pi^2 Ap\right]\right)\mathbb{1}_{(0,T]}(t) \qquad (2.86)$$
$$+\varepsilon\cosh\left(\pi^2 Ap\right)\left(\operatorname{sech}\left[\pi^2 A(T-t+p)\right] - \operatorname{sech}\left[\pi^2 A(p-t)\right]\right)\mathbb{1}_{(T,\infty)}(t) + \mathcal{O}\left(\varepsilon^2\right),$$

which can be written in the form of x_1 as a graph of x_2 at each fixed t by using (2.68) to replace p with x_2. A similar analysis leads to the representation of the stable manifold as

$$x_1^s(p,t) = 1 + \varepsilon\cosh\left(\pi^2 Ap\right)\left(\operatorname{sech}\left[\pi^2 A(p-t)\right] - \operatorname{sech}\left[\pi^2 A(T-t+p)\right]\right)\mathbb{1}_{(-\infty,0)}(t)$$
$$+\varepsilon\left(1 - \operatorname{sech}\left[\pi^2 A(T-t+p)\right]\cosh\left(\pi^2 Ap\right)\right)\mathbb{1}_{(0,T]}(t) + \mathcal{O}\left(\varepsilon^2\right). \qquad (2.87)$$

These stable and unstable manifolds are shown for two different t-values in the top two panels of Figure 2.13. As done for the "standard" double-gyre with $q(t) = \sin\omega t$ in Figure 2.10, the lower panels contain the forward- and backward-FTLE calculations for this situation. Their ridges are numerical approximations for the stable and unstable manifold, respectively. Excellent agreement between the theory and the numerics is achieved. It appears (and this can be proven; see Example 3.25) that there is one intersection between the stable and unstable manifolds, which travels downwards as t increases, towards the lower hyperbolic trajectory b_ε. The time-variation of b_ε can be obtained from (2.74) to be

$$x_1^s(0,t) = 1 + \varepsilon\, e^{\pi^2 At}\left(1 - e^{-\pi^2 AT}\right)\mathbb{1}_{(-\infty,0)}(t) + \varepsilon\left(1 - e^{\pi^2 A(t-T)}\right)\mathbb{1}_{(0,T]}(t) + \mathcal{O}\left(\varepsilon^2\right),$$

and similarly the hyperbolic trajectory a_ε has a representation

$$x_1^u(1,t) = 1 + \varepsilon\left[1 - e^{-\pi^2 At}\right]\mathbb{1}_{(0,T]}(t) + \varepsilon e^{-\pi^2 At}\left[e^{\pi^2 AT} - 1\right]\mathbb{1}_{(T,\infty)}(t) + \mathcal{O}\left(\varepsilon^2\right).$$

The variation of the x_1-coordinate of these hyperbolic trajectories is shown in Figure 2.14. The effect of the perturbation for $t \in [0,T]$ is seen to have influence beyond this time interval. The lower hyperbolic trajectory b_ε shows abrupt motion near $t = 0$ but a less abrupt settling back to zero near $t = T$ (when the perturbation is turned off). The opposite behavior is observed in a_ε. ∎

2.7 ▪ Impulsive discontinuities

The previous section outlined how stable and unstable manifolds could be interpreted when the perturbing vector field $g(x,t)$ had a finite number of jump discontinuities in time. This situation turned out to be not serious; all previous expressions related to the manifold locations remained legitimate, with the only proviso that the manifolds, when thought of as a parametrized family in time t, remained continuous across the jump discontinuities but were not C^1-smooth in t.

This section approaches a significantly more serious discontinuity in which $g(x,t)$ is permitted to have Dirac delta-type behavior at a finite number of t-values. The methods of this section are based on the development in [37]. Addressing this problem is not purely of theoretical interest; this situation could be used to model the sharp tapping of a microfluidic device, or an underwater eruption in the sea, which would thereby modify pre-existing stable and unstable manifolds. The intuition is to consider

$$\dot{x} = f(x) + \varepsilon \sum_{i=1}^{n} \delta(t - t_i) h_i(x,t), \tag{2.88}$$

where δ is the Dirac delta "function," and $\{t_1, t_2, \ldots, t_n\}$ is an increasing set of finite time values at which the impulses occur. The functions h_i are assumed smooth. Now, in comparison with (2.6), the perturbing function g is apparently supposed to be chosen as

$$g(x,t) = \sum_{i=1}^{n} \delta(t - t_i) h_i(x,t), \tag{2.89}$$

which certainly does not satisfy smoothness or boundedness, and indeed is not even a function for $t \in \mathbb{R}$. Therefore, (2.88) *is to be interpreted symbolically only*, since it does not generate a flow in the phase space in the usual sense. For example, if one starts with any initial condition $x(\beta)$ where $\beta < t_1$, this solution will evolve continuously until $x(t_1^-)$, since there are no impulses present. However, as t crosses t_1, the solution will jump to a new location $x(t_1^+)$, where the jump is apparently quantified by $\varepsilon h_1(x(t_1), t_1)$. However, this is ambiguous, since the very presence of the jump implies that $x(t_1)$ is not necessarily well-defined. Hence, (2.88) as it stands is ill-defined.

In the standard approach to *impulsive differential equations* one usually considers a smooth differential equation, to which is attached a separate equation explicitly characterizing the "kicks" which occur at discrete times [19, 58, 57, 306, 194, 245, 142, 89]. This approach has applications in the control and stabilization literature [280, 179, 214, 300, 114, 194, 89, e.g.], since one can think of using the kicks to induce trajectory jumps from undesirable to desirable locations. A typical description of this approach is to imagine trajectories which are escaping into a chaotic region by going through the point p: these can be reset by a jump to a previous (desirable) location q. Each time the trajectory once again arrives at p, an impulsive kick can be applied to make the trajectory jump back to q, resulting in a forced periodic trajectory in a region which contains no stable periodic trajectories. One usually thinks of this motion in an *autonomous* phase space, as it is a steady flow which usually governs the motion from q to p. However, the jumps imposed imply that trajectories in the autonomous phase space are reset to a new location periodically. In defining the kicks which reset trajectories, a decision regarding the above issue of $x(t_1)$ being ill-defined has been implicitly made. A subtle issue is that only *one* time direction can be considered with this approach, since the trajectory resetting equation need not be invertible. In results which attempt to characterize the stable and unstable manifold analogues using the approach of impulsive differential equations, the usual results are able to obtain existence using functional analytic methods [58, 57, 306, 142] but do not provide methods for locating the invariant manifolds at each time. As shall be shortly presented, these issues of a single time direction and inability to locate the invariant manifolds can be resolved if considering instead an *integral equation* on an appropriately restricted augmented phase space $\Omega \times \mathbb{R}$.

Time-impulsive flows have an interesting connection to Melnikov methods which have been proposed for a different scenario: when the autonomous vector field in an ordinary differential equation has discontinuities across some codimension-1 surfaces of Ω [59, 95, 230, 231, e.g.]. If this situation is also viewed in the $(n+1)$-dimensional $\Omega \times \mathbb{R}$ augmented phase space, these discontinuities appear as n-dimensional fracture surfaces. The impulsive situation that is being considered in this section is similar in that the discontinuities occur across the hypersurfaces $t = t_i$, which are also n-dimensional fracture surfaces in $\Omega \times \mathbb{R}$. Of course, the t-direction is a special one in nonautonomous phase space, which make the impulsive situation somewhat different to the autonomous discontinuous situation. Nevertheless, when viewed in $\Omega \times \mathbb{R}$ augmented phase space, the connection between these quite different issues is suggestive; there is perhaps a unified method for examining these two scenarios. The connection has further interest since several spatially discontinuous studies [59, 95, 230, 133, 429] which seek heteroclinic connections *also* utilize Melnikov-type methods.

However, the approach here is to rephrase the ill-defined equation (2.88) in terms of an integral equation

$$x(t) = x(\beta) + \int_{\beta}^{t} f(x(\xi))\, \mathrm{d}\xi + \frac{\varepsilon}{2} \sum_{i=1}^{n} u(\beta, t_i, t) \left[h_i(x(t_i^-), t_i^-) + h_i(x(t_i^+), t_i^+) \right], \tag{2.90}$$

where $x \in \Omega$, a two-dimensional open connected set, and

$$u(\beta, t_i, t) := \begin{cases} \mathbb{1}_{(\beta,t)}(t_i) & \text{if } \beta < t \\ -\mathbb{1}_{(t,\beta)}(t_i) & \text{if } \beta > t \\ 0 & \text{if else} \end{cases}. \tag{2.91}$$

The function u above "turns on" each perturbation term if t_i lies between the initial value β and the current value t but is set to zero if any two of β, t, or t_i coincide. The integral equation (2.90) models (2.88) in the sense that it is the solution obtained if approximating $\delta(t - t_i)$ as the "limit" of a square pulse of shrinking width and expanding height centred at t_i, with unit L^1-norm. In representing (2.90) a particular choice of this limiting procedure had to be made; alternative choices (e.g., pulses which are nonsymmetrical about t_i) lead to *different* integral formulations, each potentially having different solutions. The form chosen for (2.90) is the most natural choice in the sense that the limit preserves the symmetry that we "expect" from the Dirac delta distribution. Some assumptions associated with (2.90) need to be clearly stated.

Hypothesis 2.34 (conditions on f). *The function f in (2.90) satisfies $f \in \mathrm{C}^2(\Omega)$ with Df bounded in Ω. Moreover, there exists $a \in \Omega$ such that $f(a) = 0$ and $Df(a)$ has one positive and one negative eigenvalue.*

Hypothesis 2.35 (properties of impulsive perturbation). *There exists an increasing set of n time-values $\mathcal{J} := \{t_1, t_2, \ldots, t_n\}$ (which is called the* jump set*) such that the perturbation in (2.90) is smooth outside these values:*

1. *For each $i \in \{1, 2, \ldots, n\}$, $h_i(\cdot, t) \in \mathrm{C}^1(\Omega)$, with both h_i and Dh_i bounded on Ω.*
2. *For each $x \in \Omega$, and each $i \in \{1, 2, \ldots, n\}$, $h_i(x, \cdot) \in \mathrm{C}^1(\mathbb{R})$.*

Since the t-dependence of the h_i only comes into action exactly on the jump set, and since each function is assumed smooth in t, it is possible to dispense with the explicit t-dependence by defining the functions $\tilde{h}_i(x) = h_i(x, t_i)$. However, this shall not be done in what follows, in order to keep clear how the time in the original formulation (2.90) affects the solutions. Now, even though the h_is are smooth, the jump discontinuity in (2.90) at the values in $\mathcal{J}$ render its solutions—if they exist—*discontinuous.* While the existence and uniqueness of solutions of (2.90) is automatic while traversing times which do not cross $\mathcal{J}$, the potential breakdown of these properties when crossing time-values in $\mathcal{J}$ needs to be verified. This is accomplished in the following lemma.

Lemma 2.36 (existence and smoothness of jump map). *Consider the evolution of (2.90) across a time-value $t_j \in \mathcal{J}$. Under Hypotheses 2.34 and 2.35, (2.90) defines a unique solution transition on Ω for sufficiently small $|\varepsilon|$; i.e., given any $x(t_j^-) \in \Omega$, there is a unique $x(t_j^+)$ which solves (2.90), such that the associated "jump map" is in $C^1(\Omega)$. The same holds in the reverse time-direction; i.e., given any $x(t_j^+)$, there exists a unique $x(t_j^-)$ satisfying (2.90), with the mapping in $C^1(\Omega)$,*

Proof. Apply the integral formulation (2.90) from an initial value $\beta = t_j^-$ to a final value $t = t_j^+$. The first integral vanishes since it is of a bounded function over an interval which shrinks to zero. Only the terms associated with t_j survive in the summation since this is the only discontinuity, and since h_j is continuous in t,

$$x\left(t_j^+\right) = x\left(t_j^-\right) + \frac{\varepsilon}{2}\left[h_j\left(x(t_j^-), t_j\right) + h_j\left(x(t_j^+), t_j\right)\right]$$

is obtained. Now, define the mapping G_j on Ω implicitly by

$$G_j(x) - x - \frac{\varepsilon}{2}\left[h_j\left(x, t_j\right) + h_j\left(G_j(x), t_j\right)\right] = 0.$$

If $y = G_j(x)$, it is required to solve

$$y - x - \frac{\varepsilon}{2}\left[h_j(x, t_j) + h_j(y, t_j)\right] = 0.$$

Note that when $\varepsilon = 0$, a unique solution for $y(x, \varepsilon)$ is $y = x$. Now, the y-derivative of the left-hand side above is

$$d := \left(I - \frac{\varepsilon}{2}\left(\frac{\partial(h_j^1, h_j^2)}{\partial(y^1, y^2)}\right)\right),$$

where the superscripts identify components and I is the identity matrix. Each term perturbing the above from the identity is bounded by $(\varepsilon/2)\sup_i \eta_i$, where the η_is are the derivative bounds on h ensured by Hypothesis 2.35. and therefore d is a small perturbation from the identity. Hence, for small enough ε, $\det(d) \neq 0$. Thus, for any $x_0 \in \Omega$, there exists an open neighborhood $B(x_0)$ and also a small interval containing 0 (say, E), such that for $(x, \varepsilon) \in B(x_0) \times E$, y can be solved uniquely as a function of (x, ε) by the Implicit Function Theorem. This moreover establishes that y is as smooth in x as is h_j. Since this works for any $x_0 \in \Omega$, a global smooth solution $y(x, \varepsilon)$ exists on $\Omega \times E$. Now, the invertibility of this process is easy to establish as well by considering solving for x as a function of y; the argument is identical. □

Therefore, all solutions of (2.90) evolve smoothly until a transition time t_j, at which point all solutions "reset." This understanding works in either backwards or forwards time. In particular, the evolution equation (2.90) is seen to generate piecewise continuous functions $x(t)$, with jumps on $\mathcal{J}$. That is, (2.90) provides solutions for $t \in \mathbb{R} \setminus \mathcal{J}$.

When $\varepsilon = 0$, a was a saddle point, which is hyperbolic. In the smooth Melnikov analyses so far, allowing ε to be nonzero resulted in this perturbing to a trajectory $a_\varepsilon(t)$ which continued to be hyperbolic. The situation under impulsive perturbations is, however, different. Consider choosing $\beta < t_1$ in (2.90), with $x(\beta) = a$. Since a fixed point of f, a continues to be the solution until time t_1, that is, $x(t) = a$ for $t < t_1$. As t_1 is crossed, $x(t_1^+)$ will jump to a $\mathcal{O}(\varepsilon)$-close point which will generically *not* be a—and thus will not behave as a fixed point during the time interval (t_1, t_2)—which will mean that $x(t)$ will undergo a nontrivial evolution during this time interval. At t_2, $x(t)$ will once again jump, and so on, until passing the final jump time t_n. Since $x(t_n^+)$ will also not be a fixed point, the subsequent evolution will be governed by $x(t) = x(t_n^+) + \int_{t_n}^t f(x(\tau))\, d\tau$. If this trajectory just described is labeled $a_+(t)$, then it will be defined for $t \in (-\infty, T_u] \setminus \mathcal{J}$ for any finite T_u as long as the trajectory remains within Ω, and be $\mathcal{O}(\varepsilon)$-close to a in this domain of validity. Is this, then, the analogue of *the* perturbed hyperbolic trajectory of the previous section? The lack of ability to push the $\mathcal{O}(\varepsilon)$-closeness to $\mathbb{R} \setminus \mathcal{J}$ is a warning sign against this. Indeed, there is *another*, quite comparable trajectory $a_-(t)$ which arises from taking $x(\beta) = a$ for $\beta > t_n$ and evolving (2.90) *backwards* in time; this will be defined for $t \in [T_s, \infty) \setminus \mathcal{J}$ for $-T_s$ arbitrarily large but finite. The two trajectories $a_+(t)$ and $a_-(t)$ are, respectively, a's forward iterate from $-\infty$ and backward iterate from $+\infty$ under the perturbed flow and *need not coincide*. Therefore, previous results for the smooth situation, in which $a_\varepsilon(t)$ existed uniquely as a nearby hyperbolic trajectory and retained *both* its stable and its unstable manifold, fail.

How can stable and unstable manifolds be rationalized? To do so, as before, the unperturbed geometry in which a saddle point a possessing stable and unstable manifolds is assumed. In other words, when $\varepsilon = 0$, the integral equation can be written as a differential equation (2.9), for which Hypothesis 2.5 is assumed. It is clear that stable and unstable manifolds in the standard sense will *not* exist when $\varepsilon \neq 0$, since trajectories lying on these purported manifolds will get reset to new locations when t passes through points in $\mathcal{J}$. So, for example, had there been an unstable manifold at time t_1^-, all points on that manifold will jump to different locations at $t = t_1^+$. There will certainly not be any smoothness in t, implying that the usage of the word *manifolds* is illegitimate. It will be necessary to define *pseudomanifolds* in some intuitively meaningful way to discuss the evolution of flow separators in this instance. This section follows ideas developed in [28] but will need to relax assumptions on area-preservation and moreover will target the locations of each of the manifolds as opposed to the normal distance between them [28]. This latter issue of normal distances will automatically follow from what is presented here and will be important in the flux computations which will be presented subsequently in Chapter 3.

Consider solutions of (2.90) when $\varepsilon = 0$, in the augmented (x, t) phase space. This is equivalent to what has been discussed in Sections 2.2–2.5, and in particular the reader is referred to Figure 2.2. Now, the unstable manifold is a two-dimensional surface in the (x, t) space. In a time-slice t, the manifold is one-dimensional and can be parametrized by p such that the point $\bar{x}^u(p)$ describes this curve. As $p \to -\infty$, $\bar{x}^u(p) \to a$, the saddle fixed point. However, no conditions have been imposed on the

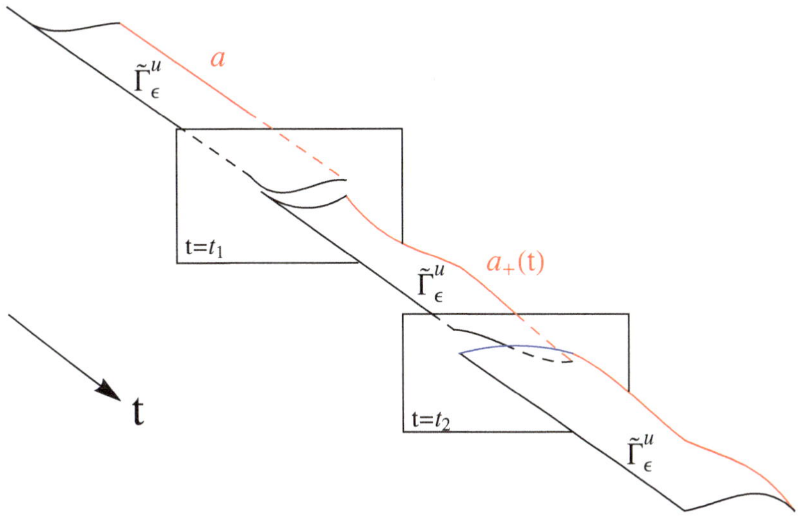

Figure 2.15. *The unstable pseudomanifold $\tilde{\Gamma}^u_\varepsilon$ of a associated with (2.90), which comprises segments of smooth surfaces with jump discontinuities at t_i, $i = 1,2,\ldots,n$. The red curve is the "hyperbolic-like" trajectory $a_+(t)$, to which trajectories on $\tilde{\Gamma}^u_\varepsilon$ are attracted in backwards time.*

unstable manifold in the opposite limit (it may escape to infinity, approach another fixed point to form a heteroclinic manifold, spiral in towards an attracting fixed point or a limit cycle, etc.). Therefore, the parametrization of the unstable manifold with respect to (p,t) will work for $p \in (-\infty, P]$ and $t \in (-\infty, T_u]$, with P and T_u being as large as one wishes, but finite.

Now, consider $\varepsilon \neq 0$. For times $t < t_1$ (when the first impulse occurs), there is no perturbation in (2.90). Thus, a remains a fixed point in the sense that a trajectory of (2.90) beginning at a remains at a until time t_1. Since there is no perturbation to the adjacent velocity field until t_1, this means that the unstable manifold therefore remains exactly as it was, until this time. Figure 2.15 shows *one branch* of this unstable manifold, which is a smooth surface for $t < t_1$. If a time-slice is taken at any time $t < t_1$, the unstable manifold curve that one gets is the *same*, since there is no nonautonomous part to the velocity. But at t_1, all trajectories on the unstable manifold undergo a jump. This is of necessity $\mathcal{O}(\varepsilon)$ by (2.90) and moreover is smooth in the spatial variable by Lemma 2.36. Therefore, the unstable manifold curve which exists at $t = t_1^-$ will get mapped smoothly to a curve at $t = t_1^+$. Intuitively, it will be the evolution of this curve that shall be defined to be the analogue of the unstable manifold of a. The point a was "at the end of the manifold" at t_1^-; this point jumps to another point at t_1^+ which, from the previous discussion, is the point $a_+(t_1^+)$. Now, until $t = t_2$, the flow will remain autonomous, and thus trajectories on this curve will evolve autonomously until t_2. However, there is a caveat here. While each trajectory evolves according to a steady velocity field, there is no reason for the curve of the unstable manifold to be itself identical to a trajectory. When $t < t_1$, the velocity field was of necessity tangential to the unstable manifold in each time-slice, since indeed the manifold could be represented as shifts of a trajectory of the steady flow. This property no longer holds when $t > t_1$; points on the "unstable manifold curve" at t_1^+ will generically have velocity components normal to the curve. Basically, this is like a curve of ink being placed in a steady flow in a way such that the velocity is not tangential to the curve, and thus the curve will evolve in a time-varying fashion. Having evolved until time t_2^-, the curve will jump to a new curve at $t = t_2^+$. This process will occur until $t = t_n^+$ (i.e., the final impulse), after which trajectories on the curve will evolve autonomously. At all instances of time not in $\mathscr{J}$, the end of this curve will be described by $a_+(t)$, shown by the red curve in Figure 2.15. Any point chosen on this t-parametrized family of curves will in backwards time asymptote to a under the flow generated by (2.90). Thus, any point chosen on the surfaces $\tilde{\Gamma}_\varepsilon^u$ shown in Figure 2.15 (illustrated only until a t-value between t_2 and t_3) will approach a as $t \to -\infty$.

Definition 2.37 (unstable pseudomanifold). *The unstable pseudomanifold of a in the augmented phase space* $\Omega \times (-\infty, T_u] \backslash \mathscr{J}$ *for any finite* T_u *is defined by*

$$\tilde{\Gamma}_\varepsilon^u := \Bigg\{ (x(\beta), \beta) \ : \ \textit{all } x(\beta) \in \Omega \textit{ for which } x(t) \to a \textit{ as } t \to -\infty \, , \qquad \textit{for each } \beta \in (-\infty, T_u] \backslash \mathscr{J} \Bigg\}, \tag{2.92}$$

where $x(t)$ *is the evolution defined in (2.90).*

Of course, there are two branches of the unstable manifold when $\varepsilon = 0$, and there are correspondingly two branches of the unstable pseudomanifold. The characterization of one of these branches (associated with the unstable branch with trajectory defined by $\bar{x}^u(t)$) is what is now sought.

Theorem 2.38 (unstable pseudomanifold normal displacement [37]). *Consider (2.90) under Hypotheses 2.34, 2.35, and 2.5. The unstable pseudomanifold of a has a parametric representation* $(x_\varepsilon^u(p,t), t)$ *with parameters* $(p,t) \in (-\infty, P] \times (-\infty, T_u] \backslash \mathscr{J}$ *for arbitrarily large but fixed P and* T_u, *such that*

$$\left[x_\varepsilon^u(p,t) - \bar{x}^u(p)\right] \cdot \hat{f}^\perp(\bar{x}^u(p)) = \varepsilon \frac{M^u(p,t)}{|f(\bar{x}^u(p))|} + \mathcal{O}\left(\varepsilon^2\right), \tag{2.93}$$

where the associated unstable Melnikov function is given by

$$M^u(p,t) = \sum_{i=1}^n \mathbb{1}_{(t_i,\infty)}(t) j_i^u(p,t) + \sum_{i=1}^{\max\{j : t_j < t\}} \int_{t_i}^t R_p^u(t-\xi) j_i^u(p,\xi) \, \mathrm{d}\xi \, , \tag{2.94}$$

in which

$$j_i^u(p,t) := f^\perp(\bar{x}^u(t_i - t + p)) \cdot h_i(\bar{x}^u(t_i - t + p), t_i) \tag{2.95}$$

and the resolvent R_p^u *is defined in terms of Laplace transforms by*

$$R_p^u(t) := \mathscr{L}^{-1}\left\{ \frac{\hat{F}_p^u(s)}{1 - \hat{F}_p^u(s)} \right\}(t), \quad \hat{F}_p^u(s) := \mathscr{L}\left\{ \mathrm{Tr} Df(\bar{x}^u(p-t)) \right\}(s). \tag{2.96}$$

Proof. See Section 2.7.1. ☐

Remark 2.39. Unlike Theorem 2.12, Theorem 2.38 only characterizes the *normal* displacement of the unstable manifold. As is apparent, the expressions are messy even under this limitation, and even numerical evaluation would not be easy in general given the difficulty of numerically computing inverse Laplace transforms. Determining formulae for the tangential displacement is likely to be nontrivial.

Remark 2.40 (extension to countable impulses). It is possible to extend the result of Theorem 2.38 to $\mathcal{J}$ having a countable number of impulses in the form $\{t_i\}_{i=1}^{\infty}$, such that $t_i \to \infty$ as $i \to \infty$. If so, the formula (2.94) has the simple modification of the upper index n in the first summation being replaced by ∞. The ability to make $\mathcal{J}$ infinite, but not bi-infinite, is related to the necessity of unidirectionality of time in impulsive differential equations [19, 58, 57, 306, 194, 245]. It will turn out that this relaxation will mean that the *stable* pseudomanifold *cannot* be defined. In the interests of being able to simultaneously define both the stable and the unstable pseudomanifolds, the approach shall therefore be restricted to having only a finite number of jump values.

Corollary 2.41 (unstable pseudomanifold under area-preservation). *Under the conditions of Theorem 2.38, consider the additional assumption that f is area-preserving. Then, (2.94) simplifies to*

$$M^u(p,t) = \sum_{i=1}^{n} \mathbb{1}_{(t_i,\infty)}(t) f^{\perp}(\bar{x}^u(t_i - t + p)) \cdot h_i(\bar{x}^u(t_i - t + p), t_i) \,. \tag{2.97}$$

Proof. Since in this case $\operatorname{Tr} Df = 0$, $R_p^u = 0$ from (2.96). Thus, from (2.94), the second summation disappears. □

Remark 2.42 (formal Melnikov computation under impulses). If $\operatorname{Tr} Df = 0$, then a *purely formal* insertion of (2.89) into the (smoothly derived) Melnikov function (2.26) recovers the result of Corollary 2.41 (see also [28]). While this formal procedure might be pursued for more general f, Theorem 2.38 shows that the result would be incorrect. A careful analysis, using the integral formulation (2.90), is therefore necessary.

So far, it is only the unstable pseudomanifold that has been addressed. The modifications for the *stable* pseudomanifold are analogous. When $\varepsilon = 0$, $\bar{x}^s(t)$ is assumed to be a trajectory on a branch of the stable manifold, such that $\bar{x}^s(t) \to a$ as $t \to \infty$. When $\varepsilon \neq 0$, a remains a fixed point for $t > t_n$, with its stable manifold well-defined; this is simply taken in backwards time across the time-discontinuities to generate the stable pseudomanifold.

Definition 2.43 (stable pseudomanifold). *The unstable pseudomanifold of a in the augmented phase space $\Omega \times [T_s, \infty) \setminus \mathcal{J}$ is defined by*

$$\tilde{\Gamma}_\varepsilon^s := \Big\{ (x(\beta), \beta) \ : \ \textit{all } x(\beta) \in \Omega \textit{ for which } x(t) \to a \text{ as } t \to \infty \,, \quad \textit{for each } \beta \in [T_s, \infty) \setminus \mathcal{J} \Big\}, \tag{2.98}$$

where $x(t)$ is the evolution defined in (2.90).

Theorem 2.44 (stable pseudomanifold normal displacement [37]). *Consider (2.90) under Hypotheses 2.34, 2.35, and 2.5. The stable pseudomanifold of a has a parametric representation* $(x_\varepsilon^s(p,t),t)$ *with parameters* $(p,t)\in[-P,\infty)\times[T_s,\infty)\setminus\mathcal{J}$ *for arbitrarily large but fixed* P *and* $-T_s$*, such that*

$$[x_\varepsilon^s(p,t)-\bar{x}^s(p)]\cdot\hat{f}^\perp(\bar{x}^s(p))=\varepsilon\frac{M^s(p,t)}{|f(\bar{x}^s(p))|}+\mathcal{O}\left(\varepsilon^2\right), \tag{2.99}$$

where the associated stable Melnikov function is given by

$$M^s(p,t)=-\sum_{i=1}^{n}\mathbb{1}_{(-\infty,t_i)}(t)j_i^s(p,t)+\sum_{i=\min\{j:t_j>t\}}^{n}\int_t^{t_i}R_p^s(t-\xi)j_i^s(p,\xi)\,\mathrm{d}\xi\,, \tag{2.100}$$

in which

$$j_i^s(p,t):=f^\perp(\bar{x}^s(t_i-t+p))\cdot h_i(\bar{x}^s(t_i-t+p),t_i) \tag{2.101}$$

and the resolvent R_p^s *is defined in terms of Laplace transforms by*

$$R_p^s(t):=\mathcal{L}^{-1}\left\{\frac{\hat{F}_p^s(s)}{1+\hat{F}_p^s(s)}\right\}(-t),\quad \hat{F}_p^s(s):=\mathcal{L}\left\{\mathrm{Tr}Df(\bar{x}^s(p+t))\right\}(s). \tag{2.102}$$

Proof. While this is analogous to the proof of Theorem 2.38, subtle issues arise in using the Laplace transform method since the relevant functions are defined on $\mathbb{R}^-$ as opposed to $\mathbb{R}^+$. The details of the proof are therefore outlined in Section 2.7.3. $\square$

Remark 2.45 (extension to countable impulses). Similar to what has been stated in Remark 2.40, it is possible to extend Theorem 2.44 to the situation in which $\mathcal{J}$ has a countable number of impulses in the form $\{t_i\}_{i=-\infty}^{n}$, such that $t_i\to-\infty$ as $i\to-\infty$. The modification to (2.100) is to replace the lower index in the first summation to $i=-\infty$.

Corollary 2.46 (stable pseudomanifold under area-preservation). *Under the conditions of Theorem 2.44, consider the additional assumption that* f *is area-preserving. Then, (2.100) simplifies to*

$$M^s(p,t)=-\sum_{i=1}^{n}\mathbb{1}_{(-\infty,t_i)}(t)f^\perp(\bar{x}^s(t_i-t+p))\cdot h_i(\bar{x}^s(t_i-t+p),t_i)\,. \tag{2.103}$$

Proof. Simply set $\mathrm{Tr}Df=0$, as in the proof of Corollary 2.41. $\square$

Example 2.47. Suppose $f(x)=(2x_1,-x_2)$. Thus, when $\varepsilon=0$, the system (2.90) is a simple linear flow with a saddle point at the origin, with unstable and stable manifolds, respectively, along the x_1 and x_2 axes. Consider only the branch of the unstable manifold lying along the positive x_1 axis. For this,

$$\bar{x}^u(t)=\begin{pmatrix}e^{2t}\\0\end{pmatrix},f(\bar{x}^u(t))=\begin{pmatrix}2e^{2t}\\0\end{pmatrix},f^\perp(\bar{x}^u(t))=\begin{pmatrix}0\\2e^{2t}\end{pmatrix}.$$

Consider a situation of only one impulse occurring at $t_1 = 0$, with corresponding $h_1(x,t) = (x_1 + x_2, x_1 \cos t)^\top$. Then,

$$j_1^u(p,t) = f^\perp \cdot h_1(\bar{x}^u(t_1 - t + p), t_1) = 2e^{2(0-t+p)}e^{2(0-t+p)}\cos 0 = 2e^{4p}e^{-4t}.$$

Now, $\operatorname{Tr} Df = 2 - 1 = 1$ (a constant) for this case. Thus, $\hat{F}_p^u(s) = 1/s$, and

$$R_p^u(t) = \mathcal{L}^{-1}\left\{\frac{1/s}{1 - 1/s}\right\}(t) = e^t,$$

which from (2.94) gives the expression

$$\begin{aligned} M^u(p,t) &= 2e^{4p}\mathbb{1}_{(0,\infty)}(t)e^{-4t} + \mathbb{1}_{(0,\infty)}(t)\int_0^t e^{t-\xi}2e^{4p}e^{-4\xi}\,\mathrm{d}\xi \\ &= 2e^{4p}\mathbb{1}_{(0,\infty)}(t)\left\{e^{-4t} + e^t \left.\frac{e^{-5\xi}}{-5}\right]_{\xi=0}^t\right\} \\ &= 2e^{4p}\mathbb{1}_{(0,\infty)}(t)\left\{\frac{4}{5}e^{-4t} + \frac{1}{5}e^t\right\}. \end{aligned}$$

Therefore, from (2.93) the component of the unstable manifold in the x_2 direction is

$$\begin{aligned} \left[x_\varepsilon^u(p,t) - \bar{x}^u(p)\right]\cdot\hat{x}_2 &= \varepsilon\frac{2e^{4p}\mathbb{1}_{(0,\infty)}(t)\left\{\frac{4}{5}e^{-4t} + \frac{1}{5}e^t\right\}}{2e^{2p}} + \mathcal{O}\left(\varepsilon^2\right) \\ &= \varepsilon\frac{e^{2p}}{5}\mathbb{1}_{(0,\infty)}(t)\left[4e^{-4t} + e^t\right] + \mathcal{O}\left(\varepsilon^2\right). \end{aligned}$$

This means that an $\mathcal{O}(\varepsilon)$-approximation to the unstable pseudomanifold is

$$\tilde{\Gamma}_\varepsilon^u = \left\{\left(\begin{pmatrix} e^{2p} + \mathcal{O}(\varepsilon) \\ \varepsilon\frac{e^{2p}}{5}\mathbb{1}_{(0,\infty)}(t)\left[4e^{-4t} + e^t\right]\end{pmatrix}, t\right) : t < T_u,\ p < P\right\},$$

where the $\mathcal{O}(\varepsilon)$-term in the x_1-component arises from the tangential movement of the pseudomanifold for which an expression is not available. This forms a (p,t) parametrization for this entity. The p can be thought of as representing a choice of trajectory in the augmented phase space, such as that shown in Figure 2.15. Indeed, an approximate expression in terms of the x_1x_2 variables is immediately obtainable by elimination of p above:

$$x_2 = \varepsilon\frac{x_1 - \mathcal{O}(\varepsilon)}{5}\mathbb{1}_{(0,\infty)}(t)\left[4e^{-4t} + e^t\right] = \varepsilon\frac{x_1}{5}\mathbb{1}_{(0,\infty)}(t)\left[4e^{-4t} + e^t\right] + \mathcal{O}\left(\varepsilon^2\right).$$

Of particular note is that for this representation of the pseudomanifold as a graph from x_1 to x_2, the lack of knowledge of the tangential component's motion does not affect the ability to give the leading-order expression, which is of a t-parametrized straight line. The slope is zero for $t < 0$ (since there is no perturbation to the steady flow), jumps to a value of $\varepsilon(4e^0 + e^0)/5 = \varepsilon$ at $t = 0^+$, and then subsequently evolves according to the above expression. The behavior of the unstable pseudomanifold in this instance is pictured in Figure 2.16(a) for $\varepsilon = 0.1$, at different values of t. Notice that control over this is lost as $t \to \infty$, in keeping with the statement in Theorem 2.38 that the results are valid for $t < T_u$.

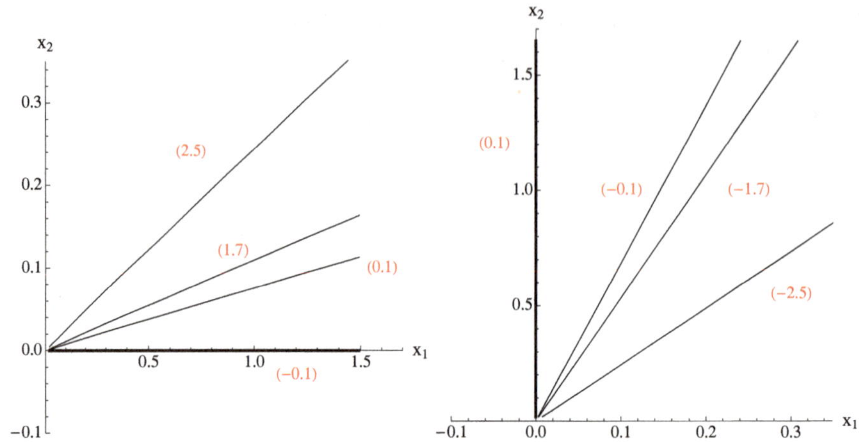

(a) Unstable pseudomanifold (b) Stable pseudomanifold

Figure 2.16. *The unstable and stable pseudomanifolds of Example 2.47 for* $\varepsilon = 0.1$, *at different t-values (shown in parantheses).*

Next, the stable pseudomanifold is addressed. Considering the unperturbed branch in the positive x_2 direction as the base manifold,

$$\bar{x}^s(t) = \begin{pmatrix} 0 \\ e^{-t} \end{pmatrix}, f(\bar{x}^s(t)) = \begin{pmatrix} 0 \\ -e^{-t} \end{pmatrix}, f^{\perp}(\bar{x}^s(t)) = \begin{pmatrix} e^{-t} \\ 0 \end{pmatrix}.$$

Therefore, from (2.101),

$$j_1^s(p,t) = f^{\perp} \cdot h_1(\bar{x}^s(t_1 - t + p), t_1) = e^{2(t-p)},$$

and from (2.102),

$$R_p^s(t) = \mathscr{L}^{-1}\left\{\frac{1/s}{1-1/s}\right\}(-t) = e^{-t}.$$

Utilizing (2.100),

$$\begin{aligned} M^s(p,t) &= \mathbb{1}_{(-\infty,0)}(t)e^{2(t-p)} - \mathbb{1}_{(-\infty,0)}(t)\int_t^0 e^{-(t-\xi)}e^{2(\xi-p)}\,\mathrm{d}\xi \\ &= \mathbb{1}_{(-\infty,0)}(t)e^{-2p}\left\{e^{2t} - e^{-t}\int_t^0 e^{3\xi}\,\mathrm{d}\xi\right\} \\ &= \mathbb{1}_{(-\infty,0)}(t)\frac{e^{-2p}}{3}\left\{4e^{2t} - e^{-t}\right\}, \end{aligned}$$

and hence from (2.99)

$$[x_\varepsilon^s(p,t) - \bar{x}^u(p)] \cdot \hat{x}_1 = \varepsilon\frac{e^{-p}}{3}\mathbb{1}_{(-\infty,0)}(t)\left[4e^{2t} - e^{-t}\right] + \mathcal{O}(\varepsilon^2).$$

The $\mathcal{O}(\varepsilon)$ parametric approximation for the stable pseudomanifold is therefore

$$\tilde{\Gamma}_\varepsilon^s = \left\{\left(\begin{pmatrix} \varepsilon\frac{e^{-p}}{3}\mathbb{1}_{(-\infty,0)}(t)\left[4e^{2t} - e^{-t}\right] \\ e^{-p} + \mathcal{O}(\varepsilon) \end{pmatrix}, t\right) : t > T_s,\ p > -P\right\},$$

which once again is a straight line at each t to leading-order. This lies in the unperturbed stable manifold direction $\hat{x}_2$ for $t > 0$ but abruptly switches to a line with reciprocal slope ε at $t = 0^-$ and evolves in backwards time subsequently. This is shown in Figure 2.16(b). ∎

Example 2.48 (double-gyre cont.). The double-gyre previously examined in Examples 2.25, 2.56, and 2.33 is reexamined, but with the term $q(t)$ therein considered an impulse. Intuitively, the idea is to consider $q(t) = \delta(t)$, the Dirac delta operating at time 0. It has already been established that $\operatorname{Tr} Df = 0$, and thus Corollaries 2.41 and 2.46 can be applied. As argued in Remark 2.42, in this area-preserving situation the formal process of simply substituting the Dirac delta inside the integral of the smooth Melnikov functions as given in Theorems 2.12 and 2.23 is legitimate. Referring to Figure 2.9, the idea in this example is to determine how using $q(t) = \delta(t)$ will change the unstable manifold of $(1,0)$, which initially lies along the line $x_1 = 1$. Using what has been already calculated in Example 2.25 in (2.67),

$$f^{\perp} \cdot g\Big|_{x_1=1} = -\frac{\pi^3 A^2}{2} \sin(2\pi \bar{x}_2)\, \delta(t)$$

with $\bar{x}_2(t) = (2/\pi)\cot^{-1} e^{\pi^2 A t}$ the relevant term inside the Melnikov integrals (2.26) and (2.62). Thus, from (2.26),

$$\begin{aligned} M^u(p,t) &= \int_{-\infty}^{p} \left(-\frac{\pi^3 A^2}{2}\right) \sin\left[2\pi \frac{2}{\pi} \cot^{-1}\left(e^{\pi^2 A \tau}\right)\right] \delta(\tau + t - p)\, \mathrm{d}\tau \\ &= -\frac{\pi^3 A^2}{2} \sin\left[4 \cot^{-1}\left(e^{\pi^2 A (p-t)}\right)\right] \mathbb{1}_{(0,\infty)}(t). \end{aligned}$$

The corresponding x_1 component, $x_\varepsilon^u(p,t;1)$, of the parametric point $x_\varepsilon^u(p,t)$ on $\tilde{\Gamma}_p^u$ satisfies

$$\begin{aligned} x_\varepsilon^u(p,t;1) - 1 &= \left[x_\varepsilon^u(p,t) - \bar{x}^u(p)\right] \cdot \hat{f}^{\perp}(\bar{x}^u(p)) \\ &= \frac{-\varepsilon \frac{\pi^3 A^2}{2} \sin\left[4\cot^{-1}\left(e^{\pi^2 A(p-t)}\right)\right] \mathbb{1}_{(0,\infty)}(t)}{|f(x^u(p))|} + \mathcal{O}\left(\varepsilon^2\right) \\ &= \frac{-\varepsilon \frac{\pi^3 A^2}{2} \sin\left[4\cot^{-1}\left(e^{\pi^2 A(p-t)}\right)\right] \mathbb{1}_{(0,\infty)}(t)}{\pi A \sin\left[\pi \frac{2}{\pi} \cot^{-1} e^{\pi^2 A p}\right]} + \mathcal{O}\left(\varepsilon^2\right) \\ &= -\varepsilon \frac{\pi^2 A}{2} \mathbb{1}_{(0,\infty)}(t) \frac{\sin\left[4\cot^{-1}\left(e^{\pi^2 A(p-t)}\right)\right]}{\sin\left[2\cot^{-1} e^{\pi^2 A p}\right]} + \mathcal{O}\left(\varepsilon^2\right). \end{aligned}$$

By writing p in terms of x_2 as given by (2.68), this enables representing the unstable manifold to leading-order as a graph from x_2 to x_1. Further analysis of this situation, in conjunction with the stable pseudomanifold of the point $(1,1)$, appears in [37]. ∎

2.7.1 ▪ Proof of Theorem 2.38 (unstable pseudomanifold)

Consider a fixed time-slice t in the augmented phase space. Suppose p is also fixed. Now, when $\varepsilon = 0$, the trajectory $\bar{x}^u(\tau - t + p)$ (with τ representing time) is a solution to (2.90) such that this passes through the point $\bar{x}^u(p)$ in the time-slice t. When $\varepsilon \neq 0$, suppose $x_\varepsilon^u(p,\tau)$ is a nearby trajectory lying on the unstable pseudomanifold. This will be represented by

$$x_\varepsilon^u(p,\tau) = \bar{x}^u(\tau - t + p) + \varepsilon x_1(p,\tau,\varepsilon), \tag{2.104}$$

where it follows that x_1 is $\mathcal{O}(\varepsilon)$ for $(p,\tau) \in (-\infty, P] \times (-\infty, T_u] \backslash \mathcal{J}$ since the effect of ε is only to introduce a finite number of $\mathcal{O}(\varepsilon)$ jumps in the solution. Moreover, as the perturbation will only begin affecting solutions for $t > t_1$, it is clear that $x_1(p,\tau,\varepsilon) = 0$ for $\tau < t_1$. Given that the behavior at time $-\infty$ is well-defined, and since τ is now being thought of as the time variable, (2.90) will be now rewritten as

$$x_\varepsilon^u(p,\tau) = a + \int_{-\infty}^{\tau} f(x_\varepsilon^u(p,\xi))\mathrm{d}\xi + \frac{\varepsilon}{2}\sum_{i=1}^{n} \mathbb{1}_{(-\infty,\tau)}(t_i)\left[h_i(x_\varepsilon^u(p,t_i^-),t_i^-) + h_i(x_\varepsilon^u(p,t_i^+),t_i^+)\right], \tag{2.105}$$

with the replacement $\beta \to -\infty$ and $t \to \tau$. When examining (2.104) and (2.105) at $\varepsilon = 0$ one observes that

$$\bar{x}^u(\tau - t + p) = a + \int_{-\infty}^{\tau} f(\bar{x}^u(\xi - t + p))\,\mathrm{d}\xi, \tag{2.106}$$

which is satisfied for *any* p and t since $\bar{x}^u(\tau - t + p)$ satisfies the differential equation

$$\frac{\partial}{\partial \tau}\bar{x}^u(\tau - t + p) = f(\bar{x}^u(\tau - t + p))$$

and moreover $\bar{x}^u(\tau - t + p) \to a$ as $\tau \to -\infty$ for any p and t. This implies that for any time-slice t chosen, and any $\bar{x}^u(p)$ chosen on the unstable manifold curve in that time-slice, the solution trajectory passing through that backwards asymptotes to a and is therefore on a's unstable manifold. What is sought in (2.104) is the $\mathcal{O}(\varepsilon)$-modification to this solution, which also approaches a as $\tau \to -\infty$. Now, substituting (2.104) into (2.105) leads to

$$\begin{aligned}
\bar{x}^u(\tau - t + p) + \varepsilon x_1(p,\tau,\varepsilon) = a &+ \int_{-\infty}^{\tau} f(\bar{x}^u(\xi - t + p) + \varepsilon x_1(p,\xi,\varepsilon))\,\mathrm{d}\xi \\
&+ \frac{\varepsilon}{2}\sum_{i=1}^{n} \mathbb{1}_{(-\infty,\tau)}(t_i) h_i\left(\bar{x}^u(t_i^- - t + p) + \varepsilon x_1(p,t_i^-,\varepsilon), t_i^-\right) \\
&+ \frac{\varepsilon}{2}\sum_{i=1}^{n} \mathbb{1}_{(-\infty,\tau)}(t_i) h_i\left(\bar{x}^u(t_i^+ - t + p) + \varepsilon x_1(p,t_i^+,\varepsilon), t_i^+\right).
\end{aligned}$$

Consider applying Taylor's theorem to the above, in expanding f and each h_i around $\bar{x}^u$. Terms beyond $\mathcal{O}(\varepsilon)$ will include $D^2 f$ and Dh_i, all of which are bounded on $\Omega \times \mathbb{R}$ by Hypotheses 2.34 and 2.35. While the Dh_i terms appear in a regular fashion, the $D^2 f$ terms appear inside an integral over an unbounded domain, and hence the fact that this term is associated with a $\mathcal{O}(\varepsilon^2)$ factor does not in and of itself guarantee that it can be written as an $\mathcal{O}(\varepsilon^2)$ factor outside the integral. However, since $x_1(p,\tau,\varepsilon)$ is zero for $\tau < t_1$, the apparently unbounded domain $(-\infty, \tau)$ of the integral reduces to a bounded domain (t_1, τ). Thus the term involving $D^2 f$ does indeed contribute a $\mathcal{O}(\varepsilon^2)$ term, and without further ado, all such higher-order terms shall be lumped together as one term, leading to

$$\bar{x}^u(\tau-t+p)+\varepsilon x_1(p,\tau,\varepsilon)=a+\int_{-\infty}^{\tau} f(\bar{x}^u(\xi-t+p))\,\mathrm{d}\xi$$
$$+\int_{-\infty}^{\tau} Df(\bar{x}^u(\xi-t+p))\varepsilon x_1(p,\xi,\varepsilon)\mathrm{d}\xi+\mathcal{O}\left(\varepsilon^2\right)$$
$$+\frac{\varepsilon}{2}\sum_{i=1}^{n}\mathbb{1}_{(-\infty,\tau)}(t_i)\left[h_i\left(\bar{x}^u(t_i^- -t+p),t_i^-\right)+h_i\left(\bar{x}^u(t_i^- -t+p),t_i^+\right)\right].$$

Now, the $\mathcal{O}(1)$ terms collectively cancel because of (2.106), and since $\bar{x}^u$ and the h_i are continuous,

$$x_1(p,\tau,\varepsilon)=\int_{-\infty}^{\tau} Df(\bar{x}^u(\xi-t+p))\,x_1(p,\xi,\varepsilon)\mathrm{d}\xi+\sum_{i=1}^{n}\mathbb{1}_{(t_i,\infty)}(\tau)h_i(\bar{x}^u(t_i-t+p),t_i)$$
$$+\mathcal{O}(\varepsilon), \tag{2.107}$$

where the indicator function has been written in an equivalent fashion. The unstable Melnikov function is now defined by

$$M^u(p,\tau,\varepsilon):=f(\bar{x}^u(\tau-t+p))^{\perp}\cdot\frac{x_\varepsilon^u(p,\tau)-\bar{x}^u(\tau-t+p)}{\varepsilon}$$
$$=f(\bar{x}^u(\tau-t+p))^{\perp}\cdot x_1(p,\tau,\varepsilon). \tag{2.108}$$

Since $M^u(p,t,\varepsilon)=f(\bar{x}^u(p))^{\perp}x_1(p,t,\varepsilon)$, the quantity $M^u(p,t,\varepsilon)$ expresses the leading-order displacement of the unstable pseudomanifold in the normal direction to the original manifold at a point $\bar{x}^u(p)$, in the time-slice t, and hence its manifestation in (2.93) is clear. However, its ε-dependence, and in particular an expression for it, is necessary.

Since $x_1(p,-\infty,\varepsilon)=0$ and $M^u(p,-\infty,\varepsilon)=0$, it is possible to rewrite (2.108) in the integral form

$$M^u(p,\tau,\varepsilon)=\int_{-\infty}^{\tau}\frac{d}{d\xi}\left[f(\bar{x}^u(\xi-t+p))^{\perp}\cdot x_1(p,\xi,\varepsilon)\right]\mathrm{d}\xi,$$

where this is legitimate for $\tau\notin\mathcal{J}$ since while $x_1(p,\xi,\varepsilon)$ is not differentiable in ξ at the jump values t_i, its temporal derivative *is* integrable. This enables the set of manipulations

$$M^u(p,\tau,\varepsilon)=\int_{-\infty}^{\tau}\frac{d}{d\xi}\left[f(\bar{x}^u(\xi-t+p))^{\perp}\right]\cdot x_1(p,\xi,\varepsilon)\mathrm{d}\xi$$
$$+\int_{-\infty}^{\tau} f(\bar{x}^u(\xi-t+p))^{\perp}\cdot\frac{d}{d\xi}[x_1(p,\xi,\varepsilon)]\,\mathrm{d}\xi$$
$$=\int_{-\infty}^{\tau}[Df(\bar{x}^u(\xi-t+p))f(\bar{x}^u(\xi-t+p))]^{\perp}\cdot x_1(p,\xi,\varepsilon)\mathrm{d}\xi+\mathcal{O}(\varepsilon)$$
$$+\int_{-\infty}^{\tau} f(\bar{x}^u(\xi-t+p))^{\perp}\left[Df(\bar{x}^u(\xi-t+p))\,x_1(p,\xi,\varepsilon)+\sum_{i=1}^{n}h_i(\bar{x}^u(t_i-t+p),t_i)\frac{d}{d\xi}\mathbb{1}_{(t_i,\infty)}(\xi)\right]\mathrm{d}\xi$$
$$=\int_{-\infty}^{\tau}\mathrm{Tr}\,Df(\bar{x}^u(\xi-t+p))f(\bar{x}^u(\xi-t+p))^{\perp}\cdot x_1(p,\xi,\varepsilon)\mathrm{d}\xi+\mathcal{O}(\varepsilon)$$
$$+\sum_{i=1}^{n}h_i(\bar{x}^u(t_i-t+p),t_i)\cdot\int_{-\infty}^{\tau} f(\bar{x}^u(\xi-t+p))^{\perp}\frac{d}{d\xi}\mathbb{1}_{(t_i,\infty)}(\xi)\mathrm{d}\xi, \tag{2.109}$$

where the second equality is obtained by using (2.107) (with its distributional derivative well-defined) and the third by collecting together the smooth terms using the same argument as in the proof of Theorem 2.12 (see (2.39) and the adjacent discussion). Now, the final (distributional) integral in (2.109) can be performed by integrating by parts:

$$
\begin{aligned}
&\int_{-\infty}^{\tau} f(\bar{x}^u(\xi-t+p))^{\perp} \frac{d}{d\xi}\mathbb{1}_{(t_i,\infty)}(\xi)\mathrm{d}\xi \\
&= f(\bar{x}^u(\xi-t+p))^{\perp}\mathbb{1}_{(t_i,\infty)}(\xi)\Big]_{\xi=-\infty}^{\tau} - \int_{-\infty}^{\tau}\frac{d}{d\xi}\left[f(\bar{x}^u(\xi-t+p))^{\perp}\right]\mathbb{1}_{(t_i,\infty)}(\xi)\mathrm{d}\xi \\
&= \left[f(\bar{x}^u(\tau-t+p))^{\perp}\mathbb{1}_{(t_i,\infty)}(\tau)-0\right] - \mathbb{1}_{(t_i,\infty)}(\tau)\int_{t_i}^{\tau}\frac{d}{d\xi}\left[f(\bar{x}^u(\xi-t+p))^{\perp}\right]\mathrm{d}\xi \\
&= f(\bar{x}^u(\tau-t+p))^{\perp}\mathbb{1}_{(t_i,\infty)}(\tau) - \mathbb{1}_{(t_i,\infty)}(\tau)\left[f(\bar{x}^u(\tau-t+p))^{\perp} - f(\bar{x}^u(t_i-t+p))^{\perp}\right] \\
&= \mathbb{1}_{(t_i,\infty)}(\tau) f(\bar{x}^u(t_i-t+p))^{\perp}.
\end{aligned}
$$

Inserting this and the fact that $M^u = f^{\perp}\cdot x_1$ into (2.109) gives the equation

$$
\begin{aligned}
M^u(p,\tau,\varepsilon) = &\int_{-\infty}^{\tau} \operatorname{Tr} Df(\bar{x}^u(\xi-t+p))M^u(p,\xi,\varepsilon)\mathrm{d}\xi + \mathcal{O}(\varepsilon) \\
&+\sum_{i=1}^{n}\mathbb{1}_{(t_i,\infty)}(\tau) f(\bar{x}^u(t_i-t+p))^{\perp}\cdot h_i(\bar{x}^u(t_i-t+p),t_i).
\end{aligned}
$$

Next, the abuse of notation of setting $M^u(p,\tau)$ as the solution to the above *with the $\mathcal{O}(\varepsilon)$-term neglected* will be made. This provides the Melnikov function as an ε-independent quantity. Since the error associated with this process is $\mathcal{O}(\varepsilon)$, by (2.108) this will only cause a $\mathcal{O}(\varepsilon^2)$ error in the normal distance measure. Furthermore, the distance we require is in the time-slice t, near the location $\bar{x}^u(p)$, and therefore is associated with $M^u(p,t)$. Replacing τ above with t then leads to the integral equation for the unstable Melnikov function:

$$
\begin{aligned}
M^u(p,t) = &\int_{-\infty}^{t} \operatorname{Tr} Df(\bar{x}^u(\xi-t+p))M^u(p,\xi)\mathrm{d}\xi \\
&+\sum_{i=1}^{n}\mathbb{1}_{(t_i,\infty)}(t) f(\bar{x}^u(t_i-t+p))^{\perp}\cdot h_i(\bar{x}^u(t_i-t+p),t_i) \\
= &\int_{-\infty}^{t} \operatorname{Tr} Df(\bar{x}^u(\xi-t+p))M^u(p,\xi)\mathrm{d}\xi \\
&+\sum_{i=1}^{n}\mathbb{1}_{(t_i,\infty)}(t) j_i^u(p,t),
\end{aligned}
\tag{2.110}
$$

where the definition (2.95) has been used. In solving (2.110), the following lemma will be useful.

Lemma 2.49. *Consider the integral equation*

$$
M(t) = j(t) + \int_{-\infty}^{t} M(\xi)F(t-\xi)d\xi,
\tag{2.111}
$$

where j is piecewise differentiable and is zero below some finite value t_1, and the kernel

$F \in C^1([0,\infty))$ satisfies $|\lim_{t\to\infty} F(t)| = F_0 < \infty$. Then, (2.111) has a solution

$$M(t) = j(t) + \int_{-\infty}^{t} R(t-\xi) j(\xi)\,d\xi \tag{2.112}$$

at values t at which j is defined, where the resolvent R is obtained from the Laplace transform $\hat{F}(s)$ of $F(t)$ by $R(t) = \mathscr{L}^{-1}\{\hat{F}(s)/[1-\hat{F}(s)]\}(t)$.

Proof. Equation (2.111) is in the form of a renewal equation [319, 64] but with an unbounded domain. The basic renewal equation solution with the infinite limit substituted is indeed (2.112) [319, 64]. However, the legitimacy of this formal process requires the conditions on F and j as given in Lemma 2.49, based on which a full proof is given in Section 2.7.2. $\square$

Now, the definitions

$$F(t) = \mathrm{Tr}\, Df(\bar{x}^u(p-t)) \quad \text{and} \quad j(t) = \sum_{i=1}^{n} \mathbb{1}_{(t_i,\infty)}(t) j_i^u(p,t)$$

are employed; it is clear that these satisfy the hypotheses of Lemma 2.49. A direct application of Lemma 2.49 on (2.110), along with the resolvent definition in (2.96), yields

$$M^u(p,t) = \sum_{i=1}^{n} \mathbb{1}_{(t_i,\infty)}(t) j_i^u(p,t) + \int_{-\infty}^{t} R_p^u(t-\xi) \sum_{i=1}^{n} \mathbb{1}_{(t_i,\infty)}(\xi) j_i^u(p,\xi)\,d\xi\,,$$

from which (2.94) arises since each of the jump functions is only turned on for ξ values greater than t_i in the integrand. Thereby, Theorem 2.38 has been proven.

2.7.2 ▪ Proof of Lemma 2.49 (integral equation for M^u)

In proving Lemma 2.49, a preliminary lemma proves convenient.

Lemma 2.50. *Let F and j satisfy the hypotheses stated in Lemma 2.49. If $w(t)$ satisfies*

$$w(t) = 1 + \int_{0}^{t} w(\xi) F(t-\xi)\,d\xi \tag{2.113}$$

for $t \geq 0$, then the solution to the integral equation (2.111) is given by

$$M(t) = \int_{-\infty}^{t} w(t-\xi) \frac{d}{d\xi}[j(\xi)]\,d\xi\,, \tag{2.114}$$

where since j is piecewise continuous the derivative in (2.114) is to be considered in a distributional sense.

Proof. After defining

$$\bar{M}(t) := \int_{-\infty}^{t} w(t-\xi) \frac{d}{d\xi}[j(\xi)]\,d\xi\,,$$

the following computation, closely following [64], is possible:

$$\begin{aligned}
\int_{-\infty}^{t} \bar{M}(\eta)F(t-\eta)\mathrm{d}\eta &= \int_{-\infty}^{t}\left[\int_{-\infty}^{\eta} w(\eta-\xi)\frac{d}{d\xi}[j(\xi)]\mathrm{d}\xi\right]F(t-\eta)\mathrm{d}\eta \\
&= \int_{-\infty}^{t} \frac{d}{d\xi}[j(\xi)]\int_{\xi}^{t} w(\eta-\xi)F(t-\eta)\mathrm{d}\eta\,\mathrm{d}\xi \\
&= \int_{-\infty}^{t} \frac{d}{d\xi}[j(\xi)]\int_{0}^{t-\xi} w(u)F([t-\xi]-u)\,\mathrm{d}u\,\mathrm{d}\xi \\
&= \int_{-\infty}^{t} \frac{d}{d\xi}[j(\xi)][w(t-\xi)-1]\,\mathrm{d}\xi \\
&= \int_{-\infty}^{t} w(t-\xi)\frac{d}{d\xi}[j(\xi)]\,\mathrm{d}\xi - \int_{-\infty}^{t}\frac{d}{d\xi}[j(\xi)]\mathrm{d}\xi \\
&= \bar{M}(t) - j(\xi)\Big]_{\xi=-\infty}^{t} = \bar{M}(t)-j(t),
\end{aligned}$$

which shows that $\bar{M}(t)$ does indeed satisfy (2.111). The interchanging of the order of integration in the second equality above was possible because $j(t)=0$ for $t<t_1$ while $F(t-\eta)$ approached a limit as $\eta\to-\infty$; the integrand was therefore absolutely integrable over the unbounded domain. The fourth equality used (2.113). □

If the solution to the auxiliary equation (2.113) can be determined, the solution to the required equation (2.111) can be written in terms of this solution using (2.114). Thus, the next step is to determine $w(t)$ which solves (2.113). Luckily, the functions associated with this are smooth, since there is no $j(t)$ in (2.113). Moreover, the integration range begins at zero. To express a solution to this, the Laplace transform of (2.113) is taken. If $\hat{w}(s) := \mathscr{L}\{w(t)\}(s)$ and $\hat{F}(s) := \mathscr{L}\{F(t)\}(s)$, then the convolution property gives

$$\hat{w}(s) = \frac{1}{s} + \hat{w}(s)\hat{F}(s).$$

Solving for $\hat{w}(s)$ gives the result

$$\hat{w}(s) = \frac{1}{s\left[1-\hat{F}(s)\right]} = \frac{1}{s} + \frac{1}{s}\frac{\hat{F}(s)}{1-\hat{F}(s)} =: \frac{1}{s} + \frac{1}{s}\hat{R}(s).$$

Inverting the Laplace transform and once again using the convolution property gives

$$w(t) = 1 + \int_{0}^{t} R(\xi)\mathrm{d}\xi.$$

Inserting the above into (2.114) then results in

$$\begin{aligned}
M(t) &= \int_{-\infty}^{t}\left(1+\int_{0}^{t-\xi} R(\eta)\mathrm{d}\eta\right)\frac{d}{d\xi}[j(\xi)]\,\mathrm{d}\xi \\
&= \int_{-\infty}^{t}\frac{d}{d\xi}[j(\xi)]\,\mathrm{d}\xi + \int_{-\infty}^{t}\left(\int_{0}^{t-\xi} R(\eta)\mathrm{d}\eta\right)\frac{d}{d\xi}[j(\xi)]\,\mathrm{d}\xi
\end{aligned}$$

$$= j(t) + \left[\left(\int_0^{t-\xi} R(\eta)\mathrm{d}\eta \right) j(\xi) \right]_{\xi=-\infty}^{t} - \int_{-\infty}^{t} [-R(t-\xi)] j(\xi)\,\mathrm{d}\xi \right]$$

$$= j(t) + \int_{-\infty}^{t} R(t-\xi) j(\xi)\,\mathrm{d}\xi\,,$$

the result required for Lemma 2.49.

2.7.3 ▪ Proof of Theorem 2.44 (stable pseudomanifold)

Many details are similar to, and with obvious modifications from, the proof of the unstable pseudomanifold expressions of Theorem 2.38 as given in Section 2.7.1. Such situations will be briefly sketched without much explanation. However, there are some issues—in particular dealing with how the Laplace transform representation is to be modified for functions with negative argument—which require more careful development.

As in Section 2.7.1, consider a fixed time-slice t and a fixed p and let τ be the time-variable. Define

$$x_\varepsilon^s(p,\tau) = \bar{x}^s(\tau - t + p) + \varepsilon x_1(p,\tau,\varepsilon), \tag{2.115}$$

where now x_1 is $\mathcal{O}(\varepsilon)$ for $(p,\tau) \in [P,\infty) \times [T_s,\infty) \setminus \mathcal{J}$. Define

$$M^s(p,\tau,\varepsilon) := f(\bar{x}^s(\tau - t + p))^{\perp} \cdot \frac{x_\varepsilon^s(p,\tau) - \bar{x}^s(\tau - t + p)}{\varepsilon} = f(\bar{x}^s(\tau - t + p))^{\perp} \cdot x_1(p,\tau,\varepsilon). \tag{2.116}$$

Now, consider using the evolution equation (2.90) with $\beta = \infty$. Not much change occurs in the derivation in Section 2.7.1, and instead of (2.110) the integral equation

$$M^s(p,t) = -\int_t^{\infty} \mathrm{Tr}Df(\bar{x}^s(\xi - t + p)) M^s(p,\xi)\,\mathrm{d}\xi - \sum_{i=1}^{n} \mathbb{1}_{(-\infty,t_i)}(t) j_i^s(p,t) \tag{2.117}$$

results for the stable Melnikov function M^s, where j_i^s is defined in (2.101). (This can be achieved by the simple strategem of replacing β with ∞ in following the derivation in Section 2.7.1, but noting from (2.91) that there needs to also be a negative sign in front of the indicator function in the perturbing term.) To solve this the following lemma, analogous to Lemma 2.49, is necessary.

Lemma 2.51. *Consider the integral equation*

$$M(t) = -j(t) - \int_t^{\infty} M(\xi) F(t-\xi)\,d\xi\,, \tag{2.118}$$

where j is piecewise differentiable and is zero above some finite value t_n, and the kernel $F \in \mathrm{C}^1((-\infty,0])$ satisfies $\left|\lim_{t\to-\infty} F(t)\right| = F_0 < \infty$. Then, (2.118) has a solution

$$M(t) = -j(t) + \int_t^{\infty} R(t-\xi) j(\xi)\,\mathrm{d}\xi \tag{2.119}$$

at values t at which j is defined, where the resolvent R is obtained from the Laplace transform $\hat{F}(s)$ of $F(-t)$ by $R(t) = \mathcal{L}^{-1}\left\{\hat{F}(s)/\left[1+\hat{F}(s)\right]\right\}(-t)$.

Proof. Analogous to Lemma 2.50, the first claim is that if $w(t)$ solves

$$w(t) = -1 + \int_0^t w(\xi) F(t-\xi)\, d\xi\,, \tag{2.120}$$

for $t \leq 0$ (with w and F being defined on $(0,\infty)$), then the solution to the integral equation (2.118) is given by

$$M(t) = -\int_t^\infty w(t-\xi) \frac{d}{d\xi}\left[j(\xi)\right] d\xi\,. \tag{2.121}$$

The proof of this is no different from that of Lemma 2.50 (save for the domain of t-validity of w) and will be skipped. The next issue is to find the solution to (2.120), for which a direct Laplace transform is not possible since the domain of validity is $t \leq 0$. Replacing t with $-t$, (2.120) can be written as

$$w(-t) = -1 + \int_0^{-t} w(\xi) F(-t-\xi)\, d\xi$$

with domain of validity now $t \geq 0$. Defining $\tilde{w}(t) = w(-t)$ and $\tilde{F}(t) = F(-t)$ enables the representation

$$\tilde{w}(t) = -1 + \int_0^{-t} \tilde{w}(-\xi) \tilde{F}(t+\xi)\, d\xi = -1 - \int_0^t \tilde{w}(\eta) \tilde{F}(t-\eta)\, d\eta$$

with the change of integration variable $\eta = -\xi$. Since each of $\tilde{w}$ and $\tilde{F}$ are defined for $t \geq 0$, it is possible to define $\hat{F}(s) = \mathscr{L}\left\{\tilde{F}(t)\right\}(s) = \mathscr{L}\left\{F(-t)\right\}(s)$ and $\hat{w}(s) = \mathscr{L}\left\{\tilde{w}(t)\right\}(s) = \mathscr{L}\left\{w(-t)\right\}(s)$. Taking the Laplace transform of the above expression gives

$$\hat{w}(s) = -\frac{1}{s} - \hat{w}(s) \hat{F}(s)\,,$$

and therefore

$$\hat{w}(s) = -\frac{1}{s} + \frac{1}{s} \frac{\hat{F}(s)}{1 + \hat{F}(s)}\,.$$

Let $\hat{R}(s) = \hat{F}(s) / \left[1 + \hat{F}(s)\right]$, with inverse Laplace transform $\tilde{R}(t)$, which is defined for $t \geq 0$. Then, applying the convolution property while inverting the Laplace transform in the above expression yields

$$\tilde{w}(t) = -1 + \int_0^t \tilde{R}(\xi)\, d\xi\,,$$

which with the replacement $t \to -t$ gives

$$w(t) = -1 + \int_0^{-t} \tilde{R}(\xi)\, d\xi\,,$$

where now $t \leq 0$. This solution for w when inserted into (2.121) yields

$$M(t) = -\int_t^\infty \left[-1 + \int_0^{-t+\xi} \tilde{R}(\eta)\, d\eta\right] \frac{d}{d\xi}\left[j(\xi)\right] d\xi$$

$$
\begin{aligned}
&= -j(t) - \int_t^{\infty} \left[\int_0^{-t+\xi} \tilde{R}(\eta)\,d\eta \right] \frac{d}{d\xi}[j(\xi)]\,d\xi \\
&= -j(t) - \left[\left(\int_0^{-t+\xi} \tilde{R}(\eta)\,d\eta \right) j(\xi) \right]_{\xi=t}^{\infty} - \int_t^{\infty} \tilde{R}(-t+\xi) j(\xi)\,d\xi \Bigg] \\
&= -j(t) + \int_t^{\infty} \tilde{R}(-t+\xi) j(\xi)\,d\xi \\
&= -j(t) + \int_t^{\infty} R(t-\xi) j(\xi)\,d\xi ,
\end{aligned}
$$

where $R(t) := \tilde{R}(-t)$ was used to express the solution in terms of a resolvent R defined for $t \leq 0$. This is the result required. □

Now, apply Lemma 2.51 to the integral equation (2.117) with the choice $M(t) = M^s(p,t)$, $F(t) = \mathrm{Tr}Df(\bar{x}^s(p-t))$ (and hence $\tilde{F}(t) = \mathrm{Tr}Df(\bar{x}^s(p+t))$, whose Laplace transform is defined for $t \geq 0$) and $j(t) = \sum_{i=1}^{n} \mathbb{1}_{(-\infty,t_i)}(t) j_i^s(p,t)$. This gives

$$
M^s(p,t) = -\sum_{i=1}^{n} \mathbb{1}_{(-\infty,t_i)}(t) j_i^s(p,t) + \int_t^{\infty} R_p^s(t-\xi) \sum_{i=1}^{n} \mathbb{1}_{(-\infty,t_i)}(\xi) j_i^s(p,\xi)\,d\xi ,
$$

which, when restricting the integral in relation to the indicator functions, gives the result of Theorem 2.44.

2.8 ▪ Finite-time invariant manifolds

The importance of finite-time issues has been discussed in Section 1.4. In this section, a discussion on how the nearly nonautonomous setting could be modified for a particular form of finite-time interpretation is provided. Assuming that the flow (2.6) is only available for a finite time duration $[-T,T]$, the various concepts associated with stable and unstable manifolds—hyperbolic trajectories and local tangent vectors—will be characterized. Thus, the system is

$$
\dot{x} = f(x) + \varepsilon g(x,t), \quad t \in [-T,T], \tag{2.122}
$$

subject to Hypothesis 2.2 and 2.5, but it is assumed that g is *only known for the finite-time interval* $[-T,T]$ but is smooth in the sense of Hypothesis 2.3 for $t \in [-T,T]$. With this understanding, and within the "nearly autonomous" framework of (2.122), the idea is to obtain reasonable definitions for hyperbolic trajectories and stable/unstable manifolds. It should be noted that the approach will be required to encapsulate the nonautonomous nature of the situation by being able to parametrize the entities as a function of t for all $t \in (-T,T)$. This is in contrast to many finite-time approaches [219, 127, 159, 281, 150, 154, 257] which perform their investigations over a time-interval $[t_i, t_f]$ by (numerically or otherwise) determining the flow map $F_{t_i}^{t_f}$ from the fixed initial time t_i to the fixed final time t_f, and then utilizing just this *one-step flow map* for their analyses. (Nonautonomy is occasionally viewed in these methods by allowing either t_i or t_f to vary, but the necessity for computing $F_{t_i}^{t_f}$ laboriously for each such change often makes this an impractical nonautonomous method.) For the definitions provided here, it will be assumed that *all* the data available from $[-T,T]$ shall be incorporated in the definitions for each t, arguably providing a more realistic viewpoint to the albeit restrictive setting of nearly autonomous flows.

Now, stable and unstable manifolds cannot be defined in the traditional way for finite-time flows, since they are defined in terms of exponential decay rates as $t \to \pm\infty$. Given the fact that there is *no* information about times outside $[-T, T]$, a practical method of defining the manifolds is to artificially extend the vector field g to $t \in \mathbb{R}$, define manifolds for this situation, and then restrict to looking at time-slices which are in the permitted $[-T, T]$ interval. The most natural way of doing this is to consider the flow

$$\dot{x} = f(x) + \varepsilon \tilde{g}(x, t), \quad t \in \mathbb{R}, \tag{2.123}$$

where

$$\tilde{g}(x,t) = \begin{cases} g(x,t) & \text{if } t \in [-T, T], \\ 0 & \text{if } t \notin [-T, T]. \end{cases} \tag{2.124}$$

This will result in a t-discontinuity in $\tilde{g}$ at $t = \pm T$, but this is not an issue in the theory of the preceding sections; as shown in Section 2.6, all derived equations remain legitimate. It should be made clear that the velocity outside the "known" interval $[-T, T]$ is *not* being set to zero but rather to the steady velocity $f(x)$, which is the dominant part of the known velocity within $[-T, T]$. This idea resonates with the concept of *transitory flow* introduced by Mosovsky and Meiss [281, 282] to examine transport occurring as a result of having two different autonomous flows in backwards and forwards time but some specified nonautonomy in $[-T, T]$ which smoothly approaches the autonomous flows as $t \to \pm T$. This also is within the framework of the scattering theory of Blazevski and collaborators [72, 73], in which stable and unstable manifolds can be thought of in terms of a diffeomorphism of a scattering map. In any case, for (2.123), the perturbed hyperbolic trajectory and its stable and unstable manifold are well-defined. By then limiting these to $t \in [-T, T]$, one can form finite-time definitions for all these entities, which are defined using all the data available for $t \in [-T, T]$ in a self-consistent way.

The change in the expression for the hyperbolic trajectory is first examined. For infinite-times, this is defined by $a_\varepsilon(t)$ as given in (2.14). The $\mathcal{O}(\varepsilon)$ projections of the vector $a_\varepsilon(t) - a$ in the directions $v_s^\perp$ and $v_u^\perp$ are those, respectively, defined by $\beta_s(t)$ and $\beta_u(t)$ in (2.15); see (2.16). The first of these, for example, can be written as

$$\beta_s(t) = -e^{\lambda_u t} \int_t^\infty e^{-\lambda_u \tau} \tilde{g}(a, \tau) \cdot v_s^\perp \, d\tau = -e^{\lambda_u t} \int_t^T e^{-\lambda_u \tau} g(a, \tau) \cdot v_s^\perp \, d\tau,$$

where it was necessary to shift the variable of integration in (2.15) to ensure that the appropriate time range, as expressed by (2.124), could be taken into account. This idea of extending g in a trivial fashion for $t \in \mathbb{R}$ enables a characterisation of finite-time hyperbolic trajectories as follows.

Definition 2.52 (finite-time hyperbolic trajectory [39]). Suppose $T < \infty$. Then, the $\mathcal{O}(\varepsilon)$-approximation to the *finite-time hyperbolic trajectory* a_ε is characterized for $t \in (-T, T)$ by

$$a_\varepsilon(t; T) := a + \varepsilon \left[\beta_u(t; T) v_u^\perp + \frac{\beta_u(t; T)(v_u \cdot v_s) - \beta_s(t; T)}{v_u^\perp \cdot v_s} v_u \right], \tag{2.125}$$

where

$$\beta_s(t; T) := -e^{\lambda_u t} \int_t^T e^{-\lambda_u \tau} g(a, \tau) \cdot v_s^\perp \, d\tau \tag{2.126}$$

and

$$\beta_u(t;T) := e^{\lambda_s t} \int_{-T}^{t} e^{-\lambda_s \tau} g(a,\tau) \cdot v_u^\perp \, \mathrm{d}\tau. \tag{2.127}$$

The intuitive meanings of $\beta_{s,u}(t;T)$ in the finite-time situation continue to be as expressed in Section 2.2; they are the $\mathcal{O}(\varepsilon)$-projections of $a_\varepsilon(t,T)$ in the directions $v_{s,u}^\perp$. Notice that as $t \to -T$, $\beta_u \to 0$, indicating that the $\mathcal{O}(\varepsilon)$ adjustment of $a_\varepsilon - a$ is purely in the v_u direction in this limit and is influenced only by the nonautonomous velocity component in the direction $v_s^\perp$, integrated over the full $(-T,T)$ time-interval. Conversely, as $t \to T$, $a_\varepsilon - a$ will lie in the v_s direction. In either case, the variation of g with time is only required at the point a, in keeping with the perturbative nature of these definitions.

Notice that the definition (2.125), in comparison with $a_\varepsilon(t)$ for the infinite-time situation as expressed in (2.14), satisfies

$$\lim_{\varepsilon \to 0} \lim_{T \to \infty} \frac{a_\varepsilon(t;T) - a_\varepsilon(t)}{\varepsilon} = 0.$$

In other words, $a_\varepsilon(t;T)$ approaches $a_\varepsilon(t)$ to leading-order in ε as the finiteness parameter T becomes infinite. In this sense, Definition 2.52 is a consistent extension of the infinite-time hyperbolic trajectory (2.14), which is an entity which—through obeying the exponential dichotomy conditions (1.22) for infinite times for the extended g—is rigorously definable as being hyperbolic.

Analyzing errors associated with Definition 2.52 is an ill-defined exercise, since a specific criterion for unequivocally identifying hyperbolic trajectories in finite-time flows does not exist. This formulation is associated with the particular extension (2.124). What happens if this is relaxed? Suppose one extends the flow (2.122) to $\mathbb{R}$ with the nonautonomous extension $\tilde{g}$ bounded; this can be thought of as any realization of a bounded stochastic perturbation, for example. That is, suppose there exists a $G > 0$ such that

$$|\tilde{g}(x,t)| + \|D\tilde{g}(x,t)\| \leq G \quad \text{for all}\, (x,t) \in \Omega \times \mathbb{R}. \tag{2.128}$$

Then, a genuine hyperbolic trajectory $a_\varepsilon(t)$ exists and is given by (2.14). Now pick any $t \in (-T,T)$. The $\mathcal{O}(\varepsilon)$-error $e_h(t)$ in using Definition 2.52 can be characterized in two linearly independent directions by

$$e_h(t) \cdot v_s^\perp := \left| \lim_{\varepsilon \to 0} \frac{a_\varepsilon(t;T) - a_\varepsilon(t)}{\varepsilon} \cdot v_s^\perp \right| \leq G e^{\lambda_u t} \int_T^\infty e^{-\lambda_u \tau} \mathrm{d}\tau = \frac{G}{\lambda_u} e^{\lambda_u (t-T)} \tag{2.129}$$

and

$$e_h(t) \cdot v_u^\perp := \left| \lim_{\varepsilon \to 0} \frac{a_\varepsilon(t;T) - a_\varepsilon(t)}{\varepsilon} \cdot v_u^\perp \right| \leq G e^{\lambda_s t} \int_{-\infty}^{-T} e^{-\lambda_s \tau} \mathrm{d}\tau = \frac{G}{-\lambda_s} e^{\lambda_s (t+T)}. \tag{2.130}$$

The error component $e_h(t) \cdot v_s^\perp$ is due to velocity uncertainties in $[T,\infty)$. Moreover, it increases from $Ge^{-2\lambda_u T}/\lambda_u$ to G/λ_u as t progresses from $-T$ to T. Definition 2.52 gives an increasingly accurate estimate for the $v_s^\perp$-component of the hyperbolic trajectory as time goes backwards. Conversely, the bound for $e_h(t) \cdot v_u^\perp$ decreases from $-G/\lambda_s$ to $-Ge^{2\lambda_s T}/\lambda_s$ and is impacted by uncertainties in $(-\infty,-T]$. In any event, an estimate for the error associated with the artificial extension of g to $\mathbb{R}$ by setting it to zero outside $[-T,T]$ is thus possible.

Definition 2.53 (finite-time local tangent vectors [39]). Suppose $T<\infty$, and $\sigma_{s,u}$ is as defined in (2.80). Then, the $\mathcal{O}(\varepsilon)$-approximation to the *finite-time local tangent vectors* to the stable and unstable directions in a time-slice $t\in(-T,T)$ are characterized by counterclockwise rotational angles

$$\theta_s(t;T):=-\varepsilon\, e^{(\lambda_u-\lambda_s)t}\int_t^T e^{(\lambda_s-\lambda_u)\tau}\sigma_s(\tau)\,d\tau \tag{2.131}$$

and

$$\theta_u(t;T):=\varepsilon\, e^{(\lambda_s-\lambda_u)t}\int_{-T}^t e^{(\lambda_u-\lambda_s)\tau}\sigma_u(\tau)\,d\tau \tag{2.132}$$

of v_s and v_u, respectively.

This is the obvious modification to Theorem 2.27. As argued for the finite-time hyperbolic trajectory definitions in comparison to the infinite-time ones, here too the statements

$$\lim_{\varepsilon\to0}\lim_{T\to\infty}\frac{\theta_i(t;T)-\theta_i(t)}{\varepsilon}=0,\quad i=s,u,$$

hold. Moreover, since $\theta_s(T;T)=0$, the proposed tangent to the nonautonomous finite-time stable manifold at $t=T$ lies in the direction of v_s. This is not merely a $\mathcal{O}(\varepsilon)$-statement but is true for any ε in the sense that $\theta_s(T^+;T)=0$. The reason is that the stable manifold in backwards time is attached to the hyperbolic trajectory (a,t) for $t>T$, and since there is no perturbation in this time-domain, the stable manifold continues to be in the direction of v_s. As time flows backwards and crosses T, this will change. At the opposite end of the time-interval, since $\theta_u(-T;T)=0$, the proposed tangent to the unstable manifold at $t=-T$ lies in the direction of v_u. Thus, the extension of the finite-time vector field $g(x,t)$ to 0 outside of $[-T,T]$ would ensure a smooth and consistent continuation of the tangent vectors. More precisely, $v_u(t)$ starts off being equal to $\mathcal{O}(\varepsilon)$ to v_u for $t\in(-\infty,-T]$ and then evolves continuously thereafter, while $v_s(t)$ is equal to $\mathcal{O}(\varepsilon)$ to v_s for $t\in[T,\infty)$ and evolves continuously from this in backwards time. One *cannot* guarantee equality to $\mathcal{O}(\varepsilon)$ of $v_u(t)$ to v_u for $t>T$ even though $\tilde{g}=0$ for $t>T$, since by $t=T$, $v_u(t)$ would have evolved to $v_u(T)$ based on g's influence in $[-T,T]$, and henceforth the tangent vector would not change from $v_u(T)$—which is generically not equal to v_u—since $\tilde{g}=0$ for $t>T$.

Next, an error estimate is established for the situation in which g is extended beyond $[-T,T]$ in a nonzero fashion but satisfying the bound (2.128). In particular, this means that G serves as a bound for the shear terms $\sigma_{s,u}$ in Definition 2.53. If the $\mathcal{O}(\varepsilon)$-errors in the rotational angles $\theta_{s,u}$ at each time $t\in(-T,T)$ are defined by $e_\theta^{s,u}(t)$, then

$$e_\theta^s(t):=\left|\lim_{\varepsilon\to0}\frac{\theta_s(t;T)-\theta(t)}{\varepsilon}\right|\le\frac{G}{\lambda_u-\lambda_s}e^{(\lambda_u-\lambda_s)(t-T)} \tag{2.133}$$

and similarly

$$e_\theta^u(t):=\left|\lim_{\varepsilon\to0}\frac{\theta_u(t;T)-\theta(t)}{\varepsilon}\right|\le\frac{G}{\lambda_u-\lambda_s}e^{(\lambda_s-\lambda_u)(t+T)}. \tag{2.134}$$

The error (2.133) in the rotation of the stable manifold at a value $t \in (-T, T)$ is caused by uncertainties of the nonautonomy in $t > T$ and moreover increases with t exponentially in the finite domain of validity $t \in (-T, T)$. If data are available over larger time-frames, that is, if T can be made larger, the bound for this error can be decreased exponentially in T. Similar statements on the error in the rotation of the unstable manifold can be made from (2.134).

Next, modifying the derived results of Theorem 2.12 for this situation gives the following definition.

Definition 2.54 (finite-time primary unstable manifold). The $\mathcal{O}(\varepsilon)$-approximation to the *finite-time primary stable manifold* for (2.122) is parametrically characterized by

$$x_\varepsilon^u(p,t;T) := \bar{x}^u(p) + \varepsilon \left[\frac{M^u(p,t;T)}{|f(\bar{x}^u(p))|} \hat{f}^\perp(\bar{x}^u(p)) + \frac{B^u(p,t;T)}{|f(\bar{x}^u(p))|} \hat{f}(\bar{x}^u(p)) \right] \quad (2.135)$$

for $(p,t) \in (-\infty, P] \times (-T, T)$ for arbitrarily large but fixed P, in which

$$M^u(p,t;T) := \int_{-T-t+p}^{p} \exp\left[\int_\tau^p \mathrm{Tr}[Df(\bar{x}^u(\xi))]\,\mathrm{d}\xi \right] f^\perp(\bar{x}^u(\tau)) \cdot g(\bar{x}^u(\tau), t+\tau-p)\,\mathrm{d}\tau \quad (2.136)$$

and

$$B^u(p,t;T) := |f(\bar{x}^u(p))|^2 \int_{-T-t+p}^{p} \frac{R^u(\tau) M^u(p,\tau+t-p;T) + f(\bar{x}^u(\tau)) \cdot g(\bar{x}^u(\tau), \tau+t-p)}{|f(\bar{x}^u(\tau))|^2} \mathrm{d}\tau, \quad (2.137)$$

where R^u is defined in (2.23).

The limits for M^u are easy to understand since the argument of $g(x,\cdot)$ must lie in $[-T, T]$ for there to be a positive contribution. The interpretation of the tangential displacement B^u is different in that a slightly different condition to (2.59) is used. In the infinite-time proof, the expression (2.56) was derived. But now this can be integrated from $-T$ to t, with the condition that $B^u(p,-T;T) = 0$ is imposed. This is possible to do since the unstable manifold is *identical* to that for $\varepsilon = 0$ for $t \in (-\infty, -T]$, since the nonautonomous contribution is zero in this domain. Performing the change of variable $\tau - t + p \to \tau$ once again leads directly to (2.137). The imposition of the condition $B^u(p,-T;T) = 0$ was possible in this finite-time situation because of the presence of a specialized time value $-T$, which was not available in the infinite-time setting, before which the unstable manifold for the $\varepsilon = 0$ flow of (2.122) can be thought of as being *identical* to the $\varepsilon \neq 0$ flow, since the vector field is zero for $t < -T$.

If $\tilde{g}$ were not zero outside $[-T, T]$ but were bounded and satisfied (2.128) instead, one can compute gross error estimates for the unstable manifold if using the above definitions. It is easy to see that the error in the normal direction is bounded by

$$\left| [x_\varepsilon^u(p,t;T) - \bar{x}^u(p)] \cdot \hat{f}(\bar{x}^u(p)) \right| \leq \varepsilon G \left| \int_{-\infty}^{-T-t+p} e^{\int_\tau^p \mathrm{Tr}[Df(\bar{x}^u(\xi))]\mathrm{d}\xi} |f(\bar{x}^u(\tau))|\,\mathrm{d}\tau \right|, \quad (2.138)$$

with the integral on the right computable for known f. Thus, a "fattened" unstable manifold can be constructed within the envelope of this error bound, to give an indication of the uncertainty of the computed finite-time unstable manifold. A similar argument leads to determining the finite-time stable manifold, in contrast to the infinite-time expressions of Theorem 2.23.

Definition 2.55 (finite-time primary stable manifold). The $\mathcal{O}(\varepsilon)$-approximation to the *finite-time primary stable manifold* for (2.122) is parametrically characterized by

$$x_\varepsilon^s(p,t;T) := \bar{x}^s(p) + \varepsilon \left[\frac{M^s(p,t;T)}{|f(\bar{x}^s(p))|} \hat{f}^\perp(\bar{x}^s(p)) + \frac{B^s(p,t;T)}{|f(\bar{x}^s(p))|} \hat{f}(\bar{x}^s(p)) \right] \quad (2.139)$$

for $(p,t) \in [-P,\infty) \times (-T,T)$ for arbitrarily large but fixed P, in which

$$M^s(p,t;T) := -\int_p^{T-t+p} \exp\left[\int_\tau^p \mathrm{Tr}[Df(\bar{x}^s(\xi))]\mathrm{d}\xi \right] f^\perp(\bar{x}^s(\tau)) \cdot g(\bar{x}^s(\tau), t+\tau-p)\,\mathrm{d}\tau \quad (2.140)$$

and

$$B^s(p,t;T) := |f(\bar{x}^s(p))|^2 \int_{T-t+p}^p \frac{R^s(\tau)M^s(p,\tau+t-p;T) + f(\bar{x}^s(\tau)) \cdot g(\bar{x}^s(\tau),\tau+t-p)}{|f(\bar{x}^s(\tau))|^2} \mathrm{d}\tau, \quad (2.141)$$

where R^s is defined in (2.60).

If permitting $\tilde{g}$ to be extended outside $[-T,T]$ in a nonzero but bounded fashion, the uncertainty in the normal direction of the stable manifold can therefore be computed by

$$\left| [x_\varepsilon^s(p,t;T) - \bar{x}^u(p)] \cdot \hat{f}(\bar{x}^s(p)) \right| \leq \varepsilon G \left| \int_{T-t+p}^\infty e^{\int_\tau^p \mathrm{Tr}[Df(\bar{x}^s(\xi))]\mathrm{d}\xi} |f(\bar{x}^s(\tau))|\,\mathrm{d}\tau \right|. \quad (2.142)$$

Example 2.56 (double-gyre cont.). Consider the double-gyre, whose infinite-time stable and unstable manifolds were characterized in Example 2.25, and the fact that finite-time leads to no rotation of the local manifold tangent vectors in Example 2.28. Suppose now that the nonautonomous component $q(t)$ is only defined for $t \in [-T,T]$. Finite-time hyperbolic trajectories and their manifolds can be defined as in Definitions 2.52, 2.54, and 2.55 by simply setting $q=0$ outside $[-T,T]$ and by retaining only the $\mathcal{O}(\varepsilon)$ terms. Since in Example 2.25 it was shown that the integrand of the $B^{s,u}$ expressions were zero, and moreover all other infinite-time calculations were performed, those results can easily be adopted in accordance with the finite-time definitions. Thus, the hyperbolic trajectory which is at $(1,1)$ when $\varepsilon=0$ progresses to

$$x_1^u(1,t) = 1 + \varepsilon\pi^2 A \exp\left(-\pi^2 At\right) \int_{-T}^t \exp\left(\pi^2 A\tau\right) q(\tau)\,\mathrm{d}\tau, \quad (2.143)$$

with its attached unstable manifold being given by x_1 as a function of x_2 by

$$x_1^u(x_2,t) = 1 - \varepsilon \frac{\pi^2 A}{2\sin(\pi x_2)} \int_{-T}^t \sin\left[4\cot^{-1}\left(\cot\frac{\pi x_2}{2} \exp\left[\pi^2 A(\tau-t)\right]\right)\right] q(\tau)\,\mathrm{d}\tau, \quad (2.144)$$

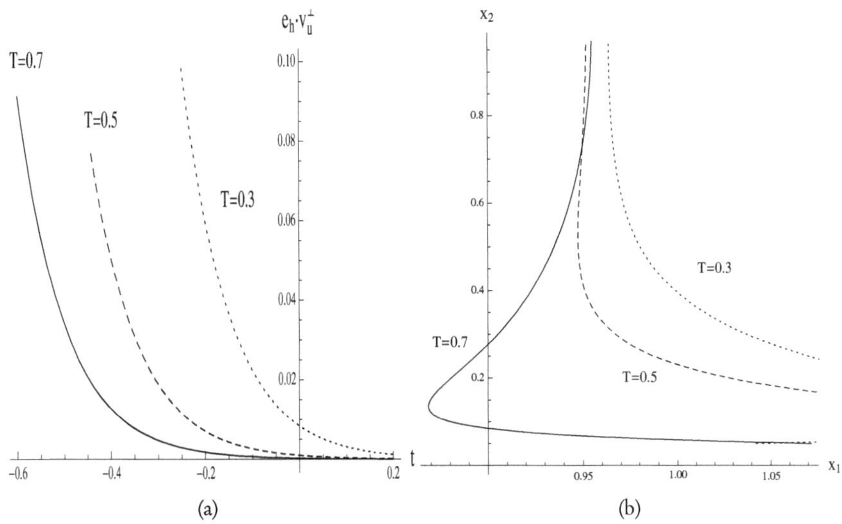

Figure 2.17. *(a) Error in hyperbolic trajectory arising from (2.143), and (b) Unstable manifold at* $t = 0$ *as given by (2.144), for the finite-time double-gyre with parameters* $A = 1$, $\omega = 2\pi$, *and* $\varepsilon = 0.1$.

by limiting (2.71) and (2.70) to the relevant range. The behavior of these expressions is numerically displayed for the standard choice $q(t) = \sin \omega t$ and parameters $A = 1$, $\omega = 2\pi$, and $\varepsilon = 0.1$ in Figure 2.17. In (a), the error of the hyperbolic trajectory in comparison to the true infinite-time trajectory as defined in (2.130) is shown as a function of t for different T-values. The error $e_h(t) \cdot v_u^\perp$ is largest at $t = -T$ and decays exponentially, and as expected, knowledge of data for smaller time-intervals results in larger errors (in this case $e_h(t) \cdot v_s^\perp = 0$). In (b) the finite-time unstable manifold computed using (2.144) is shown at $t = 0$, where now these are the leading-order unstable manifolds computed with q being set to zero outside $[-T, T]$. The result of diminished T can be quite substantial, and in particular the wiggles in the unstable manifold which exist when $T = \infty$ (see Figure 2.10) are captured less for smaller T. ∎

The approach detailed here is only one possibility in attempting to define finite-time stable and unstable manifolds. The basic difficulty in defining these for finite-time comes from the fact that the "usual" definitions for stable and unstable manifolds *locally near hyperbolic trajectories/points* rely on exponential decay estimates as $t \to \pm\infty$ [183]. The global stable/unstable manifolds are obtained by letting these local manifolds flow under the dynamics. Using exponential dichotomy conditions is intimately related, since the exponential decay along these manifolds is towards the hyperbolic trajectory. If time is clipped to a finite interval, the standard exponential decay/dichotomy conditions [115] lose their meaning, since *any* continuous function can be bounded by an exponentially decaying function with a suitably large coefficient. (One approach to circumvent this is to insist that the coefficient is exactly 1, a strong type of finite-time exponential dichotomy [127, 219, 135, 66] which will not be used here.) Thus, standard exponential dichotomies would identify *all* trajectories as hyperbolic. For such a trajectory H, it is not clear, for example, whether other trajectories which were apparently approaching H over the finite-time interval would continue to do so. So do such trajectories comprise the stable manifold of H or not?

Clearly, the ambiguity of identifying hyperbolic trajectories leads to additional ambiguity in defining stable/unstable manifolds in this instance. The approach detailed in this section provides one viewpoint in the nearly autonomous case.

Extensions of the exponential dichotomy conditions to finite-time have been attempted [219, 127, 135, 66] and continue to be developed. Basic approaches which capture exponential decay—in particular Lyapunov exponents—are often used heuristically over a finite-time interval to provide diagnostics to identify analogues of stable/unstable manifolds in finite-time. This is the FTLE approach, which serves to *define* the manifolds through the numerical procedure. One can also try to find, for example, curves which are most attracting over a fixed finite-time interval and define these to be finite-time flow barriers. In these approaches—as has been argued—the dynamics boils down to the effect of a *one-step* map from the beginning to the ending time value, which does not necessarily account for the fact that the reason for the finite-time data is because a dynamical system over a larger time-domain has been sampled only over a finite time interval. How could the information gleaned from this one-step map be extrapolated to the "real" system? The method detailed here is an attempt to go in this direction while retaining the nonautonomous nature of the flow within the given time interval. More direct results for finite-time manifolds are also available when the nearly autonomous assumption is relaxed mildly. Sandstede et al. [355] allow a nonautonomous perturbation to become unbounded as $t \to \pm\infty$ and show that as long as this rate is exponentially bounded with a sufficiently small rate constant, the Melnikov approaches can still be meaningful. The stable/unstable manifolds determined from this procedure are "fattened" ones defined for finite-times; the fattening results from the range of possibilities for extending the exponentially bounded nonautonomous velocity component [355]. Indeed, it is this understanding which enables the quantification of transport resulting from a viscous perturbation to an inviscid flow, which is addressed in detail in Section 3.8.

2.8.1 ▪ Application to data over a finite interval

The nearly autonomous hypothesis used in the majority of this chapter has a sound basis in many geophysical and microfluidic applications in which there is a dominant steady flow based on which the gross dynamical features can be identified easily. In some cases, such near autonomy arises not in the standard laboratory frame, but in a steadily moving frame as in models for the Gulf Stream [48, 314, 125, 323, 417, 225]; in such cases the same ideas can be applied to a dynamical system in the moving frame. Suppose, then, that data is available, over a finite time interval $[-T, T]$, over a discrete spatial grid, and one has the reasonable expectation that the flow is nearly autonomous in this frame of reference. Suppose two-dimensional velocity data u is available at the uniformaly spaced times t_i, $i = 1, 2, \cdots, m$, in $[-T, T]$, and at the points x_j (where x is two-dimensional, and the index j identifies points on a two-dimensional grid), and that u is expected to be nearly autonomous. Define the dominant autonomous velocity by

$$f(x_j) := \frac{1}{m} \sum_{i=1}^{m} u(x_j, t_i) \tag{2.145}$$

and the nonautonomous variation by

$$h(x_j, t_i) := u(x_j, t_i) - f(x_j). \tag{2.146}$$

The "nearly autonomous" assumption is easily verified by examining whether the vector field h is indeed much smaller than the velocities associated with f. Thus the flow that is being considered is a temporally and spatially discretised version of

$$\dot{x} = f(x) + h(x,t), \tag{2.147}$$

and consistency is achieved with the notation of this chapter with the interpretation that $h = \varepsilon g$; i.e., the small parameter ε is already present in the nonautonomous velocity h. Now, the dominant steady structures can be obtained by first examining where $f = 0$, and the hyperbolic (saddle-like) fixed points will be the subset of these such that Df has a positive and negative eigenvalue. These calculations can be done for the discretely specified f by using some interpolation method. Sketching the streamlines using standard software will enable the determination of stable and unstable manifolds emanating from hyperbolic fixed points.

Next, the issue is to determine how those autonomous entities perturb when the nonautonomous part of the velocity field is included. All formulae of the finite-time description of this section can be computed with g replaced by h, and with the prefactor ε appearing in the geometrical interpretations discarded. For example, the leading-order finite-time hyperbolic trajectory location (2.125) could be rewritten as

$$a_\varepsilon(t;T) := a + \left[\tilde{\beta}_u(t;T) v_u^\perp + \frac{\tilde{\beta}_u(t;T)(v_u \cdot v_s) - \tilde{\beta}_s(t;T)}{v_u^\perp \cdot v_s} v_u \right], \tag{2.148}$$

where

$$\tilde{\beta}_s(t;T) := -e^{\lambda_u t} \int_t^T e^{-\lambda_u \tau} h(a,\tau) \cdot v_s^\perp \, \mathrm{d}\tau \tag{2.149}$$

and

$$\tilde{\beta}_u(t;T) := e^{\lambda_s t} \int_{-T}^t e^{-\lambda_s \tau} h(a,\tau) \cdot v_u^\perp \, \mathrm{d}\tau. \tag{2.150}$$

The integrals above are easily discretized and can be computed directly from the data. The nearly autonomous finite-time unstable manifold approximation would then, from (2.135), be representible in this framework by

$$x_\varepsilon^u(p,t;T) := \bar{x}^u(p) + \left[\frac{\tilde{M}^u(p,t;T)}{|f(\bar{x}^u(p))|} \hat{f}^\perp(\bar{x}^u(p)) + \frac{\tilde{B}^u(p,t;T)}{|f(\bar{x}^u(p))|} \hat{f}(\bar{x}^u(p)) \right] \tag{2.151}$$

for $(p,t) \in (-\infty, P] \times (-T,T)$ for arbirarily large but fixed P, in which

$$\begin{aligned} &\tilde{M}^u(p,t;T) \\ &:= \int_{-T-t+p}^{p} \exp\left[\int_\tau^p \mathrm{Tr}[Df(\bar{x}^u(\xi))] \mathrm{d}\xi \right] f^\perp(\bar{x}^u(\tau)) \cdot h(\bar{x}^u(\tau), t+\tau-p) \, \mathrm{d}\tau \end{aligned} \tag{2.152}$$

and

$$\begin{aligned} &\tilde{B}^u(p,t;T) \\ &:= |f(\bar{x}^u(p))|^2 \int_{-T-t+p}^{p} \frac{R^u(\tau)\tilde{M}^u(p,\tau+t-p;T) + f(\bar{x}^u(\tau)) \cdot h(\bar{x}^u(\tau), \tau+t-p)}{|f(\bar{x}^u(\tau))|^2} \mathrm{d}\tau. \end{aligned} \tag{2.153}$$

Once again, appropriate discretizations of the integrals will be required, with interpolations of the discrete data also necessary. This requires dealing with well-established numerical data analysis methods, *without* having to do trajectory integrations which also require interpolations, in order to determine the unstable manifold (2.151) for the nearly autonomous system. The stable manifold expression is similarly modified and will not be stated for brevity. Since these calculations are those for stable and unstable manifolds for a system in which the nonautonomy is turned off outside the interval in which data is known, they automatically satisfy issues the properties outlined in Section 1.4. That is, the entities are transported in a Lagrangian fashion by the flow, the objectivity criterion is met, there is no implied time-periodicity, and time-variation is quantified within the domain in which data is available. Moreover, if one assumes that the time-variation happened to be extended outside the interval $[-T, T]$ in a bounded fashion, one can obtain error bounds for the stable and unstable manifolds, as given in (2.138) and (2.142). This approach therefore allows for a self-consistent method for computing flow barriers from nearly autonomous data over a finite-time, by explictly adapting notions of stable and unstable manifolds.

Chapter 3
Quantifying transport flux across unsteady flow barriers

Existence is no more than the precarious attainment of relevance in an intensely mobile flux of past, present and future.
—Susan Sontag

3.1 ▪ Classical results for time-periodic flow: lobe dynamics

In this chapter, the focus is on quantifying flux (i.e., transport) in a system

$$\dot{x} = F(x,t), \tag{3.1}$$

in which $x \in \Omega$, an open connected subset of $\mathbb{R}^2$, and F is assumed as smooth as required. As expressed in Chapter 1, if F has no t-dependence, the system is autonomous. Since two-dimensional, the Poincaré-Bendixson Theorem applies, and so the long-term fate of any flow trajectory is constrained: it may escape to infinity, or it may approach a fixed point, a periodic orbit, or a heteroclinic cycle [9, 183]. This will organize the phase space into regions (coherent structures), based on the basins of attraction of each entity (and of infinity), separated by flow barriers. If the initial point were on a periodic orbit which is a member of a family of nested periodic orbits, it would not be a limit cycle, and thus its basin of attraction would be itself. If so, the nested set of periodic orbits could be construed a coherent structure by itself, and its "boundary" would be a flow barrier of interest. Any flow barrier cannot vary with time since the system is autonomous, and quantification of flux is more straightforward. For example, if a stable and unstable manifold coincide (as in all the pictures in Figure 1.1) then the heteroclinic manifold created by this forms a flow barrier across which there is *zero* flux. The difficulties arise when F has genuine time-dependence, in which case stable and unstable manifolds may intersect in various ways in each time-slice, and moreover the manifolds move with time. Thus, using an Eulerian method for quantifying transport across barriers would not be meaningful. For example, if a manifold is moving in a nontangential direction, the normal component of the Eulerian velocity would indicate that there is transport across it, whereas in reality this velocity simply moves the flow barrier itself, and therefore there should be no transport. In other words, it is necessary to take into account the Lagrangian nature of particles in coming up with a definition of flux. *It is not just the velocity at present that is important, but the velocity in the past and future.*

If F's t-dependence is *periodic*, there is a rich literature on quantifying the flux as a result of a stable and unstable manifold intersecting. The key idea here is of *lobe dynamics* [259, 258, 101, 343, 421], which shall be discussed briefly in this section. Suppose F has time-periodicity T, and consider the *Poincaré map* P_τ formed by taking points in Ω at time τ to time $\tau + T$ by following the flow of (3.1). Since P_τ is a map generated from a flow, P_τ must be invertible and orientation-preserving. Without venturing into formal definitions, the intuition for these claims comes from imagining a blob of fluid A mapped by P_τ to another blob A'; then A' will get mapped to A by following the flow of (3.1) for a time $-T$ (which defines the map P_τ^{-1}). Moreover A' must be obtained from A by a deforming process which cannot reverse the orientation of nearby particles (prohibiting, for example, picking A up "out of the plane," flipping it upside down, and then putting it back on the plane). Now, since transport is to be assessed in relation to stable and unstable manifolds, the assumption will be that these exist for P_τ:

Hypothesis 3.1 (Poincaré map properties). *For some fixed $\tau \in [0, T)$, the Poincaré map P_τ associated with (3.1) possesses the following:*

(a) *Fixed points a and b (which might be the same point) such that the two eigenvalues of each of $DP_\tau(a)$ and $DP_\tau(b)$ have moduli greater than and less than one.*

(b) *A branch of the one-dimensional unstable manifold of the fixed point a transversely (nontangentially) intersects a branch of the one-dimensional stable manifold of b at a point q.*

Now, an unstable manifold is invariant. Since q lies on the unstable manifold of a, $P_\tau^n(q)$ must also lie on this same manifold for $n \in \mathbb{Z}$. Similarly, all forward and backward iterates of q must lie on b's stable manifold. This implies that the points $P_\tau^n(p)$, $n \in \mathbb{Z}$ are a countably infinite set of points which all correspond to intersections of the unstable manifold of a with the stable manifold of b. Thus, one intersection between these manifolds implies infinitely many intersections. A generic picture for such an intersection pattern is shown in Figure 3.1. This displays a *heteroclinic tangle* in which the stable and unstable manifolds intersect infinitely many times, and these intersection points accumulate towards the fixed points of P_τ. Here, only the first forward and backward iterate of q are pictured, in red. Notice that there may be other intersection points (such as that indicated by $\bar{q}$) which are not associated with iterates of q, and also that the intersection between manifolds leads to the creation of *lobes* which are bounded by a segment of each of the stable and unstable manifolds and indicated by L_i.

Under the condition that these two manifolds intersect infinitely often, how can one characterize a *transport*, or *flux*, across them? The usage of lobe areas to measure this has been advocated for some time [259, 258, 101, 343], but in what sense does this work? Following [343, 421], consider what happens to lobes under P_τ. Since q maps to $P_\tau(q)$ under the Poincaré map, it is easy to see that the nearby lobe L_4 maps to a lobe $P_\tau(L_4)$ near $P_\tau(q)$. Continuity implies that L_i, $i = 1,2,3$, get mapped, respectively, to $P_\tau(L_i)$ in Figure 3.1. It is easy to understand that the number of lobes between any adjacent iterates of q must always be four, which could be identifiable as $P_\tau^m(L_i)$, $i = 1,2,3,4$, for some $m \in \mathbb{Z}$. A natural approach to quantifying transport might be to compute the area of a lobe: an understanding that pervades the literature in many

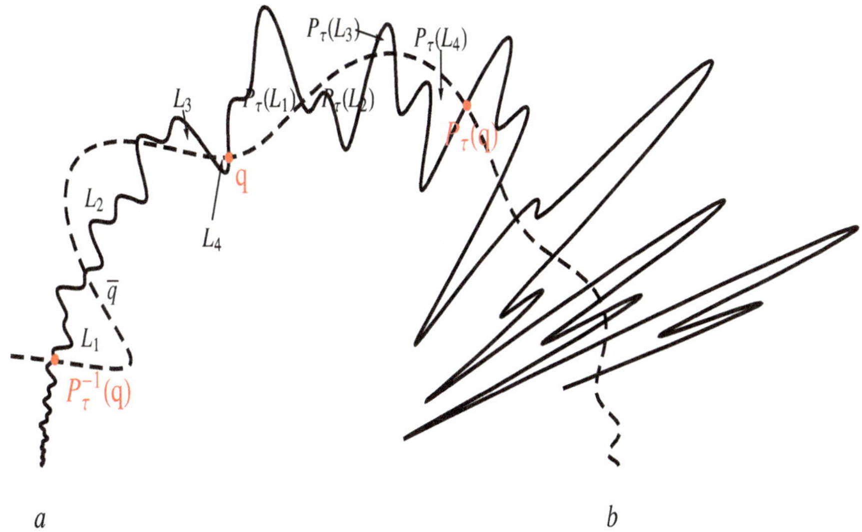

Figure 3.1. *A generic intersection pattern for a stable (solid) and an unstable (dashed) manifold associated with the Poincaré map P_τ of (3.2) when F is T-periodic in time.*

time-periodic examples. But this leads to several difficulties.

1. The area of *which* lobe? There are infinitely many lobes in Figure 3.1, all with potentially unequal areas.
2. If one specific lobe (say, L_1) out of the four present in this case is chosen as being the one of importance, even this leads to ambiguity; $P_\tau(L_1)$ and/or $P_\tau^{-1}(L_1)$ may each not have the same area as L_1. In general, the area of $P_\tau^n(L_1)$ will be different for each $n \in \mathbb{Z}$. Which specific *iterate* of the chosen lobe iterates is the right one to choose?
3. The Poincaré map takes lobes from time τ to time $\tau + T$, and hence depends on τ. The picture one gets if choosing a different τ value would be different from Figure 3.1 *in detail*, although it would need to have the same topological intersection pattern. So which τ is the "right" one to choose if trying to compute lobe areas?
4. The picture of Figure 3.1 was presumed on there being at least one intersection point q. What if there were *no* intersections between the stable and unstable manifold? This would imply that there are no lobes—but does this mean that there is no transport?

These questions indicate that *even in the time-periodic* two-dimensional situation, defining flux using lobe areas is ambiguous. However, under certain *additional* assumptions, lobe areas can be fruitfully applied.

Hypothesis 3.2 (lobe areas are applicable). *Suppose Hypothesis 3.1 is satisfied, and moreover*

(a) *the function F in (3.1) is* area-preserving *i.e., the condition* $\mathrm{Tr}\, DF = 0$ *(equivalently,* $\operatorname{div} F = 0$*) is satisfied for all $t \in \mathbb{R}$;*

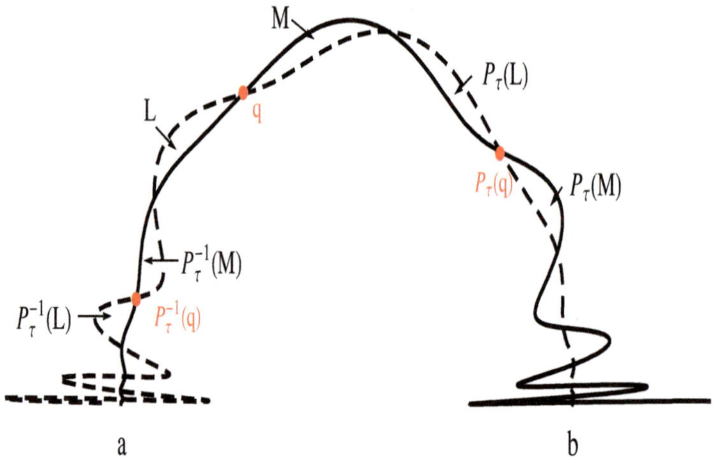

Figure 3.2. *The specific intersection pattern associated with Hypothesis 3.2.*

(b) *there are two intersection points q and $\bar{q}$ such that $P_\tau^n(q)$ and $P_\tau^n(\bar{q})$ for $n \in \mathbb{Z}$ represent the two (and only two)* distinct *heteroclinic trajectories of the Poincaré map;*

(c) *the two lobes formed between an intersection segment between q and $P_\tau(q)$ have equal areas.*

Area-preserving flows can be thought of as *incompressible* from the fluid mechanics viewpoint. This guarantees that any blob of fluid maintains the same area throughout time. The conditions on there being exactly two heteroclinic trajectories may appear somewhat unusual, but this turns out to be the generic situation in which the time-variation is *harmonic* or *sinusoidal*, as will be explained in Section 3.3 (briefly, both the sine and the cosine function have two distinct zeros in each period) . The situation arising from Hypothesis 3.2 is pictured in Figure 3.2. Here, q and its forward and backwards iterate under P_τ is pictured; continuing iteration forwards leads to a sequence of points asymptoting to b, and likewise to a under backwards iteration. The intersecting manifold segments between q and $P_\tau(q)$ reveals another intersection point, which can be thought of as the $\bar{q}$ in Hypothesis 3.2. Iterates of this, $P_\tau^n(\bar{q})$, form a heteroclinic trajectory between a and b which is distinct from $P_\tau^n(q)$; therefore, there are exactly two such heteroclinic trajectories in this instance. The lobe intersection pattern of this nature is often a signature of chaotic transport (though for the Smale-Birkhoff Theorem to guarantee chaotic transport, one needs $a = b$).

Consider the lobe labeled by L, which is bounded by a segment of each of the stable and unstable manifolds. To figure out its image under P_τ, one can use the fact that the boundary consisting of the unstable manifold must map to a segment of the unstable manifold as well, and since one endpoint of this is at q, it must map to a segment with endpoint $P_\tau(q)$. To maintain the ordering, L's image must thus be $P_\tau(L)$ as shown in Figure 3.2. Now, by the area-preserving nature of F, $P_\tau(L)$ must have the same area as L. Continuing this argument, each of the lobes $P_\tau^n(L)$ for $n \in \mathbb{Z}$ has the identical

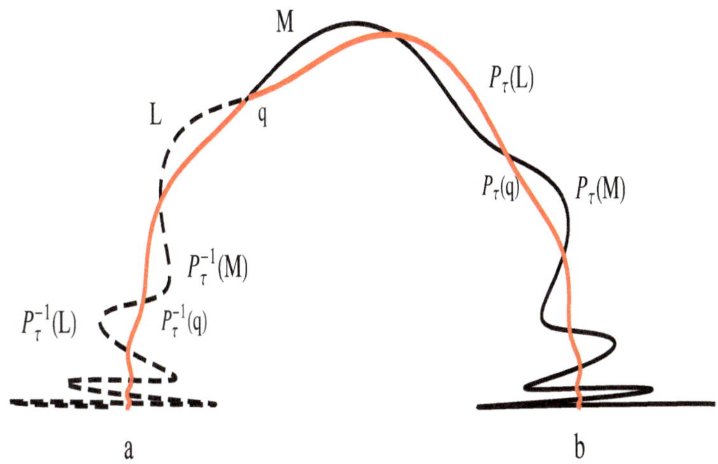

Figure 3.3. *Defining the pseudoseparatrix [thick red curve] associated with the Poincaré map of Figure 3.2.*

area. There is another set of lobes which is distinct from these, shown by $P_\tau^n(M)$ in Figure 3.2, all of which have the same area. By Hypothesis 3.2, M and $P_\tau(L)$ have equal areas, and consequently *all* the lobes in Figure 3.2 have the same area. Since the bounding points $P_\tau^n(q)$ and $P_\tau^n(\bar{q})$ of $P_\tau^n(M)$ both approach b as $n \to \infty$, the distance $|P_\tau^n(q) - P_\tau^n(\bar{q})| \to 0$, and to compensate, the lobes must elongate in the transverse direction when approaching b. This scenario is visible in Figure 3.2 as the hyperbolic fixed points a and b are approached.

The upshot of the above discussion is that all lobe areas are equal under Hypothesis 3.2, and thus the area of any *one* lobe can be used as an unambiguous measure of transport. However, what exactly does this mean? Transport across what? The fact that the stable and unstable manifold intersect in a complicated fashion means that a transport barrier cannot be defined easily, unlike in a case in which the stable and unstable manifold coincide (of which examples were given in Chapter 1). This issue was resolved by Rom-Kedar and collaborators [341, 343], and described in detail by Wiggins [421], through the definition of a *pseudoseparatrix*, which is not quite a "separatrix" since it does not exactly separate regions. The first issue is to identify a principal intersection point between the manifolds; this is a defined by Wiggins [421] as a point of intersection q such that the unstable manifold U_{aq} between a and q has no other intersection with the stable manifold S_{qb} between q and b. The union of the manifold segments U_{aq} and S_{qb}, which meet at q, is then defined to be the pseudoseparatrix, which is shown in Figure 3.3 by the thick red curve. The definition of a pseudoseparatrix is not unique since it depends on the choice of the principal intersection point at which the manifolds are to be joined together.

Once the pseudoseparatrix is defined, it is possible to rationalize transport across it per iteration of the Poincaré map P_τ. In Figure 3.3 one can have the notion of lobes which are "above" the pseudoseparatrix and those that are "below" it. What happens to these lobes under P_τ? For example, $P_\tau^{-1}(L)$ gets mapped to L, and similarly $P_\tau^{-2}(L)$

gets mapped to $P_\tau^{-1}(L)$. In either case, the lobes have *not* got transported across the pseudoseparatrix, since all these lobes are above it. On the other hand, L gets mapped to $P_\tau(L)$, which represents a transfer from above to below the pseudoseparatrix. A similar analysis of what happens to all the lobes indicates that the only lobes which cross the pseudoseparatrix under the action of P_τ are $P_\tau^{-1}(M)$ and L. The first of these crosses from below the pseudoseparatrix to M, which is above it, and the second crosses from above to below it. In effect, the *only* lobes which are associated with crossing the pseudoseparatrix are the two lobes "just before" the chosen principal intersection point q, which cross to the two lobes "just after" the intersection point. These four lobes are collectively called the *turnstile* [421, 341]. The clever definition of the pseudoseparatrix forces the understanding that these are the only lobes which participate in transport across the pseudoseparatrix. In this context, it makes sense to think of the lobe area of L (which is equal to the area of any other lobe in Figure 3.3) as a measure of the transport occurring in a situation as pictured in Figure 3.2.

This picture of *lobe dynamics*, and usage of a lobe area to quantify transport, is a pleasing one. However, it must be borne in mind that this made sense under the very restrictive assumptions of Hypothesis 3.2. In the more general time-periodic situation of Figure 3.1, it has been shown that there are several difficulties which make the appealing idea of lobe dynamics and areas inapplicable. In Section 3.5, some resolution to this is provided, along with computable transport formulae for general time-periodic situations. But Figure 3.1 is still a highly idealized situation since it depends on the presence of a Poincaré map for which time-periodicity of the flow is required. How exactly can transport be quantified in situations when all this fails? That is, how can transport be quantified in genuinely unsteady flows, where the time-variation is general? This is the subject of the next section.

3.2 ▪ Flux definition for general time-dependence

In the system

$$\dot{x} = F(x,t) \tag{3.2}$$

for $x \in \Omega$, a two-dimensional open connected surface, can transport be defined if F has *general* time-dependence?

In the time-periodic situation under several other assumptions, Section 3.1 outlined how lobe areas could be a useful method for quantifying transport. One approach for going into genuinely time-aperiodic situation would be to define a sequence of maps (rather than *one* recurring Poincaré map) and use lobe areas associated with such maps [63, 148, 264, 42, 337, e.g.]. This causes some difficulty in interpretation, since as shown in Section 3.1 repeated applications of the Poincaré map was necessary in rationalizing exactly how lobe areas were definitive entities in transport. Indeed, one aspect of such Poincaré map not emphasized in Section 3.1 was the fact that the Poincaré map P_τ depended on τ; the map took fluid particles from a time τ to a time $\tau + T$, where T was the periodicity of the flow. Thus, even in this case, there is time-dependence in the interpretation due to the presence of τ. When applying a sequence of maps to account for general time-variation, the problem of time-variation in a continuous vector field (3.2) simply gets translated to infinitely many maps—a different one for each time-instance considered. Unless there are additional time restrictions, this will not simplify the problem.

In such a situation, the attitude that *transport shall be defined as a time-varying quantity* makes sense within the time-varying situation of (3.2). But how is this to be done?

First, the transport *across some entity* must be defined. This entity presumably must be a curve in order to partition two-dimensional space, but exactly how should the curve be determined? In steady flows as shown in Figure 1.1, the stable and unstable manifolds represent a particularly important set of curves which distinguish between regions of distinct types of motion. The transport across any of these is zero. However, the transport across *any* trajectory in Figure 1.1 is also zero, and thus trying to find a curve across which transport is zero as a way of identifying flow separators, or boundaries of Lagrangian Coherent Structures, does not work. On the other hand, the method should recover the steady flow barriers of Figure 1.1 if the flow is autonomous. A *perturbation* which has a general time-variation on any of these pictures will cause transport across such flow barriers, and this will cause fluid interchange between regions which initially had distinct motions, as a result of the broken heteroclinic manifold. In other words, the intersection pattern of the stable and unstable manifolds, be it as in Figure 1.1 in which they coincide, or as in Figure 3.1, in which they intersect in some way, or in alternative situations in which they intersect a finite number of times or not at all, is the progenitor of transport between the previously coherent regions. Thus, any method that we use to choose an entity across which transport is to be assessed must have central to it the concepts of stable and unstable manifolds.

Using the language of fluid mechanics, there are apparently two possibilities that might be considered. The first is to use an *Eulerian curve*, that is, a curve which is fixed in time. This terminology relates to *Eulerian* quantities in fluid mechanics as being quantities which are fixed in time, as opposed to *Lagrangian* quantities, which are those which follow fluid particles advected by the flow (for example, the temperature at the location of a *specific* fluid particle as it is advected according to (3.2). The second type of curve is thus a *Lagrangian curve*, which is a curve which is advected by the flow of (3.2). It is the mixing together (or lack of such) of fluid particles, and of structures which display coherence under the influence of the flow, that is to be quantified. This suggests that Lagrangian curves—which incorporate transport—might be better candidates than Eulerian ones. Consider identifying a string of particles in a fluid at time zero as our Lagrangian curve. As each particle gets advected by the flow, this curve evolves, but continuity ensures that it remains a curve. It is a *material curve* of the flow (3.2), and no other fluid particles cross it while it is evolving. Thus, the Lagrangian transport across this is zero. This is so for *any* chosen string of particles, and thus using Lagrangian curves in this fashion in an unsteady flow as the entity across which flux is to be assessed is a futile exercise. The specific conclusion that can be reached is that, if using a stable or unstable manifold as a potential Lagrangian curve across which transport is to be assessed in an unsteady situation, the transport shall always be zero. In any case, choosing purely Lagrangian curves does not make sense.

Instead, consider using an Eulerian curve, which simply means a curve which is fixed in time. One can of course compute the transport across an Eulerian curve in an unsteady situation by simply quantifying the amount of fluid which crosses it per unit time; this is related to the fluid velocity normal to the curve. Such an Eulerian flux will in general change with time in unsteady flows, thereby providing a time-varying transport quantification. However, of particular interest is the transport from one coherent structure to another; this sort of issue governs, say, the heat transported from an oceanic eddy to another, or how oil which was trapped within an oceanic ring escapes to an oceanic jet. In unsteady flows such coherent entities are themselves moving around, and thus using Eulerian curves across which transport is assessed *also* makes little sense. One thought might be to move to a frame of reference in which the coherent structure is stationary and then use an Eulerian curve in that frame; this would give a reasonable way of quantifying transport into (or out of) that particular entity. But is it possible to find a frame of reference in which the entity is not moving? This is not a well-defined question since the "entity" comprises many particles, each of which is moving at different velocities.

Given the fact that neither Eulerian nor Lagrangian curves appear to be reasonable entities to consider for transport assessment, what can be done? Once again, the centrality of stable and unstable manifolds comes into play. A blob of fluid placed across a stable manifold will eventually get pulled apart in a direction transverse to the stable manifold due to the influence of the complementary *unstable* manifold emanating from the governing hyperbolic trajectory. This can be illustrated via Figure 1.4; any blob on the stable manifold will get exponentially stretched in the up-down direction once it reaches the ball of influence B_ε of the hyperbolic trajectory in which linearized motion becomes applicable. Thus, *a stable manifold can be thought of as a flow barrier in forward time*. Similarly, *an unstable manifold is a flow barrier in backward time*. These statements are true whether the flow is autonomous or nonautonomous. If a stable manifold and an unstable manifold happened to coincide to form a heteroclinic manifold, this is a happy circumstance which forms a flow barrier in *both* time directions, as happens in all the pictures in Figure 1.1. These steady pictures all possess the pleasing happenstance of possessing a codimension 1 heteroclinic manifold—these are therefore genuine flow barriers, in forward *and* backward time, which specifically partition space.

This situation is nongeneric in unsteady flows. Stable and unstable manifolds which have codimension 1, while being flow barriers in forward and backward time, respectively, may not intersect at a particular instance of time. Or their intersection may occur at lower dimensions. This is the typical expectation: for example, in all the pictures presented in the previous two sections, the manifolds were curves, and their intersections occurred at points. That is, the heteroclinic "manifold," the proposed flow barrier in forward and backward time, was a collection of points which did not *separate* space. Genuine flow barriers need to have codimension 1 and be barriers in both forward and backward time. Thus, these typically do *not* exist in the unsteady situation.

To quantify transport across a "system" consisting of a stable and unstable manifold, some assumptions will be necessary. Consider the situation of a stable and and an unstable manifold at a time instance t such that these do not coincide but are "close" to one another. This stable and unstable manifold might intersect a finite number of times, an infinite number of times, or not at all. It is this *region of intersections* which is key to transport, and a definition should be able to account for all possibilities of intersection. So, in the sense that the stable and unstable manifolds need to be considered, there must be a Lagrangian flavor to the definition of a transport flux. However, it must be specialized to include the stable and unstable manifolds. Transport, then, will be quantified in a way which accounts for the relative positioning of a stable and an unstable manifold. A particular definition—a slight generalization of that originally developed by Haller and Poje [192]—is presented; this enables the quantification of a flux "between" a stable and an unstable manifold as a time-varying entity. This definition works under the conditions.

Hypothesis 3.3 (nonautonomous pseudoseparatrix existence). *The system (3.2) possesses the following.*

1. *Specialized trajectories $a(t)$ and $b(t)$ (which may be the same) which are* hyperbolic *in the sense of exponential dichotomies;*

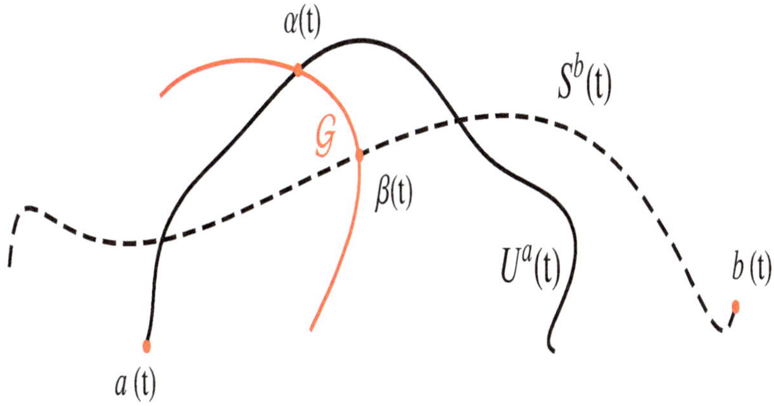

Figure 3.4. *Constructing the pseudoseparatrix $\mathscr{Q}(t)$ for the system (3.2) subject to Hypothesis 3.3 from an unstable manifold [solid], a stable manifold [dashed], and a gate curve $\mathscr{G}$ [red]; the segments between the red dots of these curves represent the three entities in (3.3).*

2. *A fixed simple smooth finite curve segment $\mathscr{G}$ in Ω, with a chosen direction of unit normal $\hat{n}$ which varies continuously along $\mathscr{G}$, such that for $t \in (T_i, T_f)$, the unstable manifold $U^a(t)$ of $a(t)$ intersects $\mathscr{G}$ at a point $\alpha(t)$, and the stable manifold $S^b(t)$ of $b(t)$ intersects $\mathscr{G}$ at a point $\beta(t)$;*
3. *If $\hat{U}^a(t)$ is the restriction of $U^a(t)$ to between the points $a(t)$ and $\alpha(t)$ (including the endpoints), and if $\hat{S}^b(t)$ is similarly the restriction of $S^b(t)$ to between the points $\beta(t)$ and $b(t)$, then $\hat{U}^a(t)$ and $\hat{S}^b(t)$ do not intersect for any $t \in (T_i, T_f)$;*
4. *If $\hat{\mathscr{G}}(t)$ is the restriction of $\mathscr{G}$ to between $\alpha(t)$ and $\beta(t)$ for $t \in (T_i, T_f)$, then $\hat{\mathscr{G}}(t)$ does not intersect with either $\hat{U}^a(t)$ or $\hat{S}^b(t)$, except at the points $\alpha(t)$ and $\beta(t)$, where it, respectively, connects up with each of these entities.*

Transport will only be defined for $t \in (T_i, T_f)$. For a reminder of the first condition, see (1.20)–(1.22). The implication is that each of $a(t)$ and $b(t)$ possess *both* a stable and an unstable manifold. When viewed in a t-slice of the augmented phase-space of (3.2), this means that there exist one-dimensional entities $U^a(t)$ and $S^a(t)$ representing the unstable and stable manifolds, respectively, associated with $a(t)$. Similarly, there are one-dimensional unstable and stable entities $U^b(t)$ and $S^b(t)$ related to $b(t)$. Such a t-slice for $t \in (T_i, T_f)$ is shown in Figure 3.4. In keeping with the ideas for the periodic case, transport is to be assessed in relation to the relative positioning of *one* stable manifold and *one* unstable manifold; in this instance, $U^a(t)$ and $S^b(t)$ at each time value $t \in (T_i, T_f)$. Thus, the transport (which shall be called the *flux*) shall be a time-dependent quantity. This transport shall be assessed across a nonautonomous version of a pseudoseparatrix, whose existence is guaranteed by Hypothesis 3.3. The pseudoseparatrix $\mathscr{Q}(t)$ in a time-slice $t \in (T_i, T_f)$ is defined by

$$\mathscr{Q}(t) := \hat{U}^a(t) \cup \hat{\mathscr{G}}(t) \cup \hat{S}^b(t). \tag{3.3}$$

This pseudoseparatrix is a combination of Lagrangian and Eulerian viewpoints, since the manifold segments are Lagrangian in nature, and $\mathscr{G}(t)$ is Eulerian. The necessity for restricting $\mathscr{G}(t)$ to $\hat{\mathscr{G}}(t)$ shows that this too is impacted at its endpoints by Lagrangian considerations. The above formalizes the process originally suggested by Haller and Poje [192] in defining a so-called *gate*; the gate here is defined by $\hat{\mathscr{G}}(t)$ in each relevant time-slice t. All entities defined here are shown in Figure 3.4. The nonintersecting conditions of Hypothesis 3.3 are very similar to the conditions for an intersection point to be a "principal" one given for the time-periodic situation of Section 3.1.

Consider the transport across the pseudoseparatrix $\mathscr{Q}(t)$ as time evolves. Since the portions $\hat{U}_a(t)$ and $\hat{S}_b(t)$ consist of parts of invariant manifolds, they are material curves which evolve such that there is *no* transport across them. On the other hand, the gate $\hat{\mathscr{G}}(t)$ consists of a fixed curve in space (it is only its endpoints which change with t), across which in general there is transport. Thus, the only instantaneous transport (instantaneous flux) that occurs across $\mathscr{Q}(t)$ is that which occurs across the gate. This is the rationalization for Haller and Poje to quantify the flux across gates, which was subsequently used in an oceanographic context [277]. This idea has been formalized through the definition of a time-varying pseudoseparatrix above. Given that the flux across $\mathscr{Q}(t)$ is only contributed to by that across $\hat{\mathscr{G}}(t)$, the following definition for the instantaneous flux is now presented.

Definition 3.4 (instantaneous flux). *The* instantaneous flux $\phi(t)$ *across a pseudoseparatrix* $\mathscr{Q}(t)$ *constructed from Hypothesis 3.3 is defined for* $t \in (T_i, T_f)$ *to be*

$$\phi(t) := \int_{\hat{\mathscr{G}}(t)} F(x,t) \cdot \hat{n} \, ds \,, \tag{3.4}$$

in which s *is the arclength parametrization along* $\hat{\mathscr{G}}(t)$ *and* $\hat{n}$ *is the unit normal vector to each point in* $\hat{\mathscr{G}}$ *chosen continuously along* $\hat{\mathscr{G}}$.

There is of course the standard uncertainty of which of the two directions to choose for the normal vector $\hat{n}$; it shall be assumed that once a choice of direction has been chosen for one point on $\hat{\mathscr{G}}$, the same direction is maintained to ensure that $\hat{n}$ varies continuously along $\hat{\mathscr{G}}$. Thus, $\phi(t)$ is unambiguously defined up to a negative sign corresponding to which direction is chosen. The above is a *signed* instantaneous flux; an *absolute instantaneous flux* could be defined via

$$\phi(t) := \int_{\hat{\mathscr{G}}(t)} |F(x,t) \cdot \hat{n}| \, ds \,, \tag{3.5}$$

in which flux across $\mathscr{G}$ in both directions is added absolutely. Choosing curves to minimize such an absolute flux is one approach for determining flow barriers in unsteady flows [44, 43], which shall not be pursued here.

For the remainder of this chapter, the focus shall be on the signed instantaneous flux as defined in (3.4). It should be noted that for arbitrarily wiggling $U^a(t)$ and $S^b(t)$ in a time-slice t, one can *always define* a finite curve segment $\hat{\mathscr{G}}(t)$ in order to satisfy Hypothesis 3.3 in some time interval around that value. The simplest choice is to use a straight line which connects the two manifolds together; as time evolves, the manifolds' positioning relative to $\hat{\mathscr{G}}$ will change but in a suitable time interval will continue to satisfy the conditions on nonintersection given in Hypothesis 3.3. Thus, the definition of a pseudoseparatrix is *always* possible for any general two-dimensional unsteady flow (3.2) "between" a one-dimensional stable and a one-dimensional unstable manifold. This is valid even under conditions such as infinite intersections between these manifolds, only finitely many intersections, or even no intersections at all. Thus, the transport relative to $U^a(t)$ and $S^b(t)$ can *always* be defined, irrespective of the nature of these two curves. Of course, $\mathcal{Q}(t)$ may only exist over a finite interval in t; this implies that the transport will only be defined for a finite-time interval.

The instantaneous flux as defined above has a pleasing physical meaning in that it corresponds to the amount of fluid per unit time, which instantaneously crosses the pseudoseparatrix at that instance in time. Since in two spatial dimensions, the "amount" here is an area of fluid; the instantaneous flux therefore has units of length-squared per unit time.

This computed transport depends on the choice of the curve $\mathcal{G}$, i.e., the choice of gate. Such a dependence is inevitable, particularly given that incompressibility has *not* been assumed for (3.2). Even for steady situations, the fact that the choice of $\mathcal{G}$ will affect the transport is easily obtained by choosing two different curves $\mathcal{G}$ and $\mathcal{G}'$ and applying the divergence theorem to the region enclosed by $\hat{\mathcal{G}}$, $\hat{\mathcal{G}}'$, the portion of $U^a(t)$ lying between $\alpha(t)$ and $\alpha'(t)$, and the portion of $S^b(t)$ lying between $\beta(t)$ and $\beta'(t)$, where the primes are the entities associated with $\mathcal{G}'$. Unless F is divergence-free, the flux across $\hat{\mathcal{G}}$ and $\hat{\mathcal{G}}'$ will be different. The dependence on $\mathcal{G}$ in the flux definition in Definition 3.4 is therefore inevitable.

Given the fact that the computed transport depends so strongly on $\mathcal{G}$, the *usefulness* of the instantaneous flux definition may be called into question. Characterizing the flux across *arbitrary* choices of $U^a(t)$ and $S^b(t)$ will always lead to this difficulty; in reality, computing such a flux will only make sense if $U^a(t)$ and $S^b(t)$ have some relationship to one another. The classical situation is if these manifold segments come from a broken heteroclinic manifold [183, 421]. That is, at a certain value of parameter, suppose (3.2) possesses a situation in which the hyperbolic trajectories $a(t)$ and $b(t)$ are connected together by a codimension 1 *heteroclinic manifold*. At this parameter value there is no flux across any choice of pseudoseparatrix $\mathcal{Q}(t)$, since the points $\alpha(t)$ and $\beta(t)$ are identical, and the gate $\hat{\mathcal{G}}(t)$ therefore vanishes. Now, suppose the parameter is changed slightly, resulting in the heteroclinic manifold splitting up into $U^a(t)$ and $S^b(t)$. For small values of the parameter change, a segment of each of these manifolds will remain close to the original heteroclinic manifold, and thus these segments will be close to one another. In this instance, as will be shown in the next section, the instantaneous flux definition works very well in quantifying the flux across this "broken separatrix."

3.3 ▪ Flux as a Melnikov integral

When lobes are applicable (i.e., the conditions of Hypothesis 3.2 are met), the lobe area forms a well-defined quantification of transport in time-periodic flows. If the system can be written as a perturbation of an autonomous flow, then there are well-known results [341, 421] which indicate that the lobe area can be expressed as an integral of a Melnikov function. In this section—following the development in [28]—it is shown that *even if lobe areas are not applicable* (by having unequal areas, say, or by not existing at all because stable and unstable manifolds do not intersect, or by being impossible to define via a Poincaré map because the flow is not time-periodic), the Melnikov function has a strong relationship to a definition for flux. Indeed, it is not an integrated version of a Melnikov function that characterizes the flux; it is the Melnikov function itself!

To explain this, the instantaneous flux definition of the previous section will be considered for the general perturbed two-dimensional system

$$\dot{x} = f(x) + \varepsilon g(x,t), \tag{3.6}$$

in which $x \in \Omega$, a two-dimensional open subset of $\mathbb{R}^2$, and $|\varepsilon| \ll 1$. This is exactly (2.6) examined in Chapter 2, in which expressions for the perturbed stable and unstable manifolds of a hyperbolic fixed point which existed when $\varepsilon = 0$ were derived. The smoothness of f and g characterized, respectively, by Hypotheses 2.2 and 2.3 are assumed to apply here as well. Now, the previous section's definition for an instantaneous flux involved arbitrary stable and unstable manifolds $S_b(t)$ and $U_a(t)$, respectively. A very useful situation would be when these two manifolds arose when $\varepsilon \neq 0$ by a heteroclinic manifold which existed at $\varepsilon = 0$ breaking apart. In this situation, the instantaneous flux would quantify the transport between coherent structures, whose "boundaries," though hard to define precisely in the unsteady context because of stable and unstable manifolds not coinciding, nevertheless are governed by how these manifolds evolve.

Hypothesis 3.5 (Heteroclinic manifold). *The system (3.6) when $\varepsilon = 0$, i.e., the system $\dot{x} = f(x)$, has the following features.*

1. *There exists a point $a \in \Omega$ such that $f(a) = 0$, and $Df(a)$ possesses eigenvalues $\lambda_s^a < 0$ and $\lambda_u^a > 0$.*
2. *There exists a point $b \in \Omega$ such that $f(b) = 0$, and $Df(b)$ possesses eigenvalues $\lambda_s^b < 0$ and $\lambda_u^b > 0$.*
3. *A branch of a's one-dimensional unstable manifold, denoted by U^a, coincides with a branch of b's one-dimensional stable manifold, denoted by S^b, thereby forming a heteroclinic manifold Γ.*
4. *The manifold Γ can be expressed by the heteroclinic trajectory $\bar{x}(t)$ of (2.9), such that $\bar{x}(t) \to a$ as $t \to -\infty$ and $\bar{x}(t) \to b$ as $t \to \infty$.*

The saddle fixed points a and b may well be one and the same, in which case the heteroclinic manifold is homoclinic. Thus, (3.6) when $\varepsilon = 0$ has a flow separator Γ, which demarcates distinct types of motion. From the pictures in Figure 1.1, Γ can therefore be thought of as a boundary between coherent structures. There is no transport between these structures when $\varepsilon = 0$.

In Figure 3.5, the heteroclinic manifold is shown by a dashed curve. When $\varepsilon \neq 0$ however, the system (3.6) becomes nonautonomous. By the analysis of Section 2.2 the saddle fixed points a and b become instead hyperbolic trajectories $a_\varepsilon(t)$ and $b_\varepsilon(t)$. They each retain their stable and unstable manifolds as well; this is a consequence of the preservation of exponential dichotomies under small perturbations. The local behavior near $a_\varepsilon(t)$ and $b_\varepsilon(t)$ can be then thought of as a "wiggled" version of the picture of Figure 2.2 when viewed in (3.6)'s augmented phase space. The two-dimensional stable and unstable manifolds in the augmented phase space can be visualized in each time-slice t as forming one-dimensional entities $U_\varepsilon^a(t)$ and $S_\varepsilon^b(t)$, which are segments of the unstable manifold of a_ε and the stable manifold of b_ε, respectively. The presence of $U_\varepsilon^a(t)$ follows from Theorem 2.12, which moreover asserts that restrictions on time and curve parametrization will be necessary. These are specifically $t \in (-\infty, T_u]$

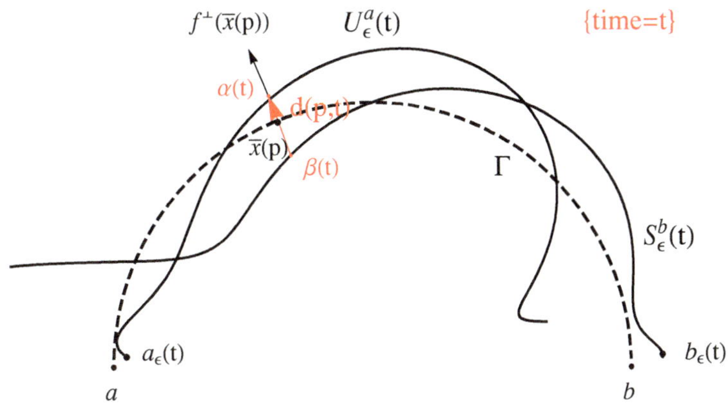

Figure 3.5. *The perturbed manifolds and their separating distance (3.7) along the normal direction $\hat{f}^\perp$ for the nonautonomously perturbed system (3.6), in a time-slice t.*

and $p \in (-\infty, P]$, for T_u and P finite. Similarly, from Theorem 2.23, $S^b_\varepsilon(t)$ exists for $t \in [T_s, \infty)$ and $p \in [-P, \infty)$. These restrictions will be implicit throughout the remainder of this chapter. Now, the primary segments of $U^a_\varepsilon(t)$ and $S^b_\varepsilon(t)$ are $\mathcal{O}(\varepsilon)$-close to Γ, as indeed characterized precisely in Theorems 2.12 and 2.23. Thus, these primary segments of the manifolds are themselves $\mathcal{O}(\varepsilon)$-close to one another, as shown by the solid curves in Figure 3.6, which is to be thought of as the picture in a time-slice t of the augmented phase space. This provides the perfect opportunity for computing the instantaneous flux across the resulting pseudoseparatrix, using the definition from Section 3.2. First, however, the related issue of determining the distance between the perturbed manifolds will be presented.

3.3.1 ▪ Distance between manifolds

Rather than fixing one specific curve $\mathcal{G}$, a *family* of curves (each of which connects together $U^a_\varepsilon(t)$ and $S^b_\varepsilon(t)$ at a different location) will be used. These curves can be parametrized by $p \in [-P, P]$ for any large finite P in the following way. The unperturbed heteroclinic manifold Γ can be parametrized in terms of the heteroclinic trajectory $\bar{x}$ of (2.9) by

$$\Gamma := \{\bar{x}(p) \, : \, p \in \mathbb{R}\} .$$

Since any shifted-time version of the heteroclinic trajectory is itself a heteroclinic trajectory, the above parametrization is not unique; for example, choosing $\tilde{p} = p - k$ (for any constant k) as a parameter also works. This turns out to not be a serious consideration, and it shall be assumed that a *specific* choice has been made for the parametrization of Γ. Now, since $\bar{x}$ is a solution to (2.9), that is, (3.6) with $\varepsilon = 0$, the vector field f is tangential to Γ. Thus, $f^\perp$, as defined by (2.13), is a normal vector to Γ. This varies continuously along Γ, always pointing across Γ in the same direction. The vector $f^\perp(\bar{x}(p))$ can be used to unequivocally define a normal direction to Γ, which is obtained by rotating f by $+\pi/2$ in the counterclockwise direction. At each point $\bar{x}(p)$ on Γ, the gate curve $\mathcal{G}(p)$ shall be defined to be the straight line crossing $\bar{x}(p)$ in the direction of $f^\perp(\bar{x}(p))$. In each time-slice t, this will intersect $U^a_\varepsilon(t)$ and $S^b_\varepsilon(t)$, at the points $\alpha(t)$ and $\beta(t)$, respectively, as shown in Figure 3.5. The distance between the two manifolds, in the time-slice t and measured along the gate $\mathcal{G}(p)$, is

$$d(p,t) := [\alpha(t)-\beta(t)] \cdot \frac{f^{\perp}(\bar{x}(p))}{|f^{\perp}(\bar{x}(p))|} =: [\alpha(t)-\beta(t)] \cdot \hat{f}^{\perp}(\bar{x}(p)), \tag{3.7}$$

where $\hat{f}^{\perp}$ indicates the unit vector in the direction of $f^{\perp}$. This is a *signed* distance with the property that if the vector drawn from the stable to the unstable manifold is in the same direction as $\hat{f}^{\perp}$, it is positive. If in the opposite direction, it is negative. It is well-known [183, 421] that this distance to leading-order (subjected additionally to a scaling) in the instance when f is area-preserving is given by a Melnikov function. This approach, however, works even when f does not have this restriction. Here, the Melnikov function shall be presented as a function of two variables (p,t), with the former representing the *p*osition and the latter the *t*ime-slice. This distance is shown by the red vector in Figure 3.5.

Theorem 3.6 (distance between manifolds [200, 32]). *Suppose Hypotheses 3.5, 2.2, and 2.3 are met. Then, signed distance between the perturbed manifolds (3.7) in (3.6) can be expressed for $(p,t) \in [-P,P] \times [T_s, T_u]$ for any positive P, T_u, $-T_s$ by*

$$d(p,t) = \varepsilon \frac{M(p,t)}{|f(\bar{x}(p))|} + \mathcal{O}(\varepsilon^2), \tag{3.8}$$

in which the Melnikov function is defined by

$$M(p,t) := \int_{-\infty}^{\infty} \exp\left[\int_{\tau}^{p} \operatorname{Tr} Df(\bar{x}(\xi))\, \mathrm{d}\xi\right] f^{\perp}(\bar{x}(\tau)) \cdot g(\bar{x}(\tau), \tau + t - p)\, \mathrm{d}\tau. \tag{3.9}$$

Proof. The formula (3.9) is well known when $\operatorname{Tr} Df = 0$; see [183, 17, 421, e.g.] and Corollary 3.14. The modification when non-area-preserving was first presented by Holmes [200], with a technical issue regarding boundary terms resolved in [32] (this is the same issue which was dealt with in the proof of Theorem 2.12). Within the context of results already developed for the unstable manifold $U_\varepsilon^a(t)$ in Theorem 2.12 and the stable manifold $S_\varepsilon^b(t)$ in Theorem 2.23, the result is easily derived. The unstable manifold's positioning in relation to the point $\bar{x}(p)$, in the direction normal to Γ, is given by (2.25) with $\bar{x}^u \to \bar{x}$. The location $x_\varepsilon^u(p,t)$ in this expression is exactly the point $\alpha(t)$. Similarly, the stable manifold location $x_\varepsilon^s(p,t) = \beta(t)$ is given by (2.61) with $\bar{x}^s \to \bar{x}$, since in this particular case, the unperturbed stable and unstable manifolds coincide. Thus, using the expressions (2.25) and (2.61) from already established theorems,

$$\begin{aligned} d(p,t) &= [\alpha(t)-\beta(t)] \cdot \hat{f}^{\perp}(\bar{x}(p)) = [x_\varepsilon^u(p,t) - x_\varepsilon^s(p,t)] \cdot \hat{f}^{\perp}(\bar{x}(p)) \\ &= \varepsilon \frac{M^u(p,t) - M^s(p,t)}{|f(\bar{x}(p))|} + \mathcal{O}(\varepsilon^2), \end{aligned}$$

which immediately gives (3.9) since $M(p,t) = M^u(p,t) - M^s(p,t)$ from the definitions (3.9), (2.26), and (2.62). □

Theorem 3.6 provides a fundamental condition for establishing whether the unstable and stable manifolds intersect near a location parametrized by (p,t). Since $f(\bar{x}(p))$

is never zero along the heteroclinic, the effective leading-order distance is encoded in $M(p,t)$. Should $M(p,t)$ have a simple zero at a specific value of (p,t), then the implicit function theorem ensures that there is an $\mathcal{O}(\varepsilon)$-close parameter value $(\tilde{p},\tilde{t})$ at which $d(p,t)$ has a zero (see, for example, [183, 17]). The presence of a zero of $M(p,t)$ is the classical method used to "prove" certain systems are chaotic using the Smale-Birkhoff Theorem [183]; for chaos, one needs additional conditions such as Γ being homoclinic (i.e., a and b being the same point) and also g being periodic in t. However, Theorem 3.6 as it stands is correct as an expression for the signed distance, without these additional conditions. It is perfectly possible, for example, for $M(p,t)$ to have only a finite number of zeros, or even no zeros at all.

Example 3.7. For a system with a general f, suppose that for some constant ω,

$$g(x,t)=f^{\perp}(x)\left[3+\cos\omega t\right], \tag{3.10}$$

for which the flow (3.6) is time-periodic. Then

$$\begin{aligned} M(p,t) &= \int_{-\infty}^{\infty} \exp\left[\int_{\tau}^{p} \operatorname{Tr} Df(\bar{x}(\xi))\,\mathrm{d}\xi\right] |f(\bar{x}(\tau))|^2 \left[3+\cos\omega(\tau+t-p)\right]\mathrm{d}\tau \\ &\geq \int_{-\infty}^{\infty} 2\exp\left[\int_{\tau}^{p} \operatorname{Tr} Df(\bar{x}(\xi))\,\mathrm{d}\xi\right] |f(\bar{x}(\tau))|^2\,\mathrm{d}\tau > 0, \end{aligned} \tag{3.11}$$

where the inequalities are possible since the integrand is strictly positive. Thus, *even in the time-periodic case*, it is possible that there are no primary intersections between the perturbed manifolds. ■

Remark 3.8. The Melnikov function, being the effective leading-order term of the distance function, is only able to quantify *primary* intersection points in the following sense. The unstable manifold emanating from $a_\varepsilon(t)$ in Figure 3.5 gets pushed out once it approaches the vicinity of $b_\varepsilon(t)$, since there is an unstable manifold (not pictured in Figure 3.5) emanating from $b_\varepsilon(t)$ as well. Consequently, the $\mathcal{O}(\varepsilon)$ distance representation becomes meaningless in these "nonprimary" parts of the manifold. In a global sense, the unstable manifold emanating from $a_\varepsilon(t)$ may get pushed outside the $\mathcal{O}(\varepsilon)$-region but then return subsequently to intersect the stable manifold emanating from $b_\varepsilon(t)$ elsewhere. The Melnikov function is not able to analyze such a possibility.

Remark 3.9 (relationship to functional-analytic approach). If the functional-analytic approach [106, 305, 61, 48, 355, 50] for determining intersections between stable and unstable manifolds were pursued, the usual result is a Lyapunov-Schmidt reduction one: the the perturbing function g would need to satisfy

$$\int_{-\infty}^{\infty} v^{\top}(\tau)\, g(\bar{x}(\tau),\tau+t)\,\mathrm{d}\tau = 0, \tag{3.12}$$

where the function v is a solution to the *adjoint equation of variations* or the *cotangent equation*

$$\dot{v} = -\left[Df(\bar{x}(t))\right]^{\top} v \tag{3.13}$$

along the unperturbed heteroclinic trajectory [106, e.g.]. This approach does not have the p-parametrization, and *the method for determining the solution to (3.13) is not always obvious.* If $\operatorname{Tr} Df = 0$, it is easily verified that $f^{\perp}(\bar{x}(t))$ works, and this leads to the standard (area-preserving) Melnikov function. The solution to (3.13) when $\mathrm{Tr}\,Df \neq 0$ is not so well known. It turns out that it is

$$v(t) = \exp\left[\int_t^0 \mathrm{Tr}\,Df\,(\bar{x}(\xi))\,\mathrm{d}\xi\right] f^{\perp}(\bar{x}(t)), \tag{3.14}$$

as can be verified by direct computation. Hence, it is clear that setting $M(0,t)=0$ in (3.9) is identical to the condition (3.12) from the functional-analytic approach.

Corollary 3.10. *Suppose the conditions of Theorem 3.6 are met. Upon defining the* scaled Melnikov function

$$\tilde{M}(p,t) := \int_{-\infty}^{\infty} \exp\left[\int_{\tau}^{0} \mathrm{Tr}\,Df\,(\bar{x}(\xi))\,\mathrm{d}\xi\right] f^{\perp}(\bar{x}(\tau))\cdot g\,(\bar{x}(\tau),\tau+t-p)\,\mathrm{d}\tau, \tag{3.15}$$

the Melnikov function (3.9) can be written as

$$M(p,t) = \tilde{M}(p,t)\exp\left[\int_0^p \mathrm{Tr}\,Df\,(\bar{x}(\xi))\,\mathrm{d}\xi\right], \tag{3.16}$$

whose simple zeros are exactly associated with simple zeros of $\tilde{M}(p,t)$.

Proof. This is a straightforward manipulation; see [35]. □

Remark 3.11. The distance function (3.8) can be written in terms of the scaled Melnikov function as

$$d(p,t) = \varepsilon\,\frac{\tilde{M}(p,t)\exp\left[\int_0^p \mathrm{Tr}\,Df\,(\bar{x}(\xi))\,\mathrm{d}\xi\right]}{|f\,(\bar{x}(p))|} + \mathcal{O}\left(\varepsilon^2\right). \tag{3.17}$$

In view of the fact that the other factors in the leading-order term are nonzero, simple zeros of the function $\tilde{M}(p,t)$ relate to intersections of the stable and unstable manifold. Thus, either the scaled or the unscaled Melnikov function could be used without prejudice when seeking intersections of manifolds for small $|\varepsilon|$.

Remark 3.12. The advantage of working with $\tilde{M}(p,t)$ rather than $M(p,t)$ in the analysis of intersections is that $\tilde{M}$, unlike M, can be expressed in terms of just *one* variable $(t-p)$. However, both quantities will be retained since, as will be shown in Theorem 3.18, M is critical to flux characterization.

Corollary 3.13 (arclength representation). *Suppose the conditions of Theorem 3.6 are met. In terms of the arclength ℓ along Γ measured from a, the Melnikov function (3.9) can be represented by*

$$M(p,t) = \int_{\Gamma} \exp\left[\int_{\tau(\ell)}^{p} \mathrm{Tr}\,Df\,(\bar{x}(\xi))\,\mathrm{d}\xi\right] g^{n}\,(\bar{x}(\tau(\ell)),\tau(\ell)+t-p)\,\mathrm{d}\ell, \tag{3.18}$$

where $\tau(\ell)$ is the inverse of the relationship given by

$$\ell(\tau) = \int_{-\infty}^{\tau} |f\,(\bar{x}(\xi))|\,\mathrm{d}\xi \tag{3.19}$$

and g^n is the "normal component" of g, i.e., the component in the direction of $\hat{f}^{\perp}$ given by $g^n := g \cdot \hat{f}^{\perp}$.

Proof. Since the rate of change of the arclength in the unperturbed flow is equal to the velocity, the expression

$$\frac{d\ell}{d\tau} = |f(\bar{x}(\tau))|$$

gives the connection between the arclength parametrization ℓ and the time-parameterization τ along Γ. Integrating from $\tau \to -\infty$ (i.e., the point a at which the the arclength is zero) gives the relationship (3.19). This is invertible because each point on Γ has a unique arclength value ℓ and also a well-defined time value given by $\bar{x}(\tau)$. Now

$$f^{\perp} \cdot g \, \mathrm{d}\tau = f^{\perp} \cdot g \frac{d\tau}{d\ell} \mathrm{d}\ell = f^{\perp} \cdot g \frac{1}{|f|} \mathrm{d}\ell = g^n \, \mathrm{d}\ell$$

as required. □

If $\operatorname{Tr} Df = 0$, (3.18) becomes

$$M(p,t) = \int_{\Gamma} g^n \left(\bar{x}(\tau(\ell)), \tau(\ell) + t - p\right) \mathrm{d}\ell\,, \tag{3.20}$$

which shows the fairly simple result that the normal component of g needs to be integrated along Γ to obtain the Melnikov function. This is intuitively a pleasing result; the component of velocity across Γ needs to be summed up along Γ through an integration with respect to arclength. However, (3.20) incorporates the time-variation within it. The *Lagrangian* variation of the trajectories that comprise the perturbed manifolds at each instance t in time are used, and thus (3.20) is not simply a frozen-time (Eulerian) calculation. The condition $\operatorname{Tr} Df = 0$ used here is actually the statement that the $\varepsilon = 0$ form of the flow (3.6) is incompressible but not necessarily the *full* flow. More familiar results are possible when $\nabla \cdot g = 0$ as well; i.e., (3.6) is fully assumed incompressible. In this case it is possible to define streamfunctions ψ_0 and ψ_1 associated, respectively, with f and g, as defined in (2.30) and (2.31) in Remark 2.16. Under this condition, the following well-known [183, 421] form for the distance function arises from Theorem 3.6.

Corollary 3.14. *Suppose the conditions of Theorem 3.6 are met, and moreover the flow (3.6) is incompressible and associated with the streamfunctions (2.30) and (2.31). Then,*

$$d(p,t) = \varepsilon \frac{M(p,t)}{|\nabla \psi_0(\bar{x}(p))|} + \mathcal{O}\left(\varepsilon^2\right),$$

in which the Melnikov function reduces to

$$M(p,t) := \int_{-\infty}^{\infty} \{\psi_0, \psi_1\}(\bar{x}(\tau), \tau + t - p) \, \mathrm{d}\tau\,. \tag{3.21}$$

Proof. The reader is referred to Remark 2.16 for the notation used in (3.21). Since f is area-preserving, $\operatorname{Tr} Df = \operatorname{div} f = 0$, resulting in the interior integral of (3.9) disappearing. The fact that $f^{\perp} \cdot g = \{\psi^0, \psi^1\}$ is a simple computation using the streamfunction definitions (2.30) and (2.31) in Remark 2.16. □

Remark 3.15. For an interesting recent computation of the Melnikov function (3.21) *numerically* using experimental velocity data, the reader is referred to [272].

Example 3.16 (Duffing oscillator (cont.)). Return to the Duffing oscillator, introduced previously in Examples 2.11, 2.21, and 2.32. The unperturbed flow (when there is no damping or forcing) is area-preserving. While the phase space here is not a physical flow space, the terminology associated with a fluid flow will be followed for convenience. Consider the situation in which the damping continues to be zero, in which case area-preservation continues to be satisfied, and the result of Corollary 3.14 can be used. The "standard" situation of imposing a time-harmonic forcing $\phi(t) = \cos(\omega t)$ will be postponed to Example 3.34, since Section 3.4's development for harmonic time-dependence can be used to great advantage. Here, suppose a "box-forcing" which only operates on the time-interval $[0, T]$ is applied, i.e., $\phi(t) = \mathbb{1}_{[0,T]}(t)$, where $\mathbb{1}$ denotes the indicator function. Since this means that

$$f = \begin{pmatrix} x_2 \\ x_1 - x_1^3 \end{pmatrix} \text{ and } g = \begin{pmatrix} 0 \\ \mathbb{1}_{[0,T]}(t) \end{pmatrix},$$

the streamfunctions can be chosen to be

$$\psi_0(x_1, x_2) = \frac{x_2^2}{2} - \frac{x_1^2}{2} + \frac{x_1^4}{4} \quad \text{and} \quad \psi_1(x_1, x_2, t) = -x_1 \mathbb{1}_{[0,T]}(t).$$

Therefore, $\{\psi_0, \psi_1\} = (x_1^3 - x_1)(0) - (x_2)(-\mathbb{1}_{[0,T]}(t)) = x_2 \mathbb{1}_{[0,T]}(t)$. Now, it has been shown in Example 2.21 that $\bar{x}_1(t) = \sqrt{2}\,\mathrm{sech}\, t$ and $\bar{x}_2(t) = -\sqrt{2}\,\mathrm{sech}\, t \tanh t$. Thus, the denominator in the distance function of Corollary 3.14 becomes

$$\begin{aligned} |\nabla \psi_0(\bar{x}_1(p), \bar{x}_2(p))| &= \sqrt{2\,\mathrm{sech}^2 p \tanh^2 p + \left(\sqrt{2}\,\mathrm{sech}\, p - 2^{2/3}\,\mathrm{sech}^3 p\right)^2} \\ &= \sqrt{2\,\mathrm{sech}^2 p \tanh^2 p + 2\,\mathrm{sech}^2 p \left(1 - 2\,\mathrm{sech}^2 p\right)^2} \\ &= \sqrt{2}\,\mathrm{sech}\, p \sqrt{2 - 5\,\mathrm{sech}^2 p + 4\,\mathrm{sech}^4 p}\,. \end{aligned}$$

The Melnikov function can be explicitly computed:

$$\begin{aligned} M(p,t) &= \int_{-\infty}^{\infty} \sqrt{2}\,\mathrm{sech}\,\tau \tanh \tau \mathbb{1}_{[0,T]}(\tau + t - p)\, \mathrm{d}\tau = \sqrt{2} \int_{p-t}^{T-t+p} \mathrm{sech}\,\tau \tanh \tau\, \mathrm{d}\tau \\ &= -\sqrt{2}\left[\mathrm{sech}\,(T + p - t) - \mathrm{sech}\,(p - t)\right]. \end{aligned}$$

Therefore, the distance function for such "box-forcing" in the Duffing oscillator is

$$d(p,t) = \varepsilon \frac{\mathrm{sech}\,(p - t) - \mathrm{sech}\,(T + p - t)}{\mathrm{sech}\, p \sqrt{2 - 5\,\mathrm{sech}^2 p + 4\,\mathrm{sech}^4 p}} + \mathcal{O}(\varepsilon^2).$$

Since $T > 0$, the only possibility for a leading-order term is when the arguments of the sech terms are negatives of each other, that is, when $t - p = T/2$. In each time-slice t, there is therefore one intersection between the stable and unstable manifolds, occurring near a p value of $t + T/2$. This intersection persists in each time-slice; indeed, the curve traced out in the augmented phase space is necessarily a homoclinic trajectory to the perturbed hyperbolic trajectory near the origin. ∎

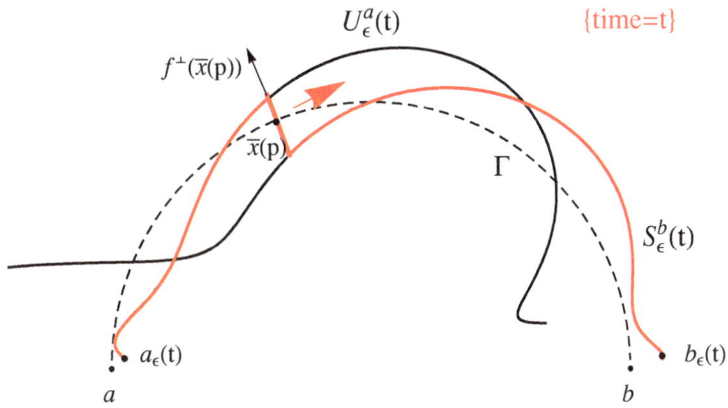

Figure 3.6. *The pseudoseparatrix $\mathscr{Q}(p,t)$ constructed according to Definition 3.17 [red curve] for the nonautonomously perturbed system (3.6), in a time-slice t, in the situation of Figure 3.5.*

3.3.2 ▪ Flux across perturbed manifolds

The separatrix Γ of (3.6) when $\varepsilon = 0$, breaks up into stable and unstable manifolds S^b_ε and U^a_ε, respectively, when $\varepsilon \neq 0$. In the previous subsection, the distance between these manifolds was characterized. Here, the question of the *flux* as a result of this breakup is considered, using the flux definition of Section 3.2 for general time-aperiodic flows. Following Hypothesis 3.3, a specific pseudoseparatrix will be defined according to the following definition.

Definition 3.17 (pseudoseparatrix for nearly autonomous flows). *The pseudoseparatrix $\mathscr{Q}(p,t)$ for the system (3.6) is defined for $(p,t) \in [-P,P] \times [T_s, T_u]$ by*

$$\mathscr{Q}(p,t) := \hat{U}^a_\varepsilon(t) \cup \hat{\mathscr{G}}(p,t) \cup \hat{S}^b_\varepsilon(t),$$

in which $\hat{U}^a_\varepsilon(t)$ is the curve segment of $U^a_\varepsilon(t)$ lying between $a_\varepsilon(t)$ and $\alpha(t)$, $\hat{S}^b_\varepsilon(t)$ is the curve segment of $S^b_\varepsilon(t)$ lying between $b_\varepsilon(t)$ and $\beta(t)$, and $\hat{\mathscr{G}}(p,t)$ is the segment of the straight line $\mathscr{G}(p)$ lying between $\alpha(t)$ and $\beta(t)$. (See Figure 3.6.)

Theorem 3.18 (instantaneous flux for nearly autonomous systems [28]). *Suppose Hypotheses 2.2 and 2.3 are satisfied for the system (3.6). The instantaneous flux $\phi(p,t)$, parametrized by gate* position *$p \in [-P,P]$ and time $t \in [T_s, T_u]$ (for any positive P, T_u, $-T_s$), across the pseudoseparatrix $\mathscr{Q}(p,t)$ in the direction of $\hat{f}^\perp$ is given by*

$$\phi(p,t) = \varepsilon M(p,t) + \mathscr{O}\left(\varepsilon^2\right), \tag{3.22}$$

in which the Melnikov function M is defined by (3.9).

Proof. By the construction of the pseudoseparatrix, the instantaneous flux is specified according to Definition 3.4. Thus, it is only the flux across the gate $\hat{\mathscr{G}}(p,t)$ that needs to be quantified in (3.4). Now, the velocity at $\bar{x}(p)$ when $\varepsilon = 0$ is given by $f(\bar{x}(p))$, which is perpendicular to the gate at the point $\bar{x}(p)$. To leading-order in ε, this is the same velocity at all points on $\hat{\mathscr{G}}(p,t)$. When $\varepsilon \neq 0$, this velocity acquires a further $\mathcal{O}(\varepsilon)$-correction, but to leading-order, the normal velocity at *all* points on $\hat{\mathscr{G}}(p,t)$ is $f(\bar{x}(p))$. Since the normal vector $\hat{n}$ on $\hat{\mathscr{G}}(p,t)$ is $\hat{f}(\bar{x}(p))$ at all points, using (3.4),

$$\begin{aligned}\phi(p,t) &= \int_{\hat{G}(p,t)} \{|f(\bar{x}(p))| + \mathcal{O}(\varepsilon)\}\, ds = \{|f(\bar{x}(p))| + \mathcal{O}(\varepsilon)\} \int_{\hat{G}(p,t)} ds \\ &= \{|f(\bar{x}(p))| + \mathcal{O}(\varepsilon)\} d(p,t) = \{|f(\bar{x}(p))| + \mathcal{O}(\varepsilon)\}\left\{\varepsilon \frac{M(p,t)}{|f(\bar{x}(p))|} + \mathcal{O}\left(\varepsilon^2\right)\right\} \\ &= \varepsilon M(p,t) + \mathcal{O}\left(\varepsilon^2\right)\end{aligned}$$

by using (3.8). □

Theorem 3.18 states that the leading-order flux is *exactly* the Melnikov function, with the second argument of the Melnikov function representing precisely time. The first argument, p, represents the location at which the gate is positioned in constructing the pseudoseparatrix. This interpretation depends only on the flow being of the form (3.6); that is, it needs to be nonautonomously perturbed. There is no necessity for time-periodicity or the more restrictive time-harmonicity in the perturbation g. Furthermore, unlike in the lobe dynamics interpretation of [341, 421], it is *not* necessary to integrate the Melnikov function in order to obtain a measure of the flux. Such an integral does make sense when computing lobe areas (an issue which shall be returned to), but Theorem 3.18 offers the opportunity of using the second argument of the Melnikov function precisely as time, in representing explicitly the leading-order instantaneous flux and its temporal variation.

An important feature of the instantaneous flux is that the direction of the flux is incorporated within the definition. For the unperturbed heteroclinic manifold Γ as pictured in Figures 3.5 and 3.6, the vector field f is from left to right along the dashed curve, since Γ is the unstable manifold of a. Thus, $\hat{f}^{\perp}$ is in the upwards direction at all points on Γ. For the situation pictured in Figure 3.5, the distance $d(p,t)$ is therefore positive; the vector from the stable manifold to the unstable manifold drawn at this point $\bar{x}(p)$ lies in the same direction as $\hat{f}^{\perp}(\bar{x}(p))$. Therefore, from (3.8), $M(p,t)$ is positive, which implies from (3.22) that the instantaneous flux is positive. Now, examining Figure 3.6, the direction of flow across the gate is what is shown by the red arrow in this case, since f is from left to right. But this implies transport across the pseudoseparatrix from the lower part to the upper part. In other words, fluid crosses the previously impermeable Γ in the direction of $\hat{f}^{\perp}(\bar{x}(p))$ at this instance in time. It is easy to see that if the stable manifold $S_{\varepsilon}^{b}(t)$ is above the unstable manifold $U_{\varepsilon}^{a}(t)$ with respect to positioning along $\hat{f}^{\perp}(\bar{x}(p))$ in Figure 3.6 instead, the flow would be from above the pseudoseparatrix to below it, i.e., in the direction of $-\hat{f}^{\perp}(\bar{x}(p))$. But this would correspond to both $d(p,t)$ and $M(p,t)$ being negative.

Remark 3.19 (the sign of the instantaneous flux). The above arguments show that the instantaneous flux $\phi(p,t)$ measures the instantaneous transport *in the direction* $\hat{f}^{\perp}$ across the pseudoseparatrix. The Melnikov function $M(p,t)$ therefore provides the *leading-order instantaneous signed flux* across Γ in the direction of $\hat{f}^{\perp}$, at time t, with gate location $\bar{x}(p)$.

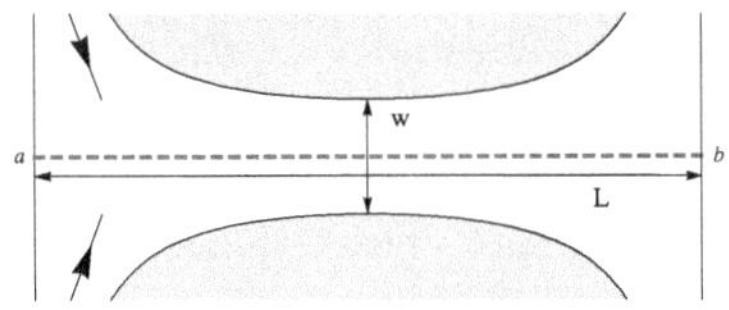

Figure 3.7. *The T-mixer geometry associated with the flow (3.23), with the unperturbed heteroclinic manifold as the dashed line. [31] Reproduced with permission from The American Physical Society.*

Remark 3.20. In view of the fact that the component of the perturbing velocity g in the direction of f has no contribution to the Melnikov integral (3.9), leading-order flux can be generated only by velocity perturbations *normal* to the unperturbed heteroclinic manifold Γ.

Example 3.21 (T-mixer). Consider the flow

$$\dot{x}_1 = x_1(L - x_1), \qquad \dot{x}_2 = -x_2(L - 2x_1) \tag{3.23}$$

in the channel $0 < x_1 < L$, with sides $|x_2| < wL^2/[8x_1(L - x_1)]$, which is a kinematic model for a "T-mixer" inlet geometry [244, 430, 242, 172, 376, 31]. Two fluids arrive along different channels and then flow together in a channel in a main channel as shown in Figure 3.7. For devices with small dimensions (micro- or nanofluidic channels), the two fluids tend not to mix across the dashed red line, which is a heteroclinic trajectory going from a to b for the velocity field f as given in (3.23). Suppose transport across this is to be obtained by imposing a transverse time-periodic velocity $\varepsilon g(x_1, x_2, t) = \varepsilon(0, 1)\cos\omega t$. This is a simple model for *shaking* the microfluidic device in the transverse direction, back and forth, with frequency ω. Additional simplifications occur here since $\operatorname{Tr} Df = 0$ and the arclength along the heteroclinic is precisely its x_1-coordinate. So in computing the resulting flux, several of the different forms of $M(p, t)$ that have been derived could be used. In using these, the unperturbed heteroclinic trajectory is needed; this is $\bar{x}_1(t) = L/\left(1 + e^{-Lt}\right)$, $\bar{x}_2(t) = 0$ [31]. From the original Melnikov definition (3.9), the resulting leading-order flux is

$$\begin{aligned} M(p,t) &= \int_{-\infty}^{\infty} x_1(\tau)[L - x_1(\tau)]\cos[\omega(\tau + t - p)]\, d\tau \\ &= \int_{-\infty}^{\infty} \dot{x}_1(\tau)\cos[\omega(\tau + t - p)]\, d\tau \\ &= \cos[\omega(t-p)] \int_{-\infty}^{\infty} \frac{L^2 e^{-L\tau}}{(1 + e^{-L\tau})^2} \cos(\omega\tau)\, d\tau \\ &\quad - \sin[\omega(t-p)] \int_{-\infty}^{\infty} \frac{L^2 e^{-L\tau}}{(1 + e^{-L\tau})^2} \sin(\omega\tau)\, d\tau \\ &= -\pi\omega \operatorname{cosech} \frac{\pi\omega}{L} \cos[\omega(t-p)]\,. \end{aligned} \tag{3.24}$$

Consider any fixed p location; for example, $p = 0$ would correspond to the centerpoint $(x_1, x_2) = (L/2, 0)$ under the chosen time-parametrization of the unperturbed heteroclinic trajectory. For a pseudoseparatrix constructed with a gate at this point, the instantaneous flux varies sinusoidally with time, with an amplitude

$\pi\omega \operatorname{cosech} \pi\omega/L$. The sinusoidal variation means that at some instances in time, the stable manifold will be above the unstable manifold. However, as t progresses, the leading-order instantaneous flux will be zero instantaneously as the two manifolds cross each other at the gate. For t-values slightly larger, the stable manifold will be below the unstable one, and thus the sign of the flux will flip. This process will continue periodically in time and will be essentially the same for any choice of gate location p. The implications from this is that during each period, there will be a flux of the lower fluid into the upper regions, followed by flux of the upper fluid into the lower regions. If the two fluids were dyed with different colors, one would see blobs of one color being "spat" into the other region, followed by blobs of the other color being spat back, with this periodically repeating. Eventually, these smaller fluid blobs will be subject to elongation (as they approach the right hyperbolic trajectory $b_\varepsilon(t)$, occurring somewhere near b), and consequently diffusion will come into play. ∎

Remark 3.22. Let $\rho_0(x)$ be an equilibrium scalar density associated with the steady flow ((3.6) when $\varepsilon = 0$). That is, it is a density of something (chemical quantity, heat, etc.), which after advection by f has settled into an equilibrium state (this would be an *invariant density* of (3.6) with $\varepsilon = 0$). Suppose $\rho(x,t)$ is a scalar density advected by the perturbed flow (3.6). The quantity ρ could represent, for example, a pollutant/heat/salinity concentration, or a vorticity density. This will obey $\rho(x,t) = \rho_0(x) + \mathcal{O}(\varepsilon)$ in the when the flow is perturbed by setting $\varepsilon \neq 0$. The transport of ρ across the pseudoseparatrix can also only occur when ρ is carried by particles crossing the gate at each instance in time. Therefore, the instantaneous signed flux of ρ across the pseudoseparatrix (in the direction specified by $\hat{f}^\perp$) is

$$\text{Instantaneous Flux of Scalar} = \varepsilon \rho_0(\bar{x}(p)) M(p,t) + \mathcal{O}\left(\varepsilon^2\right). \tag{3.25}$$

Remark 3.23. In view of the calculation in the proof of Corollary 3.13, and the previous remark, the instantaneous flux of a scalar density $\rho(x,t)$ can be written as

$$\phi(p,t) = \varepsilon \rho_0(\bar{x}(p)) \int_0^L e^{\int_p^{\tau(\ell)} \operatorname{Tr} Df(\bar{x}(\xi))\,\mathrm{d}\xi} g^n\left(\bar{x}(\tau(\ell)), \tau(\ell)+t-p\right) \mathrm{d}\ell + \mathcal{O}\left(\varepsilon^2\right), \tag{3.26}$$

where L is the length of the heteroclinic manifold Γ which is parametrized by arclength ℓ and $\tau(\ell)$ is the inverse of the relationship (3.19). (If a straightforward flux of fluid is needed, the ρ_0 term above can be neglected.)

Remark 3.24. A direct *Eulerian* definition for flux of a scalar $\rho(x,t)$ across the heteroclinic manifold Γ due to the presence of the time-varying velocity perturbation $\varepsilon g(x,t)$ would take the form

$$\phi_E(t) = \varepsilon \int_0^L \rho(\bar{x}(\tau(\ell)), t)\, g^n\left(\bar{x}(\tau(\ell)), t\right) \mathrm{d}\ell \tag{3.27}$$

at each instance in time t. Eulerian flux definitions remain useful for capturing precisely the transport across a fixed entity in space [44, 43, cf.] in a particular frame of reference (e.g., the earth's) but do *not* carry information on the origins of the fluid parcels instantaneously crossing the entity. The Eulerian definition (3.27) can be seen to differ from the instantaneous flux (3.26) in several ways. In determining the instantaneous flux, note that it is not the normal component of g at that instance in time t that needs to be used but this component at a time value which is shifted by $p-\tau(\ell)$. This reflects how the Lagrangian motion of the fluid parcels are incorporated into the flux definition, which is therefore a *Lagrangian flux*. This is further evidenced by the time-evolving compressibility term of the interior integral in (3.26) and the fact that the Eulerian definition (3.27) does not have the $\mathcal{O}\left(\varepsilon^2\right)$-error term in the Lagrangian one, since it is only the instantaneous velocity which matters. The Eulerian definition moreover has no dependence on the position of a gate.

Example 3.25 (double-gyre cont.). The double-gyre whose hyperbolic trajectory location and stable/unstable manifolds were addressed in Example 2.25 will be revisited. Here, the focus shall be on quantifying the flux between the right and left gyres of Figure 2.9, which is zero when there is no perturbation. Under the influence of ε as given in (2.64) with $q(t)$ being a general bounded function, flux occurs across the red heterolinic line in Figure 2.64. In computing the leading-order flux as given in Theorem 3.18, the Melnikov function (3.9) needs to be computed. However, the integrand of this has already been evaluated for general $q(t)$ in Example 2.25 in (2.67), with the x_2-coordinate's variation given immediately below. Thus, using (3.9), the $\mathcal{O}(\varepsilon)$-flux function is

$$\begin{aligned} M(p,t) &= -\frac{\pi^3 A^2}{2}\int_{-\infty}^{\infty} \sin\left[4\cot^{-1}\left(\exp\left[\pi^2 A\tau\right]\right)\right] q(\tau+t-p)\,\mathrm{d}\tau \\ &= -\pi^3 A^2 \int_{-\infty}^{\infty} \tanh\left(\pi^2 A\tau\right)\operatorname{sech}\left(\pi^2 A\tau\right) q(\tau+t-p)\,\mathrm{d}\tau, \end{aligned} \tag{3.28}$$

analogously to the alternative representations in (2.69) and (2.70). The flux function clearly depends on the combination $(t-p)$ as opposed to independently varying with t and p; varying the location of the gate is akin to examining the flow in a different time-slice. If the standard double-gyre choice $q(t)=\sin\omega t$ is made, then splitting the integral exactly as in Example 3.21 gives

$$\begin{aligned} M_p(p,t) &= \pi^3 A^2 \sin[\omega(t-p)]\int_{-\infty}^{\infty} \tanh\left(\pi^2 A\tau\right)\operatorname{sech}\left(\pi^2 A\tau\right)\sin(\omega\tau)\,\mathrm{d}\tau \\ &= \omega\operatorname{sech}\frac{\omega}{2\pi A}\sin[\omega(t-p)], \end{aligned} \tag{3.29}$$

where the integral can be explicitly performed in this instance and the subscript in M_p represents "periodic." Paying attention to the sign convention used for the instantaneous flux, the expression (3.29) is the instantaneous volume flux from the right to the left gyre in Figure 2.9, with a gate at a location parametrized by p. Observe moreover the similarity of (3.29) with the computation for the T-mixer as given in (3.24); these sorts of expressions are fairly generic under time-harmonic perturbations to heteroclinics and shall be investigated further in Section 3.4. The flux is sinusoidally varying, with fluid sloshing back and forth from one gyre to the other. One question that might be asked is whether there is an optimum frequency for the transport of fluid between the gyres. Such optimizing flux issues are the subject of Chapter 4, but in this case a straightforward answer can be given. The amplitude of the sinusoidal flux in (3.29) is maximized when ω satisfies the equation

$$\tanh\frac{\omega}{2\pi A} = \frac{2\pi A}{\omega},$$

whose numerical solution is $\omega/A \approx 7.53781$. There have been many studies which compute in various ways the mixing occurring in the double-gyre [364, 156, 150, 391, e.g.]; the focus is usually numerical since there are few analytical tools available. The above analysis—albeit perturbative in ε—provides a method of quantifying mixing from one gyre to the other as precisely a transport of fluid across the gyre boundaries. If transfer between gyres is the focus of a mixing study, issues such as the prediction for an ω for optimum transport above, or the lobe area analysis in Example 3.27, can be addressed from this viewpoint.

Next, $q(t) = \mathbb{1}_{[0,T]}(t)$, as investigated in Example 2.25, is examined. In this aperiodic but simple instance, the leading-order flux in (3.28) is explicitly statable using the already established calculations in Example 2.25, leading to

$$M_b(p,t) = \pi^3 A^2 \left(\operatorname{sech}\left[\pi^2 A(T+p-t)\right] - \operatorname{sech}\left[\pi^2 A(p-t)\right]\right), \tag{3.30}$$

where the subscript in M_b represents "box-forcing." Suppose a gate has been chosen, and thus p is fixed. The time-varying behavior of the flux is straightforward; for $t < p+T/2$, there is transport from the right to the left gyre, but for $t > p+T/2$, fluid transport occurs in the opposite direction. In either case, the transport decays to zero as $|t| \to \infty$. There is instantaneously no leading-order flux when $t = p + T/2$, which is the instance at which the stable and unstable manifolds intersect on the gate (this is when $t - p = T/2$, the midpoint of the interval of nonzero q). This is the only zero of the flux function, or alternatively of the distance function. Compare the Melnikov function in (3.30) to that of the Duffing oscillator in Example 3.16; quite different problems often have essentially the same solutions due to the exponentially decaying structure of the integrand of the Melnikov function. As another time-aperiodic situation, consider $q(t) = \tanh\left(\pi^2 A t\right)$, which is chosen since when inserted into (3.28), the integral can once again be computed explicitly. Doing so, the expression

$$M(p,t) = \pi^2 A\left[1 - \tanh\left(\pi^2 A(p-t)\right) \operatorname{cosech}\left(\pi^2 A(p-t)\right)\right]$$

results. This is a nonnegative and even function of $(p-t)$ which touches zero only at $p-t=0$ (and does so tangentially) and approaches 1 as $p-t \to \pm\infty$. The implication is that for $t \neq p$, the flux is always positive, that is, from the right cell to the left cell. At $t = p$, the flux is zero to first-order in ε, but higher-order terms may mean that the flux takes on a nonzero value. The fact that the zero $t = p$ is not a simple zero of $M(p,t)$ leads to the difficulty of not being able to definitively say whether there is an intersection of the perturbed stable and unstable manifolds at the position p in the time-slice t, since, once again, higher-order terms become necessary, this time in (3.8). Nevertheless, from a flux perspective, this issue is not particularly important, since the flux is basically positive and $\mathcal{O}(\varepsilon)$ at all values other than $p = t$, at which it is at most $\mathcal{O}\left(\varepsilon^2\right)$. ∎

The main result in this section is that the Melnikov function $M(p,t)$ is the leading-order term in the flux function $\phi(p,t)$. How smooth is the flux with time? Since $M(p,t)$ involves integrating combinations of f and g, the expectation is that it is, if anything, smoother than those functions. As shown in Corollary 3.10, $M(p,t)$ is $\tilde{M}(p,t)$ multiplied by a p-dependent factor which is smooth in p, and $\tilde{M}(p,t)$ has the p and t appearing not independently but as one variable $(t-p)$. Therefore, when examining smoothness of $M(p,t)$, it suffices to focus on one of M's arguments. A natural approach might be to think of p as fixed (i.e., the gate location chosen) and then view the flux variation with time t. Under this interpretation, the smoothness of the leading-order flux can be characterized in terms of the smoothness of g, as follows.

Theorem 3.26 (see [28]). *Consider the flow (3.6) subject to Hypotheses 2.2 and 2.3. For any fixed $p \in \mathbb{R}$, consider the leading-order flux function $M(p,t)$ as a function of time t. Then the following hold:*

(a) *The leading-order flux function $M(p,\cdot) \in L^\infty(\mathbb{R})$.*

(b) *If there is a $k \in [1,\infty)$ such that $g(x,\cdot) \in L^k(\mathbb{R})$ for all $x \in \Omega$, then $M(p,\cdot) \in L^k(\mathbb{R})$.*

(c) *If there exist $r \in \{0,1,2,\ldots\}$ and $k \in [1,\infty)$ such that $(\partial^r g)/(\partial t^r) \in \mathrm{L}^k$ for all $x \in \Omega$ and $(\partial^r g)/(\partial t^r)$ satisfies Hypothesis 2.3, then $M(p,\cdot) \in W^{r,k}(\mathbb{R})$.*

(d) *If the conditions of* (c) *above hold and moreover $kr > 1$, then $M(p,\cdot) \in C^0(\mathbb{R})$.*

(e) *If there exists a $k \in \mathbb{N}$ such that $g(x,\cdot) \in W^{k,\infty}(\mathbb{R})$ for all $x \in \Omega$, then $M(p,\cdot) \in C^k(\mathbb{R})$.*

Proof. This is Proposition 3.1 in [28]; the proof is furnished via a collection of lemmas in that article and will be skipped. □

3.3.3 ▪ Relationship to lobe areas

Lobe areas were developed as measures of transport principally in time-periodic flows in which lobe dynamics across some entity could be identified in a precise way [343, 341, 421]. Motivated by these elegant theories, many studies continue to use lobe areas as measures of transport. Even in time-periodic flows, for the usage of a lobe area as a measure of transport, one needs to have additional conditions, as specified in Hypothesis 3.2. Under general time-dependence and compressibility, additional difficulties arise. Even if there are lobes present in some time-slice, their interpretation for transport is not obvious. Unlike in time-peroidic situations in which the presence of a Poincaré map allows for determining where such a lobe may go after an iteration, there is no longer any favored time-interval to consider for transport assessment. Time needs to be considered as a *continuous* entity under this situation, and then such lobes will gradually deform and move. There may also be many, differently sized lobes in a time-slice. So if using such lobes as measures of transport, several questions arise. Which lobe is to be used? Exactly in what way is that lobe "transported," allowing for an interpretation for its area to be a measure of transport? What can one do with the difficulty faced in compressible flows in which the lobe areas will change in different time-slices? And so on.

Nevertheless, in general time-aperiodic situations, in fixed time-slice, there can be lobes. Should the area of such a lobe be important in an application, it *can* be computed using the theory derived in the previous subsections. (For such a computation in a time-aperiodic setting using numerically generated data related to oceanography, the reader is referred to [277].) While the current situation is time-aperiodic and potentially compressible, the original method developed which computes an integral of the Melnikov function [258, 228, 341, 421] is essentially still applicable in computing a lobe area (whether or not this area can be interpreted in transport). This will be briefly presented.

Consider a fixed time-slice t. Suppose in this time-slice there is a specific lobe generated by the intersection of stable and unstable manifolds. There may of course be several, but the focus shall be on determining the lobe area of this specific lobe, in this specific time-slice. Now, this lobe is defined by the intersection of the stable and unstable manifold at two points. Each intersection point is associated with a p-value; call these two values p_1 and p_2. Thus, the intersections occur near the points $\bar{x}(p_1)$ and $\bar{x}(p_2)$, and (in order to have a primary lobe) it shall be assumed that there are no intersection points in between these values. Now, for intersections to occur, the Melnikov function (3.9) must have zeros at the values p_1 and p_2, at this value of time t. Suppose these p-values are, respectively, associated with the arclength parameter values ℓ_1 and ℓ_2, where the arclength ℓ is defined in (3.19), i.e.,

$$\ell(p) = \int_{-\infty}^{p} |f(\bar{x}(\xi))| \, d\xi ,$$

and in particular $\ell(p_i) = \ell_i$ for $i = 1,2$. If $p(\ell)$ is the inverse of the above relationship, the lobe area is easily computed by

$$\text{Lobe Area} = \int_{\ell_1}^{\ell_2} d(p(\ell),t)\, d\ell + \mathcal{O}\left(\varepsilon^2\right) = \varepsilon \int_{p_1}^{p_2} |M(p,t)| \, dp + \mathcal{O}\left(\varepsilon^2\right), \qquad (3.31)$$

using a simple change of variables from ℓ to p, and utilizing (3.8) and the fact that $\ell'(p) = |f(\bar{x}(p))|$. This well-known geometric idea [341, 421] can alternatively be obtained by an action formalism [258, 228].

While the formula (3.31) is valid in the compressible, general time situation, it should be remembered that the Melnikov function definition is that of (3.9) and that its interpretation for time-continuous transport is unclear. For an incompressible general time situation in which a lobe area is computed using (3.31) as by Miller et al. [277], one possibility of relating this to transport is to track the evolution of the lobe and divide by the time taken when the lobe is deemed to have "gone to the other side." This is an ambiguous and difficult exercise [277]. In contrast, the instantaneous flux as given precisely by the Melnikov function is more natural, *and* is valid under compressibility, general time-dependence, and even in the absence of lobes.

How does the usage of the Melnikov function as the flux work in the classical time-harmonic setting in which lobe dynamics can be used to good effect, as described in Section 3.1? (So the setting is of incompressible flow, with all lobes having equal areas.) Imagine viewing a picture such as Figure 3.2 and drawing a gate at some fixed location. Rather than following the approach of Section 3.1, which considered what happened after one period of the Poincaré map, think of time now as continuous. As time progresses, the lobes and intersections will "move to the right" since intersection points, being on the stable manifold, will gradually approach the (time-varying) hyperbolic trajectory $b(t)$. Thus, the lobes will flow through the gate. At the instances when the stable and unstable manifolds intersect exactly on the gate at an intersection point p_1, the instantaneous flux will be zero. A moment afterwards, it shall take some sign (suppose it is positive, for the sake of argument). Then, until the next intersection point p_2 passes through, the instantaneous flux will be positive. During this time interval, the fluid which is inside the lobe subtended by the (moving) intersection points p_1 and p_2 will pass from one side of the pseudoseparatrix to the other. This is of course exactly the area of the lobe and can be computed by integrating the Melnikov function as in (3.31). But this continuous-time viewpoint tells us exactly *how* this area of fluid is passing through, and not simply a "time-averaged" snapshot as is implicit in the Poincaré map approach for this situation detailed in Section 3.1. Now, after p_2 has gone through the gate, the Melnikov function will become *negative* in this classical situation. Until the next intersection point (call it p_3) passes through, there is flux in the *opposite* direction. Thus, during the time between p_2 and p_3 passing through the gate, an amount of fluid, once again given by the lobe area, transports in the opposite direction. Notice that the times between p_1 and p_3 passing through the gate corresponds to exactly one iteration of the Poincaré map. This gives the continuous-time interpretation of, for example, the fact that lobes L and M in Figure 3.2 transport in opposite directions during one iteration of the Poincaré map. The continuous approach is more informative in that it tells us the time-variation of the flux and how the fluid will initially slosh in one direction across the pseudoseparatrix, and then in the other direction, during a time-interval corresponding to the period of the Poincaré map.

Thus, the continuous-time approach, and its representibility in terms of the Melnikov function (with the t therein directly being the time-variation of the flux), is a more general way to interpret transport. Moreover, in general time-dependent systems with no preferred time-scales, it very naturally retains the time-variation of the system and is able to encapsulate compressibility and any intersection pattern between the stable/unstable manifolds.

Example 3.27 (double-gyre cont.). The double-gyre whose flux was computed in Example 3.25 is re-examined. The flux function was computed in (3.29) for the periodic forcing $q(t) = \sin \omega t$. The area of a lobe generated by the intersecting stable and unstable manifolds in a time-slice t with adjacent zeros of $M(p,t)$ identified by p_1 and p_2 is, from (3.31),

$$\begin{aligned}\text{Lobe Area} &= \varepsilon \omega \operatorname{sech} \frac{\omega}{2\pi A} \int_{p_1}^{p_2} \left|\cos\left[\omega(t-p)\right]\right| \mathrm{d}p + \mathcal{O}\left(\varepsilon^2\right) \\ &= \varepsilon \omega \operatorname{sech} \frac{\omega}{2\pi A} \int_{-\pi/(2\omega)}^{\pi/(2\omega)} \cos(\omega p)\, \mathrm{d}t + \mathcal{O}\left(\varepsilon^2\right) \\ &= 2\varepsilon \operatorname{sech} \frac{\omega}{2\pi A} + \mathcal{O}\left(\varepsilon^2\right), \end{aligned} \tag{3.32}$$

since the integral of $\left|\cos\left[\omega(t-p)\right]\right|$ between adjacent zeros is identical for any choice of such zeros, and therefore any convenient shift can be employed. Notice from this that all lobe areas are *identical to leading-order*, whatever time-slice one chooses. Moreover, they exhibit the standard behavior as characterized in [342]: larger frequencies lead to smaller lobe areas. ∎

3.4 ▪ Flux function for time-harmonic perturbations

The previous sections dealt with quantifying flux in nonautonomously perturbed systems (3.6), for which the Melnikov function (3.9) was shown to quantify the instantaneous leading-order signed flux to leading-order in ε. The perturbing velocity $g(x,t)$ had a general bounded form. A situation which has been much addressed is when $g(x,t) = \tilde{g}(x)\cos(\omega t)$ for some constant $\omega > 0$; this shall be defined to be *time-harmonic* or *time-sinusoidal*. This has been a popular area of study for quite some years, with the principal tool being the Poincaré map which samples the flow at time intervals of $2\pi/\omega$ [183, 421, 17, 341, 342, 354, e.g.]. It turns out that this well-understood situation can be recast in the time-continuous sense, expanding on ideas developed in [26, 28, 35]. In addition, unlike in the standard approaches [183, 421, 17, 341, 342, 354, e.g.], area-preservation shall *not* be assumed. Explicit expressions can be derived for the distance between perturbed manifolds in a time-slice t, the flux across a heteroclinic separator, and the lobe areas, and these can all be related to one another in pleasing ways using the ideas of the previous sections.

The system of study in this section shall therefore be

$$\dot{x} = f(x) + \varepsilon \tilde{g}(x) \cos(\omega t), \tag{3.33}$$

where $|\varepsilon|$ is small, $\omega > 0$ is the perturbing frequency, and x is two-dimensional. (The minor modification of including an additional phase factor in the cosine is described in [35], but this shall be dispensed with here since it can be neglected via a simple time shift.) Hypotheses 2.2 and 2.3 shall also be assumed. The second of these translates to $\tilde{g}$ and $D\tilde{g}$ being continuous and bounded. The unperturbed ($\varepsilon = 0$) flow is assumed to satisfy Hypothesis 3.5; i.e., there is a heteroclinic manifold Γ from a point a to b, which can be parametrized by $\bar{x}(t)$ for $t \in \mathbb{R}$ corresponding to a solution to (3.33) when $\varepsilon = 0$. Now, the crucial quantity which characterizes how Γ splits when $\varepsilon \neq 0$ is the Melnikov function, for which a very straightforward representation is possible. The Fourier transform convention

$$\mathscr{F}(h)(\omega) := H(\omega) := \int_{-\infty}^{\infty} h(t) e^{-i\omega t} \, \mathrm{d}t \tag{3.34}$$

for functions $h \in \mathrm{L}^1(\mathbb{R})$ shall be useful in presenting the results.

Theorem 3.28 (Melnikov function for harmonic perturbations). *Consider (3.33) under Hypotheses 2.2, 2.3, and 3.5. Then, the Melnikov function defined in (3.9) can be represented by*

$$M(p,t) = \exp\left[\int_0^p \mathrm{Tr}\, Df(\bar{x}(\xi)) \, \mathrm{d}\xi\right] |\Lambda(\omega)| \cos\left[\omega(t-p) - \mathrm{Arg}\left[\Lambda(\omega)\right]\right], \tag{3.35}$$

where

$$\Lambda(\omega) := \mathscr{F}\left\{\exp\left[\int_t^0 \mathrm{Tr}\, Df(\bar{x}(\xi)) \, \mathrm{d}\xi\right] f^{\perp}(\bar{x}(t)) \cdot \tilde{g}(\bar{x}(t))\right\}(\omega). \tag{3.36}$$

Proof. The proof extends results available for area-preserving situations [26]. Using Corollary 3.10, the Melnikov function $M(p,t)$ in (3.9) can be written as

$$M(p,t) = \tilde{M}(p,t) \exp\left[\int_0^p \mathrm{Tr}\, Df(\bar{x}(\xi)) \, \mathrm{d}\xi\right],$$

where $\tilde{M}$ is the scaled Melnikov function as given in (3.15). Applying the fact that $g(x,t) = \tilde{g}(x)\cos(\omega t)$, this becomes

$$\tilde{M}(p,t) = \int_{-\infty}^{\infty} \exp\left[\int_\tau^0 \mathrm{Tr}\, Df(\bar{x}(\xi)) \, \mathrm{d}\xi\right] f^{\perp}(\bar{x}(\tau)) \cdot \tilde{g}(\bar{x}(\tau)) \cos\left[\omega(\tau + t - p)\right] \mathrm{d}\tau.$$

Now define the function

$$\lambda(t) := \exp\left[\int_t^0 \mathrm{Tr}\, Df(\bar{x}(\xi)) \, \mathrm{d}\xi\right] f^{\perp}(\bar{x}(t)) \cdot \tilde{g}(\bar{x}(t)) \tag{3.37}$$

and let $\Lambda(\omega)$ be its Fourier transform with respect to t. This is well-defined since λ has exponential decay in both the limits $t \to \pm\infty$, as can be ascertained using the identical argument associated with (2.77) in Section 2.4.1. Note that

$$|\Lambda(\omega)| = \left| \int_{-\infty}^{\infty} \lambda(t) e^{-i\omega t} \, \mathrm{d}t \right| = \sqrt{\left(\int_{-\infty}^{\infty} \lambda(t) \cos(\omega t) \, \mathrm{d}t \right)^2 + \left(\int_{-\infty}^{\infty} \lambda(t) \sin(\omega t) \, \mathrm{d}t \right)^2} .$$

Thus, using elementary trigonometry,

$$\begin{aligned}
\tilde{M}(p,t) &= \int_{-\infty}^{\infty} \lambda(\tau) \cos[\omega(\tau + t - p)] \, \mathrm{d}\tau \\
&= \cos[\omega(t-p)] \int_{-\infty}^{\infty} \lambda(\tau) \cos(\omega\tau) \, \mathrm{d}\tau - \sin[\omega(t-p)] \int_{-\infty}^{\infty} \lambda(\tau) \sin(\omega\tau) \, \mathrm{d}\tau \\
&= \mathrm{Re}[\Lambda(\omega)] \cos[\omega(t-p)] + \mathrm{Im}[\Lambda(\omega)] \sin[\omega(t-p)] \\
&= |\Lambda(\omega)| \left(\frac{\mathrm{Re}[\Lambda(\omega)]}{|\Lambda(\omega)|} \cos[\omega(t-p)] + \frac{\mathrm{Im}[\Lambda(\omega)]}{|\Lambda(\omega)|} \sin[\omega(t-p)] \right) \\
&= |\Lambda(\omega)| (\cos(\mathrm{Arg}[\Lambda(\omega)]) \cos[\omega(t-p)] + \sin(\mathrm{Arg}[\Lambda(\omega)]) \sin[\omega(t-p)]) \\
&= |\Lambda(\omega)| \cos[\omega(t-p) - \mathrm{Arg}[\Lambda(\omega)]] .
\end{aligned} \tag{3.38}$$

This shows that the scaled Melnikov function is itself a simple harmonic function of $(t-p)$, with the identical frequency as the perturbing velocity field. Applying the connection between the scaled and the unscaled Melnikov functions leads to the result. □

Remark 3.29. The time-variation of the Melnikov function (i.e., the leading-order flux function) in this instance where the perturbing velocity field is time-harmonic is itself time-harmonic with the identical frequency. However, there is a phase shift, and the amplitude depends on the gate position p.

Remark 3.30. If $\Lambda(\omega) \neq 0$ and $\omega \neq 0$, Theorem 3.28 immediately gives the fact that there are infinitely many intersections between the perturbed stable and unstable manifolds, occurring when

$$t - p = \frac{2m+1}{2\omega}\pi - \frac{\mathrm{Arg}[\Lambda(\omega)]}{\omega} \qquad (m \in \mathbb{Z}). \tag{3.39}$$

In a fixed time-slice t, for example, the above gives the p values (corresponding to being near the point $\bar{x}(p)$ in the direction of $f^{\perp}(\bar{x}(p))$) near which intersections occur. This is usually thought of as a signature of chaotic transport. Each such intersection—indexed by $m \in \mathbb{Z}$—when continued in t, describes a heteroclinic trajectory in the augmented phase space of (3.33) which forwards asymptotes to the perturbed hyperbolic trajectory $(b_\varepsilon(t), t)$ and backwards asymptotes to $(a_\varepsilon(t), t)$.

Remark 3.31 (proof of chaos). The expression (3.35) circumvents contour-integration or numerical methods usually used to prove that a system is chaotic by computing the Melnikov function in time-harmonic homoclinic situations [183, 17, 421, 341, 417, 178, 354, e.g.]. Indeed, a proof of chaos in the homoclinic instance via the Smale-Birkhoff Theorem [183] is automatic if $\omega \neq 0$ and $\Lambda(\omega) \neq 0$.

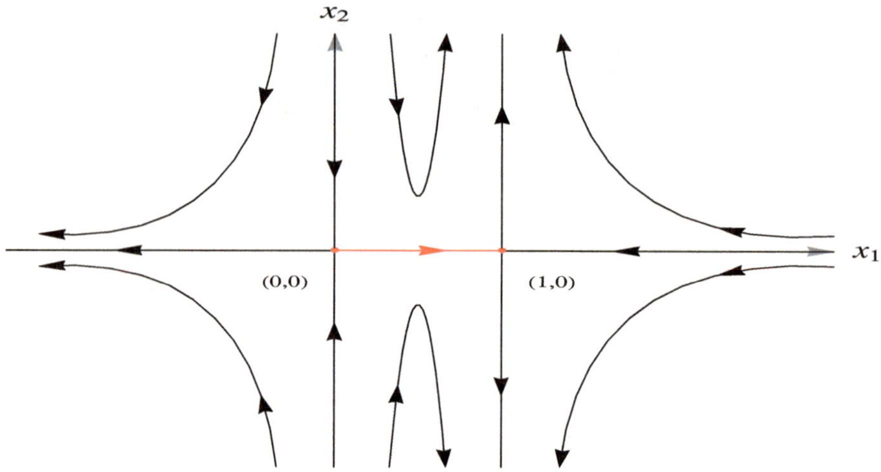

Figure 3.8. *Phase portrait of (3.40) with $\varepsilon = 0$, showing the heteroclinic manifold [red] across which flux is to be assessed.*

Example 3.32. Consider the system

$$\frac{d}{dt}\begin{pmatrix} x_1 \\ x_2 \end{pmatrix} = \begin{pmatrix} x_1x_2 + x_1 - x_1^2x_2 - x_1^2 \\ 2x_1x_2 - x_2 \end{pmatrix} + \varepsilon \begin{pmatrix} x_1x_2 - x_2^2 \\ x_1^2 \end{pmatrix} \cos(\omega t) . \tag{3.40}$$

When $\varepsilon = 0$, this has critical points at $(x_1, x_2) = (0,0)$ and $(1,0)$, which are saddle points, and the phase portrait is shown in Figure 3.8. The goal is to use Theorem 3.28 to evaluate the flux generated across the heteroclinic between these two points (shown in red in Figure 3.8) when $\varepsilon \neq 0$, at each instance in time t. Since the flux definition takes flow in the direction of $f^{\perp}$ as positive, this would indicate that flux occuring across the red curve in the direction of $+x_2$ would be considered positive. It is left as an exercise to obtain the fact that the heteroclinic, as a solution to (3.40) with $\varepsilon = 0$, is expressible as

$$\bar{x}_1(t) = \frac{e^t}{1+e^t}, \quad \bar{x}_2(t) = 0$$

under the symmetric choice of time-parametrization (i.e., $\bar{x}_1(0) = 0.5$). In preparation for using Theorem 3.28, the identification

$$f^{\perp} = \begin{pmatrix} 0 \\ \dot{\bar{x}}_1 \end{pmatrix} = \begin{pmatrix} 0 \\ e^t/(1+e^t)^2 \end{pmatrix} \quad \text{and} \quad g = \begin{pmatrix} 0 \\ \bar{x}_1^2 \end{pmatrix} = \begin{pmatrix} 0 \\ e^{2t}/(1+e^t)^2 \end{pmatrix}$$

along the heteroclinic is made. Moreover, $\operatorname{Tr} Df = x_2 - 2x_1x_2 + x_1$, but since $x_2 = 0$ on the heteroclinic,

$$\int_t^0 \operatorname{Tr} Df(\bar{x}(\xi))\, d\xi = \int_t^0 \bar{x}_1(\xi)\, d\xi = \log\left(1+e^{\xi}\right)\Big|_t^0 = \log \frac{2}{1+e^t}.$$

Therefore, from (3.36), $\Lambda(\omega)$ is the Fourier transform of the function $2e^{3t}/(1+e^t)^5$, which is

$$\Lambda(\omega) = \frac{\pi\omega\left(1+\omega^2\right)}{12} \operatorname{cosech}(\pi\omega)(2 - i\omega),$$

enabling the representation

$$M(p,t)=\frac{(1+e^p)\pi\omega\left(1+\omega^2\right)\sqrt{4+\omega^2}\,\mathrm{cosech}\,(\pi\omega)}{24}\cos\left[\omega(t-p)+\tan^{-1}\frac{\omega}{2}\right] \quad (3.41)$$

for $\omega \neq 0$. A specific choice of p would identify where on the red line in Figure 3.8 the gate is to be located; for example, $p=0$ would locate it symmetrically at $x_1 = e^0/(1+e^0)=1/2$. The flux is therefore harmonic, while its amplitude is easily read-off from (3.41). ∎

It should be noted that, unlike in most applications in the literature, Theorem 3.28 and the "proof of chaos" argument of Remark 3.31 does *not* require that the vector field be area-preserving. In this sense, the previous example is unusual in the literature. Should area-preservation occur, an even easier description of the Melnikov function is possible in the form of the following corollary.

Corollary 3.33 ([26, 27]). *In addition to the conditions of Theorem 3.28, suppose f is area-preserving. Then*

$$M(p,t)=|\Lambda(\omega)|\cos\left[\omega(t-p)-\mathrm{Arg}\left[\Lambda(\omega)\right]\right]. \quad (3.42)$$

Proof. Since $\mathrm{Tr}\,Df=0$, this is obvious. Note also that the definition for $\Lambda(\omega)$ simplifies in this case, with the exponential term dropping out. □

Example 3.34 (Duffing oscillator cont.). The Duffing oscillator examined in Examples 2.11 and 2.21 is revisited. Consider (2.19) with no damping, but with the forcing $\varepsilon\cos(\omega t)$ with $|\varepsilon|\ll 1$. This is a classical situation often considered in establishing chaotic motion in the system [183]. The quantities f, $\mathrm{Tr}\,Df$ and $\bar{x}$ have already been computed under Example 2.21. The perturbation here is, however, $\varepsilon g(x,t)=(0,\varepsilon\cos(\omega t))$, and thus $\tilde{g}=(0,1)$ leading to

$$\Lambda(\omega)=\mathcal{F}\left\{\begin{pmatrix}\bar{x}_1^3(t)-\bar{x}_1(t)\\ \bar{x}_2(t)\end{pmatrix}\cdot\begin{pmatrix}0\\1\end{pmatrix}\right\}(\omega)=\mathcal{F}\left\{-\sqrt{2}\,\mathrm{sech}\,t\tanh t\right\}(\omega).$$

This Fourier transform is easily computable to be $\Lambda(\omega)=i\sqrt{2}\pi\omega\,\mathrm{sech}(\pi\omega/2)$. Therefore, from Corollary 3.33,

$$M(p,t)=\sqrt{2}\pi\omega\,\mathrm{sech}\frac{\pi\omega}{2}\cos\left[\omega(t-p)-\frac{\pi}{2}\right]=\sqrt{2}\pi\omega\,\mathrm{sech}\frac{\pi\omega}{2}\sin\left[\omega(t-p)\right].$$

Knowing this explicit expression for the Melnikov function immediately permits the conclusion of the presence of simple intersections between the stable and unstable manifold, and—since the system is periodic—these intersections get mapped to one another under the Poincaré map. All these zeros are visible in the Melnikov function above, occurring when $t-p=m\pi/\omega$ for $m\in\mathbb{Z}$. Thus, the Duffing system is chaotic for small $|\varepsilon|$. Indeed, the distance between the manifolds can be characterized by (3.8) to be

$$d(p,t)=\varepsilon\frac{\pi\omega\,\mathrm{sech}\frac{\pi\omega}{2}\sin\left[\omega(t-p)\right]}{\mathrm{sech}\,p\sqrt{2-5\,\mathrm{sech}^2 p+4\,\mathrm{sech}^4 p}}+\mathcal{O}\left(\varepsilon^2\right),$$

but a quantification of "flux across manifolds" will not be attempted here since it has no real meaning here since the phase space is not physical space. ∎

Example 3.35 (Rossby waves). The next example arises from the atmospheric sciences and oceanography, in which meandering jet structures (i.e., jet streams in the atmosphere, oceanic jets like the Gulf Stream and Kuroshio Current) are important. Rossby waves in a meandering jet occur because of the planetary vorticity, and their theory provides the dynamical connection between the wavenumbers of the meandering jet and its phase speed [331, 310, 314, 352, 324, 348, 56, 298, 354]. Here, x_1 and x_2 shall be used to represent locally eastward ("zonal") and northward ("meridional") directions, respectively. A Rossby wave is then given by the streamfunction [314, 310]

$$\psi_0(x_1,x_2,t) = A\sin\left[k_0(x_1 - c_0 t)\right]\sin\left[l_0 x_2\right], \tag{3.43}$$

in which $c_0 = -\beta/(k_0^2 + l_0^2)$ is the speed at which the Rossby wave is traveling zonally and β is the Coriolis parameter which encodes the earth's rotation and local curvature effect on the wave [310, 22, 314] and is given by $(2\Omega/R)\cos\phi$, where R is the earth's radius, Ω its angular rotation around its axis, and ϕ is the latitude at the location of interest. If $0 < \beta < 2A(k_0^2 + l_0^2)$, the kinematic structure is of a meandering jet with adjacent eddy (or vortex) structures [314, 225, 417, cf.]. Following Pierrehumbert [314], set $k_0 = 1$ and $l_0 = 1$ (this is effectively a nondimensionalization of the (x,y)-variables in terms of the zonal and meridional lengthscales of this Rossby wave) and $A = 1$ (which essentially corresponds to a nondimensionalization of the velocity scale). Next, consider a frame moving with the Rossby wave, by setting $x_1 - c_0 t \to x_1$. With an abuse of notation, the same notation (x_1, x_2) shall be used for the *nondimensional* variables in this moving frame of reference. The fluid particle trajectories in this frame are therefore the solution to

$$\dot{x}_1 = -\frac{\partial \psi_0}{\partial x_2}, \quad \dot{x}_2 = \frac{\partial \psi_0}{\partial x_1}, \tag{3.44}$$

where

$$\psi_0(x_1,x_2) = \sin x_1 \sin x_2 - \frac{\beta x_2}{2} \tag{3.45}$$

is now independent of time. (The relationship between the streamfunction and the velocity given in (3.44) is that common to geophysics; standard fluid mechanics usually puts the negative sign with the x_2 equation rather than the x_1 one.) Note moreover that $0 < \beta < 2$ and $c_0 = -\beta/2$. In these coordinates, the meandering jet with adjacent flanking vortices (or eddies) is shown in Figure 3.9; for one Rossby wave of this form, there is no transport between the jet and the vortex. The bounding flow barrier to be examined (shown in red) is the heteroclinic manifold connecting $(\sin^{-1}\beta/2, 0)$ to $(\pi - \sin^{-1}\beta/2, 0)$; this is moreover part of the level curve $\psi_0 = 0$. The question is to determine the transport impact of other Rossby waves, i.e., those with different wavenumbers (k,l), perturbing this dominant Rossby wave [352, cf.]. Noting that a Rossby wave with wavenumber (k,l) will possess a *different* speed of $c_{kl} = -\beta/(k^2 + l^2)$ [314, 310], the new flow in the frame of reference of the dominant Rossby wave pictured in Figure 3.9 shall be

$$\dot{x}_1 = -\frac{\partial \psi}{\partial x_2}, \quad \dot{x}_2 = \frac{\partial \psi}{\partial x_1} \tag{3.46}$$

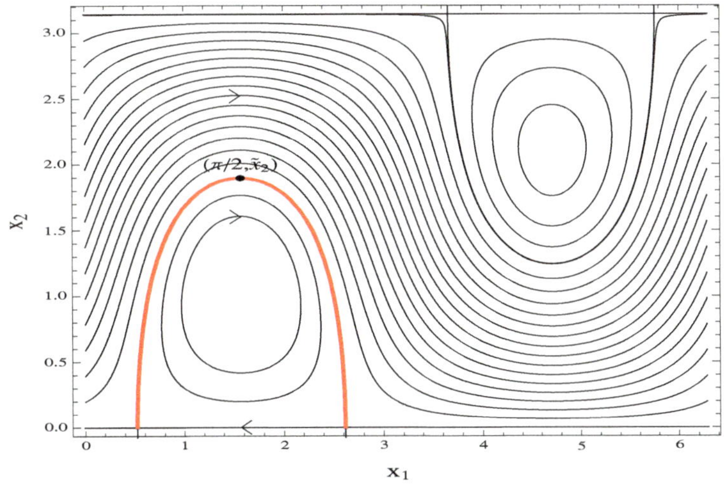

Figure 3.9. *Dominant Rossby wave in the moving frame for Example 3.35.*

with streamfunction

$$\psi(x_1,x_2,t) = \psi_0(x_1,x_2) + \psi_1(x_1,x_2,t) := \psi_0(x_1,x_2) + \varepsilon \sin[k(x_1 - c_{kl}t)]\sin(lx_2), \tag{3.47}$$

where $|\varepsilon| \ll 1$ [225, 417, 125, 264, cf.]. The full streamfunction is therefore

$$\begin{aligned}\psi(x_1,x_2,t) &= \sin(x_1)\sin(x_2) - \frac{\beta x_2}{2} + \varepsilon \sin\left[k(x_1 + (-c_{kl} - \frac{\beta}{2})t)\right]\sin(lx_2)\\ &= \sin(x_1)\sin(x_2) - \frac{\beta x_2}{2} + \varepsilon \sin\left[k\left(x_1 + \beta\left(\frac{1}{k^2+l^2} - \frac{1}{2}\right)t\right)\right]\sin(lx_2). \end{aligned}\tag{3.48}$$

Now, one can compute the fluid flux for this situation, which is given by the relevant Melnikov function. The pseudoseparatrix in this instance shall be thought of as being associated with a gate at the topmost location of the heteroclinic manifold in Figure 3.9. Thus, the gate is located at $(\pi/2, \tilde{x}_2)$, where (since located on the curve $\psi_0 = 0$) $\tilde{x}_2$ is the solution to the transcendental equation

$$2\sin\tilde{x}_2 = \beta\tilde{x}_2, \tag{3.49}$$

which lies in the domain $\tilde{x}_2 \in (0,\pi)$. Given that $f^\perp$ for this situation points from the vortex to the jet at points on the heteroclinic, flux from the vortex to the jet is considered positive. Since the flow is incompressible, Corollary 3.14 is valid in determining the flux function; since moreover harmonic, Theorem 3.28 is valid. Combining, the (fluid) flux takes the form

$$\text{Fluid Flux}(t) = \varepsilon M_{kl}(t) + \mathcal{O}(\varepsilon^2), \tag{3.50}$$

in which the Melnikov function is

$$M_{kl}(t) = A_{kl}\sin\left[k\left(\frac{\pi}{2} + \beta\left(\frac{1}{k^2+l^2} - \frac{1}{2}\right)t\right) - \phi_{kl}\right]. \tag{3.51}$$

If $k^2 + l^2 \neq 2$, the flux is sinusoidally oscillating between the vortex and the jet with frequency

$$\omega_{kl} = |k|\beta\left|\frac{1}{k^2+l^2} - \frac{1}{2}\right|, \tag{3.52}$$

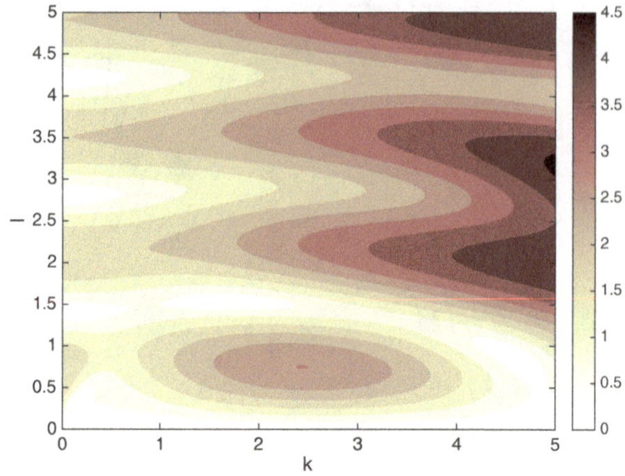

Figure 3.10. *The amplitude of the $\mathcal{O}(\varepsilon)$ flux function, (3.53), for $\beta = 0.7$, over a range of (k, l) values.*

and the flux has an amplitude A_{kl} and phase ϕ_{kl}. At times at which the flux is positive, there is fluid flux from the eddy to the jet, and at times at which it is negative, the instantaneous flux is in the opposite direction.

The "size" of the flux is encoded within A_{kl}. By performing the calculations (not shown) associated with the Melnikov function given in Theorem 3.28 with $\mathrm{Tr}\, Df = 0$, it turns out [21] that it can be given by

$$A_{kl} = 2\sqrt{l^2 I_s^2 + k^2 I_c^2}, \tag{3.53}$$

where

$$I_s := \int_0^{\tilde{x}_2} \cos[k\xi(x_2)] \cos[l x_2]\, \mathrm{d}x_2,$$

$$I_c := \int_0^{\tilde{x}_2} \cos[k\xi(x_2)] \sin[l x_2] \frac{\beta(x_2 \cot x_2 - 1)}{\sqrt{4\sin^2 x_2 - \beta^2 x_2^2}}\, \mathrm{d}x_2,$$

$$\xi(x_2) := \frac{\pi}{2} - \sin^{-1}\frac{\beta x_2}{2\sin x_2} + 2\beta\left(\frac{1}{k^2+l^2} - \frac{1}{2}\right)\int_{x_2}^{\tilde{x}_2} \frac{\mathrm{d}u}{\sqrt{4\sin^2 u - \beta^2 u^2}}.$$

Since, as argued previously, the Melnikov function provides an appropriate measure of the flux, it is clear that understanding how its amplitude A_{kl} behaves with respect to the perturbing wavenumbers (k, l) provides information on which wavenumbers offer the greatest and the least transport between the vortex and the jet. Using the above formulae, where the integrals and (3.49) require standard numerical evaluation, a contour plot of this amplitude is provided in Figure 3.10 over a range of wavenumbers. There are clearly wavenumbers associated with (local) maxima of the flux. Moreover, certain wavenumber perturbations generate minimal flux.

The amplitude A_{kl} has a relationship to the lobes generated when viewing the perturbed system in terms of a Poincaré map in the standard approach. Such lobes have equal areas, and following the development of Section 3.3.3, the area of each lobe is

$$\text{Lobe Area} = \frac{2|\varepsilon| A_{kl}}{\omega_{kl}} + \mathcal{O}(\varepsilon^2). \tag{3.54}$$

This is valid if $k \neq 0$ and $k^2+l^2 \neq 2$, since if these equalities hold, no lobes are created. This is clear from (3.51); if either $k = 0$ or $k^2 + l^2 = 2$, then the flux function is independent of time; it turns out this is actually zero [21]. ∎

3.5 ▪ Fourier flux formulae for general time-periodicity

The previous section outlined how to compute the flux in situations in which the perturbation was time-harmonic. While it is this situation which is often referred to in the literature as *time-periodic*, time-periodicity is more general than being simply a cosine function in time multiplying a function which includes the spatial dependence. Thus, in this section a genuinely time-periodic perturbation, in its most general sense, is addressed. The corresponding flow takes the form

$$\dot{x} = f(x) + \varepsilon g(x,t), \tag{3.55}$$

in which the function g, in addition to the smoothness requirements outlined in Hypothesis 2.3, satisfies the condition

$$g(x,t+T) = g(x,t) \quad \text{for } (x,t) \in \Omega \times \mathbb{R}, \tag{3.56}$$

for some constant $T > 0$ which is the *period* of g. The time-harmonic forms examined in the previous section in which $g(x,t) = \tilde{g}(x)\cos(\omega t)$ are clearly a special case of this, with g taking on a specific separable form and in which $T = 2\pi/\omega$. The presence of the period T would enable viewing (3.55) in terms of a Poincaré map which samples the flow at time intervals $0, \pm T, \pm 2T, \pm 3T, \ldots$. Should there be a heteroclinic manifold in the $\varepsilon = 0$ system which breaks apart into a stable and unstable manifold, a typical picture associated with such a Poincaré map would then be Figure 3.1 as opposed to Figure 3.2.

From the continuous perspective, that is, in viewing (3.55) in conjunction with $\dot{t} = 1$, the pictures associated with these two figures can be construed as $t =$ constant intersections with the augmented phase space. For general periodicity, as shown in Figure 3.1 the lobe sizes can be quite varied, and indeed, it is also possible that there are no intersections between the stable and unstable manifolds (see Example 3.7). Therefore, under general time-periodicity (in contrast to time-harmonicity) the usage of lobes as a flux measure requires some clarification. The instantaneous flux, on the other hand, is valid, since it is developed from general time-dependent perturbations. Under the continuous viewpoint, the flux across the heteroclinic separatrix has the distinguishing characteristic of being periodic in time, but apart from that, the flux function can take on very general forms. The periodicity enables the flux to be represented very easily in terms of Fourier ideas, leading to computable formulae which converge rapidly [26]. The development in this section is an extension of these previous results [26] to non-area-preserving f.

Given the periodicity hypothesis on g, it can be represented for each x in terms of a complex Fourier series

$$\frac{g(x,t^-)+g(x,t^+)}{2} = \sum_{n=-\infty}^{\infty} g_n(x) \exp\left(\frac{2\pi i n t}{T}\right),$$

where $g(x,t^-) := \lim_{\tau \uparrow t} g(x,\tau)$ and $g(x,t^+) := \lim_{\tau \downarrow t} g(x,\tau)$, and the functions $g_n : \Omega \to \mathbb{C}^2$ satisfy

$$g_n(x) = \frac{1}{T}\int_0^T g(x,t) \exp\left(-\frac{2\pi i n t}{T}\right) dt,$$

which are bounded in Ω by Hypothesis 2.3. Define for each $n \in \mathbb{Z}$ the function

$$\lambda_n(t) := \exp\left[\int_t^0 \operatorname{Tr} Df(\bar{x}(\xi))\,\mathrm{d}\xi\right] f^{\perp}(\bar{x}(t)) \cdot g_n(\bar{x}(t)), \tag{3.57}$$

where, as before, $\bar{x}(t)$ represents the heteroclinic trajectory of the $\varepsilon = 0$ system, whose destruction leads to a flux across the now distinct stable and unstable manifolds at each instance in time. The exponential decay of $\lambda_n(t)$ as $t \to \pm\infty$ has essentially been established previously (see the derivation of (2.77) in Section 2.4.1), and thus its Fourier transform $\Lambda_n(\omega)$ is well defined for each n.

Theorem 3.36 (Fourier representation of Melnikov function). *The Melnikov function (3.9) associated with (3.55) has a complex Fourier series representation*

$$M(p,t) = \exp\left[\int_0^p \operatorname{Tr} Df(\bar{x}(\xi))\,\mathrm{d}\xi\right] \sum_{n=-\infty}^{\infty} d_n \exp\left(\frac{2\pi i n (p-t)}{T}\right), \tag{3.58}$$

where the Fourier coefficients satisfy

$$d_n = \Lambda_n\left(-\frac{2\pi n}{T}\right), \tag{3.59}$$

in which Λ_n is the Fourier transform of λ_n using definition (3.34).

Proof. By Corollary 3.10, the Melnikov function can be written as

$$M(p,t) = \exp\left[\int_0^p \operatorname{Tr} Df(\bar{x}(\xi))\,\mathrm{d}\xi\right] \tilde{M}(p,t),$$

where the scaled Melnikov function can be written as

$$\begin{aligned}
\tilde{M}(p,t) &:= \int_{-\infty}^{\infty} \exp\left[\int_\tau^0 \operatorname{Tr} Df(\bar{x}(\xi))\,\mathrm{d}\xi\right] f^{\perp}(\bar{x}(\tau)) \cdot g(\bar{x}(\tau), \tau + t - p)\,\mathrm{d}\tau \\
&= \int_{-\infty}^{\infty} \exp\left[\int_\tau^0 \operatorname{Tr} Df(\bar{x}(\xi))\,\mathrm{d}\xi\right] \sum_{n=-\infty}^{\infty} f^{\perp}(\bar{x}(\tau)) \\
&\qquad \cdot g_n(\bar{x}(\tau)) \exp\left(\frac{2\pi i n (\tau - t + p)}{T}\right) \mathrm{d}\tau \\
&= \sum_{n=-\infty}^{\infty} \exp\left(\frac{2\pi i n (p-t)}{T}\right) \int_{-\infty}^{\infty} \lambda_n(\tau) \exp\left(\frac{2\pi i n \tau}{T}\right) \mathrm{d}\tau,
\end{aligned}$$

from which the result follows. The dominated convergence theorem has been used to interchange the infinite summation and improper integral. $\square$

Remark 3.37. Theorem 3.36 builds on previous approaches to write periodic Melnikov functions in terms of Fourier series for *Hamiltonian* systems [105, 126, 26]. In this non-area-preserving case, the Melnikov function is not actually periodic in both p and t, but once the scaling prefactor (dependent on p) is removed, one can take advantage of the fact that the remaining scaled Melnikov function $\tilde{M}$ is periodic in

$(t-p)$, and hence a Fourier representation works. Its Fourier coefficients d_n can be read immediately from (3.59).

Remark 3.38. The d_n are given by the Fourier transforms of the λ_n, evaluated at $-2\pi n/T$. Since the Fourier transform of each λ_n must decay at $\pm\infty$, the coefficients become increasingly small. Hence, as shown in [28], keeping a small number of terms in the Fourier expansion (3.58) is usually adequate. This also indicates that while g may have jump discontinuities in t (Hypothesis 2.3 only required that g be piecewise smooth in t), the flux/Melnikov function will be smooth; a precise quantification of smoothness is possible [29] using elementary Fourier transform properties.

It has been argued in Sections 3.1 and 3.2 that in compressible flows with general time-dependence, thinking of transport as a continuous time-varying function makes the most sense. In time-periodic situations such as is considered in this section, one might draw the situation such as Figure 3.1 in a particular time-slice and attempt to quantify transport as iterations of the Poincaré map P of period T. The lobe areas are not good indicators of transport for *compressible* flow since their iterates will in general have variable area; i.e., $P(L)$ need not have the same area as L. For time-periodic *incompressible* flows, however, one might quantify transport as the lobe areas of a turnstile region which transport fluid across the pseudoseparatrix in either direction. The turnstile lobes turn out to be more complicated in general than for the time-sinusoidal situation, in which there were exactly four lobes, all with the same areas, which participated in the exchange of fluid across the pseudoseparatrix (see Section 3.1). By carefully evaluating how these lobes—of arbitrary number and size in general—transport across a carefully defined pseudoseparatrix, one can continue to use the Poincaré map idea to quantify lobe transport in each direction, and also the directional imbalance in lobe transport, in terms of the Fourier coefficients d_n. The interested reader may find a considerable discussion on these topics in [26], but this section shall retain the generality of compressible flows, and thus determining $M(p,t)$ shall be deemed a sufficient quantification of transport, and lobes will be ignored.

Corollary 3.39. *Under the conditions of Theorem 3.36, suppose that* $g(x,t)$ *is* separable *in that it can be represented by*

$$g(x,t)=\tilde{g}(x)\theta(t),$$

where $\tilde{g}:\Omega\to\mathbb{R}^2$ *and* $\theta:\mathbb{R}\to\mathbb{R}$. *If* θ*'s complex Fourier coefficients are given by* $\{c_n\}_{n\in\mathbb{Z}}$, *then the complex Fourier coefficients in (3.58) can be represented by*

$$d_n=c_n\Lambda\left(-\frac{2\pi n}{T}\right),$$

where Λ *is defined in (3.36) and is the Fourier transform of* $\lambda(t)$ *defined in (3.37).*

Proof. In this case, $g_n(x)=\tilde{g}(x)c_n$ for $n\in\mathbb{Z}$, and thus from (3.57) $\lambda_n(t)=c_n\lambda(t)$ using the definition (3.37). Therefore, $\Lambda_n=c_n\Lambda$, and the result follows. □

It will be left as an exercise to verify that the Melnikov function for harmonic perturbations (Theorem 3.28) can be derived as a special case of the above results. Another situation which can be easily addressed is when the perturbation consists of one harmonic mode plus a constant term, as follows.

Proposition 3.40. *Suppose*

$$g(x,t) = \tilde{g}(x)\left[a_0 + a_1 \cos\omega t + b_1 \sin\omega t\right]$$

for some real coefficients a_0, a_1, *and* b_1, *positive frequency* ω, *and smooth function* $\tilde{g}$: $\Omega \to \mathbb{R}^2$ *as addressed in Section 3.4. If* Λ *is as defined in (3.36), then*

$$M(p,t) = \exp\left[\int_0^p \operatorname{Tr} Df\left(\bar{x}(\xi)\right) \mathrm{d}\xi\right]\left[a_0\Lambda(0) + \operatorname{Re}\left\{(a_1 + i b_1)\Lambda(\omega) e^{i\omega(t-p)}\right\}\right]. \tag{3.60}$$

Proof. This extends a result available in the area-preserving situation [26]. In this case, $T = 2\pi/\omega$, and the separable characterization of Corollary 3.39 is applicable with $c_0 = a_0$ and $c_{\pm 1} = (a_1 \mp i b_1)/2$, with all other c_ns equaling zero. The only nontrivial complex Fourier coefficients in the scaled Melnikov function are therefore

$$\begin{aligned} d_0 &= c_0\Lambda(0) = a_0\Lambda(0), \\ d_1 &= c_1\Lambda(-\omega) = \frac{a_1 - i b_1}{2}\Lambda(-\omega) \quad \text{and} \\ d_{-1} &= c_{-1}\Lambda(\omega) = \frac{a_1 + i b_1}{2}\Lambda(\omega). \end{aligned}$$

Since $d_{-1}^{*} = d_1$ here, substituting into (3.58) yields the desired result. □

For the situation of Proposition 3.40, is the flux unidirectional, or does fluid slosh back and forth across the broken heteroclinic? In other words, are there any primary intersections between the stable and unstable manifolds? This can be answered directly by observing that there are essentially two terms in (3.60): a constant term and a term which has harmonic dependence on $(t-p)$. If the coefficient of the harmonic term is less than the constant term, then $M(p,t)$ can never be zero. Simple algebra shows that if $|\Lambda(\omega)|\sqrt{a_1^2 + b_1^2} < |a_0\Lambda(0)|$, the flux is unidirectional. If this inequality is reversed, there is transport back and forth with time. The condition for unidirectionality can also be written as

$$\frac{|\Lambda(\omega)|}{|\Lambda(0)|} < \frac{|a_0|}{\sqrt{a_1^2 + b_1^2}}, \tag{3.61}$$

where the left-hand side quantifies the ratio between a flux response with frequencies ω and 0, and the right-hand side shows the reciprocal relationship between the *imposed* frequencies.

Example 3.41 (Taylor-Green flow). The Taylor-Green flow (planar cellular flow) consists of a square array of counterrotating vortices and is a common test bed for dynamical analyses [5, 27, 1, 26, 54, 69, e.g.]. It is kinematically equivalent to the double-gyre flow with $\varepsilon = 0$. The form considered in this example is

$$\dot{x}_1 = -\sin(2\pi x_1)\sin(2\pi x_2), \qquad \dot{x}_2 = -\cos(2\pi x_1)\cos(2\pi x_2), \tag{3.62}$$

which satisfies $\operatorname{Tr} Df = 0$ and is thus area-preserving. The line segment $x_1 = 0$, $1/4 \le x_2 \le 3/4$ forms a heteroclinic separatrix between adjacent cells,[6] which can

[6] See Figure 6.1 in [26].

be parametrized by $\bar{x}_1(t)=0$ and

$$\bar{x}_2(t)=\begin{cases} \frac{1}{2\pi}\cos^{-1}[-\operatorname{sech}(2\pi t)] & \text{if } t\leq 0,\\ 1-\frac{1}{2\pi}\cos^{-1}[-\operatorname{sech}(2\pi t)] & \text{if } t>0,\end{cases}$$

and hence $f(\bar{x}(t))=(0,\operatorname{sech}(2\pi t))$. The flux function shall be computed for a variety of time-periodic perturbations to the flow. In view of the fact that $\lambda_n(t)$ depends on $f^{\perp}\cdot g$, it is only the component of the perturbation $\varepsilon g(x,t)$ in the x_1-direction which contributes to the flux. Accordingly, the analysis shall be confined to the flows of the form

$$\dot{x}_1=-\sin(2\pi x_1)\sin(2\pi x_2)+\varepsilon g^1(x_1,x_2,t), \qquad \dot{x}_2=-\cos(2\pi x_1)\cos(2\pi x_2),$$

where g^1 is periodic in t with period T. For simplicity, the situations where g^1 is separable are first examined. In this simplified setting

$$\lambda(t)=f^{\perp}(\bar{x}(t))\cdot\tilde{g}(\bar{x}(t))=-\operatorname{sech}(2\pi t)\,\tilde{g}^1(0,\bar{x}_2(t)).$$

As a first case, suppose $\tilde{g}^1=-K$, a constant. Then the Fourier coefficients of the scaled Melnikov function of Theorem 3.36 are

$$d_n=\frac{K c_n}{2}\operatorname{sech}\left(\frac{\pi n}{2T}\right),$$

where the c_n are the Fourier coefficients of $\theta(t)$. For the one-mode situation as in Proposition 3.40, the condition for unidirectionality (3.61) becomes

$$\operatorname{sech}\frac{\omega}{4}<\frac{|a_0|}{\sqrt{a_1^2+b_1^2}}=:R.$$

The right-hand side above represents the relative amplitudes R of the constant term in comparison to the harmonic term; if this is large, one would expect the harmonic wiggles to not be sufficiently large to cause zeros. This is indeed the case, but the relative size depends on the frequency ω. If $R>1$ (i.e., if the harmonic amplitude is smaller than that of the constant in the perturbation θ), the inequality is met for all ω, and there can be no intersections. If $R<1$, intersections cannot occur for large enough $|\omega|$ but will occur for small $|\omega|$. A more general time-variation is achieved when θ is the $(2\pi/\omega)$-periodic square wave which in a base period is defined by

$$\theta(t)=\begin{cases} 1 & \text{if } 0<t<\frac{\pi}{2\omega} \text{ or } \frac{3\pi}{2\omega}<t<\frac{2\pi}{\omega},\\ -1 & \text{if } \frac{\pi}{2\omega}<t<\frac{3\pi}{2\omega}.\end{cases}$$

By Corollary 3.39, the scaled Melnikov function has coefficients

$$d_0=0 \quad\text{and}\quad d_n=\left(\frac{K}{2}\right)\frac{1-(-1)^n}{n\pi}\operatorname{sech}\frac{n\omega}{4}\sin\frac{n\pi}{2} \quad\text{for } n\neq 0,$$

and therefore the leading-order instantaneous flux is

$$M(p,t)=\frac{2K}{\pi}\sum_{j=1}^{\infty}\frac{(-1)^{j+1}}{2j-1}\operatorname{sech}\left[\frac{(2j-1)\omega}{4}\right]\cos[(2j-1)\omega(t-p)].$$

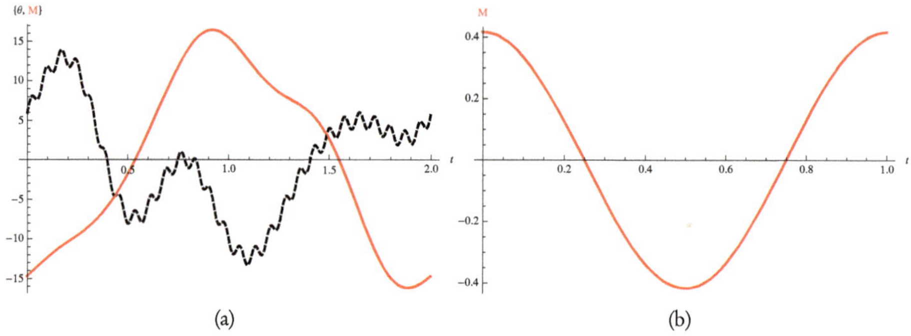

Figure 3.11. *Flux functions (solid red curves) for Taylor-Green flow: (a) using (3.63), with θ shown by the dashed curve, and (b) using (3.64).*

Within a particular period $[0, 2\pi/\omega]$, M as a function of $(t-p)$ clearly has zeros at $\pi/(2\omega)$ and $3\pi/(2\omega)$. The fact that these are the only zeros is easily confirmed (see Figure 6.4 in [26]), as well as the fact that keeping only seven terms in the summation guarantees about six-digit accuracy [26], despite θ's Fourier series having slow convergence.

As the next modification, suppose θ is the 2-periodic function

$$\theta(t) = 8\cos(\pi t) + 5\sin(3\pi t) - 2\cos(4\pi t) + \sin(27\pi t), \tag{3.63}$$

which is shown in its fundamental domain $[0, 2)$ by the dashed curve in Figure 3.11 in the left panel. The corresponding Melnikov function, computed via the the Fourier coefficients in the usual fashion, is also displayed in this figure. While θ has six zeros, the Melnikov function has only two, implying that the flux flips sign twice times during each period. The greater smoothness of M in comparison to θ occurs because the process of computing the flux function from the temporal perturbation θ can be thought of as a low-pass filter operation, in view of the fact that higher-frequency terms are damped (see Corollary 3.39).

Next, an example in which the time-dependence cannot be separated out is considered. Consider the 1-periodic function

$$g^1(x_1, x_2, t) = \cos[2\pi(x_2 - t)] - 3\sin[5\pi(x_1 - 2t)], \tag{3.64}$$

which combines a horizontal and a vertical traveling wave with unequal speeds. The only terms which survive in g^1's Fourier expansion are

$$g_{\pm 1}(x_1, x_2) = \frac{1}{2}\exp(\mp i\, 2\pi x_2), \quad g_{\pm 5}(x_1, x_2) = \mp\frac{3i}{2}\exp(\mp i\, 5\pi x_1).$$

The only nontrivial λ_n are

$$\lambda_{\pm 1}(t) = \frac{1}{2}\operatorname{sech}(2\pi t)[1 \mp \tanh(2\pi t)], \quad \lambda_{\pm 5}(t) = \pm\frac{3i}{2}\operatorname{sech}(2\pi t),$$

with associated Fourier transforms

$$\Lambda_{\pm 1}(\omega) = \frac{\omega}{8\pi}\left[\operatorname{cosech}\left(\frac{\omega}{4}\right) \mp \operatorname{sech}\left(\frac{\omega}{4}\right)\right], \quad \Lambda_{\pm 5}(\omega) = \pm\frac{3i}{4}\operatorname{sech}\left(\frac{\omega}{4}\right).$$

The Melnikov function can be exactly obtained from Theorem 3.36 to be

$$M(p,t)=\frac{1}{2}\left(\operatorname{cosech}\frac{\pi}{2}+\operatorname{sech}\frac{\pi}{2}\right)\cos(2\pi[p-t])-\frac{3}{2}\operatorname{sech}\frac{5\pi}{2}\sin(10\pi[p-t]),$$

which is shown in the right panel of Figure 3.11 for the choice $p=0$ (gate at location $(x_1,x_2)=(0,1/2)$) for one period. ∎

3.6 ▪ Flux for impulsive perturbations

The transport quantified so far in this chapter in terms of a flux depends on the smoothness of the perturbation g. Can a flux be defined if g, as in Section 2.7, had impulses in time? This would appear to be pertinent in several applications. Consider, for example, the flow of some quantity (a pollutant, plankton distribution, heat) in the ocean, which is impacted by an explosion (an underwater volcanic eruption, an earthquake, or a missile, satellite, or asteroid falling into the ocean). This might be modeled by a spatially dependent impulse in the velocity field. As another example consider a microfluidic device in which mixing is to be enhanced by the procedure of tapping the device sharply and perhaps doing this many times. This will impact the interior velocity field in an impulsive fashion, and the resulting velocities will have an influence on fluid mixing within the device. Thus, understanding fluid transport under the influence of impulses is not merely of academic interest.

The basic unperturbed setting will be that of Section 3.3; that is, there will be an unperturbed steady flow in $\Omega\subset\mathbb{R}^2$ which possesses a heteroclinic manifold. The intention is to quantify the transport across the *broken* heteroclinic manifold under the influence of an *impulsive* perturbation. As has been developed in Section 2.7, this requires thinking of the system in terms of integral as opposed to the more standard approach of impulsive differential equations which augments the differential equation with a set of trajectory resetting expressions [19, 58, 57, 306, 194, 245, 133]. Thus, consider (2.90) once again:

$$x(t)=x(\beta)+\int_\beta^t f(x(\xi))\,\mathrm{d}\xi+\frac{\varepsilon}{2}\sum_{i=1}^n u(\beta,t_i,t)\left[h_i\left(x(t_i^-),t_i^-\right)+h_i\left(x(t_i^+),t_i^+\right)\right], \tag{3.65}$$

where $x\in\Omega$, a two-dimensional open connected set, $\mathcal{J}=\{t_1,t_2,t_3,\dots,t_n\}$ is an increasing set of time values at which the impulses occur, and f and the h_i satisfy the smoothness assumptions of Hypotheses 2.34 and 2.35. As for Section 2.7, the development of this section is based on [37]. Now when $\varepsilon=0$, the system is smooth and thereby can be represented in the form (2.9), i.e., $\dot{x}=f(x)$. A heteroclinic manifold Γ connecting the saddle fixed points a and b (which may be the identical point) is then assumed; this is formalized in Hypothesis 3.5.

Section 2.7 has established that when $\varepsilon\neq 0$ (but is small), the unstable manifold of a can be rationalized instead as an *unstable pseudomanifold*, as illustrated in Figure 2.15. At each time $t\in\mathbb{R}\setminus\mathcal{J}$, the unstable pseudomanifold is a curve in Ω which evolves smoothly in time, but as each time value $t_j\in\mathcal{J}$ is crossed, this curve "resets" itself to a new curve. This therefore defines the unstable pseudomanifold $\tilde{\Gamma}_\varepsilon^u(a)$ of a, which satisfies the property that any $(x(\beta),\beta)\in\Omega\times\mathbb{R}\setminus\mathcal{J}$ that lie on $\tilde{\Gamma}_\varepsilon^u(a)$, if used as an initial condition to (3.65), generates a trajectory $x(t)$ defined for $t\in(-\infty,T_u]\setminus\mathcal{J}$ for any large finite T_u such that $x(t)\to a$ as $t\to-\infty$ (see Definition 2.37). In exactly the same

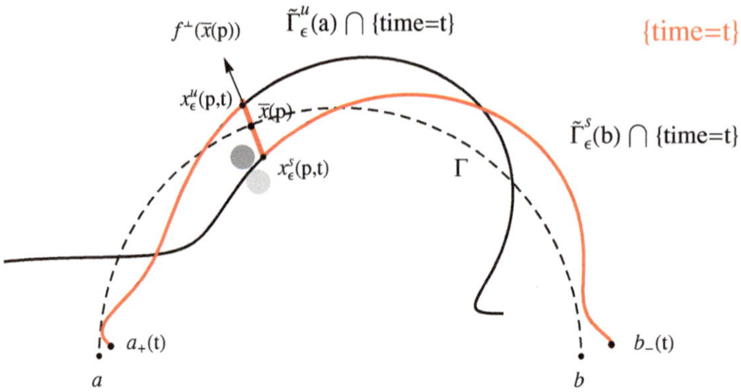

Figure 3.12. *The pseudoseparatrix $\mathcal{Q}(p,t)$ constructed analogously to Definition 3.17 (red curve) using* pseudomanifolds *for the nonautonomously perturbed system (3.65), in a time-slice $t \in [T_s, T_u] \setminus \mathcal{J}$.*

way, b will possess a stable pseudomanifold $\tilde{\Gamma}^s_\varepsilon(b)$ such that any $(x(\beta),\beta) \in \Omega \times \mathbb{R} \setminus \mathcal{J}$ that lie on $\tilde{\Gamma}^s_\varepsilon(b)$, if used as an initial condition to (3.65), generates a trajectory $x(t)$ defined for $t \in [T_s, \infty) \setminus \mathcal{J}$ for any large finite $-T_s$ such that $x(t) \to b$ as $t \to \infty$, as characterized in Definition 2.43). So if T_u and $-T_s$ are sufficiently large, there are t-values in $[T_s, T_u] \setminus \mathcal{J}$ at which *both* $\tilde{\Gamma}^u_\varepsilon(a)$ and $\tilde{\Gamma}^s_\varepsilon(b)$ exist. The reader is referred to Section 2.7 for a description of all hypotheses, notation, and results on such pseudomanifolds, based on which the results of this section rests.

Now, each of the stable and unstable pseudomanifold curves in the t-slice are $\mathcal{O}(\varepsilon)$-close to Γ, and therefore $\mathcal{O}(\varepsilon)$-close to each other. On the other hand, they need not coincide; they may intersect one another (either a finite or an infinite number of times), or not intersect at all. Determining an intersection, and thus a persistent heteroclinic trajectory, relates exactly to determining simple zeros of the Melnikov function $M(p,t) = M^u(p,t) - M^s(p,t)$ with $M^{u,s}$ defined in (2.94) and (2.100), as has been shown in [37]. Finding heteroclinic connections through the usage of a Melnikov function in this situation in which the $\Omega \times \mathbb{R}$ augmented phase space is fractured by the codimension-1 surfaces $t = t_i$ has an interesting connection to determining heteroclinics using a Melnikov approach in a situation in which the autonomous velocity field is discontinuous [59, 95, 230, 133] since such discontinuity surfaces can once again be thought of as codimension-1 fracture surfaces in $\Omega \times \mathbb{R}$. In any event, this section shall not be too concerned with the heteroclinic persistence issue, to which the reader is referred to [37], but rather on how one might define a flux.

The relative closeness of $\tilde{\Gamma}^u_\varepsilon(a)$ and $\tilde{\Gamma}^s_\varepsilon(b)$ in the time-slice t provides exactly the geometry required for defining a flux across the (now broken) separatrix Γ. The picture is therefore exactly the same as that of Figure 3.6 in the smooth situation but requires a more careful interpretation, necessitating the usage of the slightly different Figure 3.12, drawn in a time-slice in $[T_s, T_u] \setminus \mathcal{J}$. Exactly as described in Definition 3.17, the pseudoseparatrix $\mathcal{Q}(p,t)$ in the time-slice t will be formed by proceeding now from $a_+(t)$ along the unstable pseudomanifold $\tilde{\Gamma}^u_\varepsilon(a)$ until it intersects the normal vector to Γ drawn at the location $\bar{x}(p)$, and similarly from $b_-(t)$ along the stable pseudomanifold $\tilde{\Gamma}^s_\varepsilon(b)$ until it too intersects the same normal vector, and then connecting these two curves along the normal vector with a gate.

Does it still make sense to only compute the flux through the gate in this impulsive instance? This requires understanding why the remaining portions of $\mathcal{Q}$, i.e., the portions of the pseudomanifolds, can legitimately be thought of as flow barriers. Consider fluid particles lying on $\tilde{\Gamma}_\varepsilon^u(a)$. In backwards time, all of these particles will get tracked back to the (nonpseudo) unstable manifold of a which coincides with the pseudomanifold when $t < t_1$; see Figure 2.15. Think of two fluid particles, A and B, which are near to $\tilde{\Gamma}_\varepsilon^u(a)$ in some time-slice, but which lie on opposite sides of $\tilde{\Gamma}_\varepsilon^u(a)$, and consider their fate according to the flow (3.65) in backwards time. Now, the continuity of the mapping process as established in Lemma 2.36 ensures that these particles will continue to lie on opposite sides as they are tracked backwards in time. Once in the domain $t < t_1$, when considering their evolution in backwards time, each will get stripped away from a due to the influence of the exponential stretching associated with a's stable manifold. However, since A and B will lie on opposite sides of $\tilde{\Gamma}_\varepsilon^u(a)$, they will get pulled out in *opposite* directions, exponentially. Thus, $\tilde{\Gamma}_\varepsilon^u(a)$ forms a separating curve in each time-slice in the sense that particles on its two sides will be pushed away from $\tilde{\Gamma}_\varepsilon^u(a)$ in backwards time, thereby experiencing exponential repression from a in opposite directions. Particles lying exactly on $\tilde{\Gamma}_\varepsilon^u(a)$, on the other hand, will approach a. Thus, $\tilde{\Gamma}_\varepsilon^u(a)$ is a flow barrier in backwards time. The same argument, of course, works for $\tilde{\Gamma}_\varepsilon^s(b)$ in forwards time. In particular, when $\varepsilon = 0$, since these two pseudomanifolds coincide to form Γ, this is a flow barrier in both backwards and forwards time. This interpretation has not changed in thinking of the perturbation as impulsive.

Now consider the part of $\tilde{\Gamma}_\varepsilon^u(a)$ chosen to form $\mathcal{Q}$ as pictured in Figure 3.12. Consider two groups of particles—the dark and light blobs—*below* $\tilde{\Gamma}_\varepsilon^u(a)$ in Figure 3.12. In their backwards time evolution, they belong to the same "grouping" since they are on the same side of $\Gamma_\varepsilon^u(a)$. They both collapse towards $a_+(t)$ with exponential attraction. Thus, they are similar in backward time. But what happens in forward time? In their continuing evolution, the darker blob will shortly pass through the gate in Figure 3.12, while the lighter blob will not. In the long term the blobs will be on opposite sides of $\tilde{\Gamma}_\varepsilon^s(b)$ and thus will undergo exponential separation, away from $b\_(t)$ as $t \to \infty$. The two blobs evolved from particles which were very near each other in the past, and indeed approached each other with an exponential attraction rate. Going through the gate means that they will be later *repulsed* exponentially. By crossing the gate in Figure 3.12, the dark blob will no longer be repulsed away from b in the lower direction (as it would if $\varepsilon = 0$) but will instead be repulsed from b in the *upper* direction. Thus, it is these blobs of fluid which cross through the gate which represent a transfer of fluid from regions which have the long-term fate of being repelled from b in the lower direction, to the region which is eventually repelled from b in the upper direction. In this sense, the flux of particles across the gate is a sensible measure of transport across the previously impermeable Γ.

Computing the leading-order term of this instantaneous flux is no different from the smooth case. Once again, the tangential movements of the stable and unstable manifolds (pseudomanifolds in this case) become irrelevant, which is fortunate since Section 2.7 was only able to derive the normal displacements. The instantaneous flux, specifically as an area of fluid per unit time crossing the pseudoseparatrix in the direction of $\hat{f}^\perp$, in relation to a gate chosen at $\bar{x}(p)$, is therefore exactly as in the development of Theorem 3.18. The statement associated with the impulsive flow (3.65) can therefore be written as the following theorem.

Theorem 3.42 (instantaneous flux for impulsive systems [37]). *Suppose Hypotheses 3.5, 2.34, and 2.35 are satisfied for the system (3.65). The instantaneous flux $\phi(p,t)$, parametrized by gate position $p \in [-P,P]$ and time $t \in [T_s, T_u] \setminus \mathcal{J}$, across the pseudoseparatrix $\mathcal{Q}(p,t)$ in the direction of $\hat{f}^{\perp}$, is given by*

$$\phi(p,t) = \varepsilon M(p,t) + \mathcal{O}\left(\varepsilon^2\right), \tag{3.66}$$

in which the Melnikov function M is defined for $p \in [-P,P]$ and $t \in [T_s, T_u] \setminus \mathcal{J}$ by

$$M(p,t) = \sum_{i=1}^{n} \left[j_i(p,t) + \int_{t_i}^{t} R_p(t-\xi) j_i(p,\xi) \, \mathrm{d}\xi \right], \tag{3.67}$$

where

$$j_i(p,t) = f^{\perp}(\bar{x}(t_i - t + p)) \cdot h_i(\bar{x}(t_i - t + p), t_i), \; i = 1,2,3,\cdots,n, \tag{3.68}$$

and the resolvent $R_p(t)$ is defined on $\mathbb{R} \setminus \{0\}$ by

$$R_p(t) = \begin{cases} \mathcal{L}^{-1}\left\{ \frac{\hat{F}_p^u(s)}{1-\hat{F}_p^u(s)} \right\}(t) & \text{if } t > 0, \\ \mathcal{L}^{-1}\left\{ \frac{\hat{F}_p^s(s)}{1+\hat{F}_p^s(s)} \right\}(-t) & \text{if } t < 0, \end{cases} \tag{3.69}$$

with

$$\hat{F}_p^u(s) = \mathcal{L}\{\mathrm{Tr}\, Df(\bar{x}(p-t))\}(s) \quad \text{and} \quad \hat{F}_p^s(s) = \mathcal{L}\{\mathrm{Tr}\, Df(\bar{x}(p+t))\}(s). \tag{3.70}$$

Proof. When $\varepsilon = 0$, the trajectory along the unstable manifold of a ($\bar{x}^u$) and the trajectory along the stable manifold of b ($\bar{x}^s$) are identical. This shall be called $\bar{x}(t)$, which satisfies the condition that $\bar{x}(t) \to a$ as $t \to -\infty$ and $\bar{x}(t) \to b$ as $t \to \infty$. Now, consider the point $x_\varepsilon^u(p,t)$ which lies on $\tilde{\Gamma}_\varepsilon^u(a)$ but is along the normal vector to $\bar{x}(p)$. That is, this is the point at which $\tilde{\Gamma}_\varepsilon^u(a)$ and the gate meet in defining the pseudoseparatrix $\mathcal{Q}$, as shown in Figure 3.12. From Theorem 2.38, its displacement from $\bar{x}(p)$ along the normal direction $\hat{f}^{\perp}(\bar{x}(p))$ is given by $\varepsilon M^u(p,t)/|f(\bar{x}(p))| + \mathcal{O}\left(\varepsilon^2\right)$, where

$$M^u(p,t) = \sum_{i=1}^{n} \mathbb{1}_{(t_i,\infty)}(t) j_i(p,t) + \sum_{i=1}^{\max\{j : t_j < t\}} \int_{t_i}^{t} R_p^u(t-\xi) j_i(p,\xi) \, \mathrm{d}\xi,$$

in which

$$j_i(p,t) = f^{\perp}(\bar{x}(t_i - t + p)) \cdot h_i(\bar{x}(t_i - t + p), t_i),$$

and $R_p^u : \mathbb{R}^+ \to \mathbb{R}$ is defined by

$$R_p^u(t) = \mathcal{L}^{-1}\left\{ \frac{\hat{F}_p^u(s)}{1-\hat{F}_p^u(s)} \right\}(t), \quad \hat{F}_p^u(s) := \mathcal{L}\{\mathrm{Tr}\, Df(\bar{x}(p-t))\}(s).$$

Similarly, from Theorem 2.44, the point $x_\varepsilon^s(p,t)$ in Figure 3.12 has a displacement from $\bar{x}(p)$ in the normal direction given by $\varepsilon M^s(p,t)/|f(\bar{x}(p))| + \mathcal{O}\left(\varepsilon^2\right)$, where

$$M^s(p,t) = -\sum_{i=1}^{n} \mathbb{1}_{(-\infty,t_i)}(t) j_i(p,t) + \sum_{i=\min\{j : t_j > t\}}^{n} \int_{t}^{t_i} R_p^s(t-\xi) j_i(p,\xi) \, \mathrm{d}\xi,$$

where $R_p^s : \mathbb{R}^- \to \mathbb{R}$ is defined by

$$R_p^u(t) = \mathscr{L}^{-1}\left\{\frac{\hat{F}_p^s(s)}{1+\hat{F}_p^s(s)}\right\}(-t), \quad \hat{F}_p^s(s) := \mathscr{L}\left\{\mathrm{Tr}\, Df\left(\bar{x}(p+t)\right)\right\}(s).$$

Therefore,

$$[x_\varepsilon^u(p,t) - x_\varepsilon^s(p,t)]\cdot\hat{f}^\perp(\bar{x}(p)) = \varepsilon\frac{M^u(p,t)-M^s(p,t)}{|f(\bar{x}(p))|} + \mathcal{O}\left(\varepsilon^2\right) =: \varepsilon\frac{M(p,t)}{|f(\bar{x}(p))|} + \mathcal{O}\left(\varepsilon^2\right)$$

with the definition $M(p,t) = M^u(p,t) - M^s(p,t)$. This is exactly the framework for rationalizing the flux as given in Theorem 3.18, and thus the flux $\phi(p,t)$ satisfies (3.66). It now remains to compute $M(p,t)$ for $p \in [-P,P]$ and $t \in [T_s, T_u] \setminus \mathscr{J}$, which can be accomplished by the expressions for $M^{u,s}$ above:

$$\begin{aligned}
M(p,t) &= M^u(p,t) - M^s(p,t) \\
&= \sum_{i=1}^{n}\left[\mathbb{1}_{(t_i,\infty)}(t) j_i(p,t) + \mathbb{1}_{(-\infty,t_i)}(t) j_i(p,t)\right] \\
&\quad + \sum_{i=1}^{\max\{j:t_j<t\}} \int_{t_i}^{t} R_p^u(t-\xi) j_i(p,\xi)\,\mathrm{d}\xi - \sum_{i=\min\{j:t_j>t\}}^{n} \int_{t}^{t_i} R_p^s(t-\xi) j_i(p,\xi)\,\mathrm{d}\xi \\
&= \sum_{i=1}^{n} j_i(p,t) + \sum_{i=1}^{\max\{j:t_j<t\}} \int_{t_i}^{t} R_p(t-\xi) j_i(p,\xi)\mathrm{d}\xi - \sum_{i=\min\{j:t_j>t\}}^{n} \int_{t}^{t_i} R_p(t-\xi) j_i(p,\xi)\mathrm{d}\xi \\
&= \sum_{i=1}^{n} j_i(p,t) + \sum_{i=1}^{\max\{j:t_j<t\}} \int_{t_i}^{t} R_p(t-\xi) j_i(p,\xi)\mathrm{d}\xi + \sum_{i=\min\{j:t_j>t\}}^{n} \int_{t_i}^{t} R_p(t-\xi) j_i(p,\xi)\mathrm{d}\xi \\
&= \sum_{i=1}^{n} j_i(p,t) + \sum_{i=1}^{n} \int_{t_i}^{t} R_p(t-\xi) j_i(p,\xi)\,\mathrm{d}\xi,
\end{aligned}$$

since $R_p(t-\xi)$ is equal to $R_p^u(t-\xi)$ in the first integrand since $t > \xi$ and to $R_p^s(t-\xi)$ in the second since $t < \xi$. □

Remark 3.43 (flux function for impulses is continuous). Even though $M(p,t)$ is defined for $t \notin \mathscr{J}$, the expression (3.67) indicates that for any $t_i \in \mathscr{J}$, $\lim_{t\uparrow t_i} M(p,t) = \lim_{t\downarrow t_i} M(p,t)$. Thus, the $\{t_i\}$ consist of removable singularities; if "filled in," M would be continuous in t. The reason for this is that when crossing a jump value t_i, both $x_\varepsilon^u(p,t)$ and $x_\varepsilon^s(p,t)$ get reset according to the same jump map, which according to Lemma 2.36 is continuous. Their relative distance to $\mathcal{O}(\varepsilon)$ turns out to be preserved during this jump map; (3.67) implicitly establishes this fact. This "continuity" of the Melnikov function in t was also observed in an early attempt [28] to rationalize flux under impulses (but restricted to area-preservation).

Example 3.44. Consider the unperturbed system

$$\frac{d}{dt}\begin{pmatrix} x_1 \\ x_2 \end{pmatrix} = \begin{pmatrix} 2x_1 - 2x_1^2 \\ 4x_1x_2 - x_2 \end{pmatrix}. \tag{3.71}$$

It will be left as an exercise to determine that the phase space behavior here is qualitatively identical to that shown in Figure 3.8, and as in Example 3.32, the intention is to quantify the flux across the red flow barrier going from $a \equiv (0,0)$ to $b \equiv (1,0)$. However, in this instance an impulse perturbation will be imposed, and the flux will be associated with a pseudoseparatrix formed from the relevant pseudomanifolds. It shall be noted that $\mathrm{Tr}\, Df = 2 - 4x_1 + 4x_1 - 1 = 1$ for the vector field (3.71) and that the relevant heteroclinic and its velocity evolution is given by

$$\bar{x}(t) = \begin{pmatrix} \bar{x}_1(t) \\ \bar{x}_2(t) \end{pmatrix} = \begin{pmatrix} \frac{e^{2t}}{1+e^{2t}} \\ 0 \end{pmatrix}, \; f(\bar{x}(t)) = \begin{pmatrix} \frac{2e^{2t}}{(1+e^{2t})^2} \\ 0 \end{pmatrix}, \; f^{\perp}(\bar{x}(t)) = \begin{pmatrix} 0 \\ \frac{2e^{2t}}{(1+e^{2t})^2} \end{pmatrix}.$$

Now if a general impulsive perturbation of the form (3.65) is imposed, (3.68) from Theorem 3.42 becomes

$$j_i(p,t) = \frac{2e^{2(t_i - t + p)}}{\left(1 + e^{2(t_i - t + p)}\right)^2} h_{i,2}\left(\frac{e^{2(t_i - t + p)}}{1 + e^{2(t_i - t + p)}}, 0, t_i \right),$$

where $h_{i,2}$ is the second component of the vector h_i. Since $\hat{F}^{u,s}(s) = 1/s$, evaluating both parts of (3.69) gives $R_p(t) = e^t$ for $t \neq 0$. Equation (3.67) then gives the general formula for the flux term under n impulses. To be concrete, take a situation in which $n = 2$, with $t_1 = -1$, $t_2 = 1$, $h_{1,2}(x_1, x_2, t) = x_2 t$, and $h_{2,2}(x_1, x_2, t) = x_1$. Then $j_1(p,t) = 0$ and

$$j_2(p,t) = \frac{2e^{2(1-t+p)}}{\left(1 + e^{2(1-t+p)}\right)^2} \frac{e^{2(1-t+p)}}{1 + e^{2(1-t+p)}} = \frac{2e^{4(1-t+p)}}{\left(1 + e^{2(1-t+p)}\right)^3}.$$

Therefore, from (3.67),

$$\begin{aligned} M(p,t) &= \frac{2e^{4(1-t+p)}}{\left(1 + e^{2(1-t+p)}\right)^3} + \int_1^t e^{t-\xi} \frac{2e^{4(1-\xi+p)}}{\left(1 + e^{2(1-\xi+p)}\right)^3} \, d\xi \\ &= \frac{2e^{4(1-t+p)}}{\left(1 + e^{2(1-t+p)}\right)^3} + \frac{e^{t-p-1}}{4} \left[\frac{5e^{3+3p+t} + 3e^{1+p+3t}}{(e^{2+2p} + e^{2t})^2} - 3\cot^{-1}(e^p) \right. \\ &\quad \left. + 3\tan^{-1}\left(e^{t-p-1}\right) - \frac{1}{2}\operatorname{sech}(p)(1 + \tanh(p)) \right], \end{aligned} \tag{3.72}$$

an explicit expression for the leading-order flux computed by positioning a gate at any position along the former heteroclinic. The behavior of the Melnikov function in (3.72) is shown in Figure 3.13. Positive M relates to flow from below to above Γ. In (a), M's t-variation is shown for several different gate choices p ($p = 0$ would be the midpoint, $x_1 = 1/2$). The flow profile through the leftmost ($p = -2$) gate is seen to gradually flow through the next gate, but with additional accummulating effects. As time progresses, it appears that the flux increases without bound. The theory does not preclude this, since M represents an $\mathcal{O}(\varepsilon)$-term of the flux only for $t \in [T_s, T_u]$. There is, however, a more dominant reason for the increase in *positive* time but not in *negative* time: it is because $\mathrm{Tr}\, Df > 0$ in this case, and the flow is expanding in forward time. The rapid growth in the flux in forwards time, and indeed the rapid decay in backwards time, is therefore no surprise. Figure 3.13(b) shows M's variation with p, at several t-values. Each curve shows the instantaneous flux at varying gates

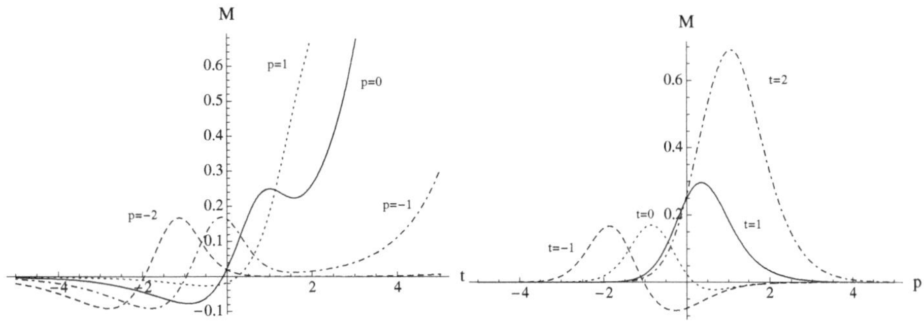

(a) Fixed p (b) Fixed t

Figure 3.13. *The variation of the Melnikov function (3.72) for Example 3.44 with (p, t), demonstrated in each panel with one of the parameters fixed.*

spanning the domain $x_1 \in (0, 1)$, at a fixed instance in time. The peak flux location is initially towards the left but moves towards the right as time progresses. ■

Corollary 3.45 (instantaneous flux for impulses under area-preservation [28, 37]). *Suppose the conditions of Theorem 3.42 are met, and moreover suppose that* $\mathrm{Tr}\, Df = 0$. *Then, the leading-order flux term in (3.67), i.e., the Melnikov function, simplifies to*

$$M(p,t) = \sum_{i=1}^{n} j_i(p,t). \tag{3.73}$$

Proof. In this case, both $\hat{F}_p^u$ and $\hat{F}_p^s$ are zero, and therefore $R_p = 0$. □

Remark 3.46 (formal flux computation for impulses). It has already been argued in Remark 2.42 that the pseudomanifold formulae for impulses are equivalant to those obtained from the *smooth* Melnikov development by the formal substitution of Dirac delta impulses into the relevant theorems (Theorems 2.12 and 2.23), *in the situation in which* $\mathrm{Tr}\, Df = 0$. A simple extension of this is that the *flux formula* in Corollary 3.45 can also be obtained by this process from Theorem 3.18 if $\mathrm{Tr}\, Df = 0$. Tellingly, however, the formal process does *not* work in attempting to derive the formulae in Theorem 3.42 if $\mathrm{Tr}\, Df \neq 0$, necessitating the laborious integral equation approach for determining the normal distances between the pseudomanifolds.

Example 3.47 (double-gyre cont.). Once again, consider the double-gyre previously detailed in Examples 2.25, 2.56, 2.28, 2.33, 2.48, 3.25, and 3.27. Here, the function $q(t)$ shall be taken as $\delta(t)$, as was done in Example 2.48 in determining pseudomanifolds, but the focus shall be the flux across the flow barrier shown in Figure 2.9. Remark 3.46 can be used in this instance since $\mathrm{Tr}\, Df = 0$ for the double-gyre. That is, the result shown in (3.73) could equivalently be found by simply inserting the perturbation, with its Dirac delta impulse, into (3.9). A quicker method to proceed is to use the calculation of the integrand already determined in Example 2.48 in its computation of

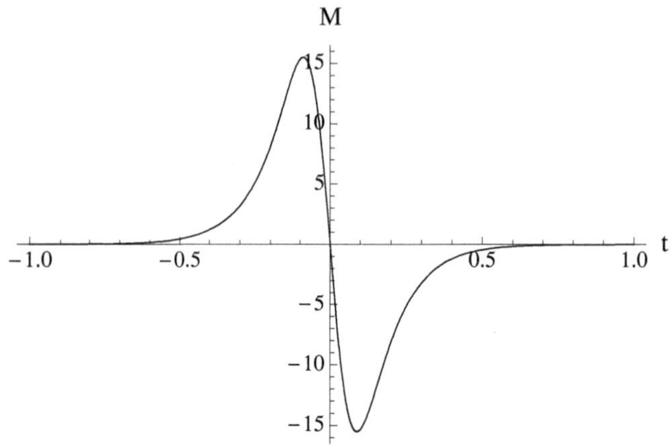

Figure 3.14. *Flux from the left to the right cell in the double-gyre of Figure 2.9 occurring due to an impulsive perturbation associated with $q(t)=\delta(t)$: here, $A=1$, and the gate at $(1,1/2)$ is chosen as described in Example 3.47.*

M^u; therefore, the result will be

$$M(p,t) = -\frac{\pi^3 A^2}{2}\sin\left[4\cot^{-1}\left(e^{\pi^2 A(p-t)}\right)\right], \tag{3.74}$$

which is shown in Figure 3.14 with $A=1$ and $p=0$. Thus, for the gate chosen at the center $(1,1/2)$ of the flow barrier in Figure 2.9, there is a fairly sharp flux to the right cell just before $t=0$ and subsequently a sharp flux to the left just after this t-value. Indeed, the time-scale over which this occurs has size, from (3.74), of $A^{-1}\pi^{-2}$. The flux dies down as $t\to\pm\infty$ and moreover, in spite of the impulsive discontinuity in the perturbation, is smooth (see Remark 3.43). ∎

3.7 ▪ Flux for nonheterolinic flow barriers

In some systems of interest, the transport is to be assessed across a flow barrier *which is not a heteroclinic manifold.* The simplest examples arise in microfluidics in which two different fluids are flowing in a laminar fashion parallel to each other, with a flow interface between them. This interface is itself a flow trajectory but may not be a heteroclinic manifold. Indeed, from the dynamical systems perspective there is nothing distinguished about this trajectory; it is important only because of its characterization as separating fluids; it is a *nonheteroclinic flow barrier.* The material in this section is based on [40]. Thus, in this section, the flow considered is still (3.6), i.e.,

$$\dot{x} = f(x) + \varepsilon g(x,t), \tag{3.75}$$

with the smoothness assumptions of Hypotheses 2.2 and 2.3 included. However, the conditions on the specific flow barrier will be different from Hypothesis 3.5 and will instead satisfy the following hypothesis.

Hypothesis 3.48 (nonheteroclinic flow barrier).

1. The system (3.75) when $\varepsilon=0$ possesses a *separating* trajectory $\bar{x}(t)$ for $t\in\mathbb{R}$, which is the flow interface between two different fluids.

2. The perturbing function g is such that there exist finite p_i and p_f such that $p_i < p_f$ and

$$g(\bar{x}(p),t)=0 \quad \text{for all } p\in(\infty,p_i]\cup[p_f,\infty) \text{ and } t\in\mathbb{R},$$

with $g(\bar{x}(p),t)$ being smooth and bounded for $(p,t)\in(p_i,p_f)\times\mathbb{R}$.

The fact that the perturbation is confined over a specific range of the flow barrier will enable the determination of the perturbed flow barrier, in a way which is not very different from the approach to the heteroclinic barrier case. The motivation for this is once again from microfluidics, in which a common methods to attempt to break the flow barrier is to introduce a velocity perturbation (via cross-channel pumping, boundary vibration, or electromagnetic means) which is limited to a certain region. To set up notation, the reader is referred to Figure 3.15(a), drawn when $\varepsilon = 0$. At a general point $\bar{x}(p)$ on the flow barrier, consider setting up a gate surface perpendicular to the barrier, i.e., in the direction of $f^{\perp}(\bar{x}(p))$ at a general p-value.

Definition 3.49 (steady stable and unstable nonheteroclinic flow barriers). The *unstable* (U) and *stable* (S) *flow barriers* for $\varepsilon = 0$ are, respectively, defined by

$$U := \bigcup_{p\in(-\infty,P_u]} \bar{x}(p) \quad \text{and} \quad S := \bigcup_{p\in[P_s,\infty)} \bar{x}(p), \tag{3.76}$$

where P_u and P_s are any finite values such that $(p_i,p_f)\subset[P_s,P_u]$.

The steady stable and unstable flow barriers overlap in the region $p\in[P_s,P_u]$. Now suppose $\varepsilon\neq 0$. Spatially, the perturbation will only apply in the middle region of the flow interface, and (in a sense to be made precise) one expects the *unsteady* versions of the stable and unstable flow barriers to be $\mathcal{O}(\varepsilon)$-close to (3.76).

Definition 3.50 (unsteady stable and unstable flow barriers). The *unstable flow barrier* for (3.75) when $\varepsilon\neq 0$ at a general time-slice t is defined by

$$\begin{aligned} U_\varepsilon(t) := \bigcup_{p\in(-\infty,P_u]} & \{x_\varepsilon^u(p,t) : x_\varepsilon^u(p,t) \text{ satisfies (3.75)} \\ & \text{with } x_\varepsilon^u(p,t-p+p_i)=\bar{x}(p_i)\}, \end{aligned} \tag{3.77}$$

where the phrase "satisfies (3.75)" means that at each fixed p, $x_\varepsilon^u(p,t)$ as a function of t obeys (3.75). Similarly, the *stable flow barrier* when $\varepsilon\neq 0$ in a time-slice t is defined by

$$\begin{aligned} S_\varepsilon(t) := \bigcup_{p\in[P_s,\infty)} & \{x_\varepsilon^s(p,t) : x_\varepsilon^s(p,t) \text{ satisfies (3.75)} \\ & \text{with } x_\varepsilon^s(p,t-p+p_f)=\bar{x}(p_f)\}. \end{aligned} \tag{3.78}$$

Remark 3.51 (flow barriers are streaklines). The unstable flow barrier (3.79) is a *streakline* [3] of particles flowing through the point $\bar{x}(p_i)$. Similarly, the stable flow barrier (3.81) is a streakline of particles flowing through the point $\bar{x}(p_f)$.

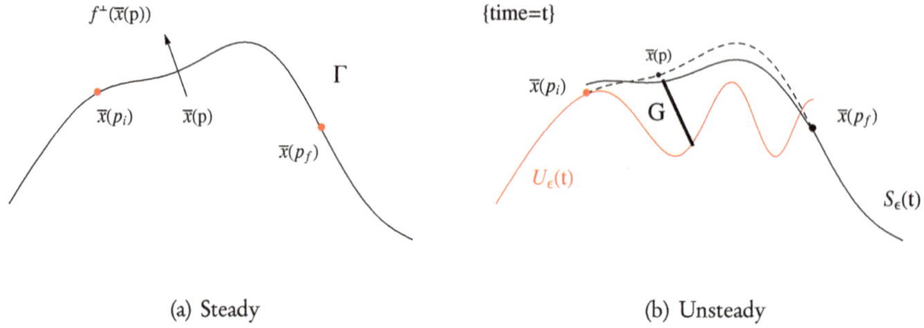

Figure 3.15. *Nonheteroclinic flow barriers are streaklines: (a) steady, and (b) unsteady (stable in black, unstable in red, steady barrier is dashed), as relevant to Definitions 3.49 and 3.50.*

To clarify Remark 3.51 further, let t be fixed; the intention is to characterize the curve $x_\varepsilon^u(p,t)$ which is parametrized by p. The structure of this curve is shown in red in Figure 3.15(b) in the time-slice t. A point parametrized by p in $x_\varepsilon^u(p,t)$ is to be chosen so that it is $\mathcal{O}(\varepsilon)$-close to the point $\bar{x}(p)$ on the unperturbed flow barrier. But $\bar{x}(p)$ would have gone through the point $\bar{x}(p_i)$ at a time $p-p_i$ previous to t, i.e., at a time $t-p+p_i$, when $\varepsilon=0$. Thus, the t-evolution of $x_\varepsilon^u(p,t)$—which is required to remain close to the trajectory passing through $\bar{x}(p)$ at time t when $\varepsilon=0$—will be made subject to the "initial condition" as given in (3.77). When thinking of this at varying p, this generates a curve of points which had each passed through the point $\bar{x}(p_i)$ at a time $p-p_i$ prior to the present instance. Note that in the argument here it has been assumed, consonant with Figure 3.15(b), that $p>p_i$. However, actually the definition in (3.77) includes more points corresponding to $p\leq p_i$: these are points that pass through the point $\bar{x}(p)$ in the *future*. Thus, this is identical to the part of the unsteady flow barrier $\bar{x}(p)$ for $p\leq p_i$. The full unsteady flow barrier, then, consists of the red curve in Figure 3.15(b). The term *streakline* is from fluid mechanics terminology [3] for trajectories passing through a fixed point, which in this case is $\bar{x}(p_i)$.

Analogously, the stable flow barrier definition (3.81) consists of streaklines associated with the point $\bar{x}(p_f)$. The points on $S_\varepsilon(t)$ pass through $\bar{x}(p_f)$ in both backward and forward time, depending on whether $p\lesseqgtr p_f$. This is shown at an instance in time t by the solid black curve in Figure 3.15(b), where the dashed curve is the unperturbed (steady) flow barrier for reference. What is important to note from Definition 3.50 is that for each fixed p, the t-variation of points is *exactly* the time-evolution according to (3.75).

The unsteady flow barriers can in these instances be quantitatively characterized exactly as they were done in the heteroclinic situation. Basically, Theorems 2.12 and 2.23 can be reproduced in an appropriately modified fashion, as shown in the following theorem.

Theorem 3.52 (nonheteroclinic flow barriers [40]). *Consider (3.75) using Hypotheses 2.2, 2.3, and 3.48. Points x_ε^u on the unstable flow barrier satisfy*

$$[x_\varepsilon^u(p,t)-\bar{x}(p)]\cdot\hat{f}^\perp(\bar{x}(p))=\varepsilon\frac{M^u(p,t)}{|f(\bar{x}(p))|}+\mathcal{O}\left(\varepsilon^2\right) \tag{3.79}$$

for $p \in (-\infty, P_u]$ *for any finite* P_u *and* $t \in \mathbb{R}$, *where*

$$M^u(p,t) := \mathbb{1}_{[p_i,P_u]}(p) \int_{p_i}^{\min(p,p_f)} \exp\left[\int_\tau^p \mathrm{Tr}[Df(\bar{x}(\xi))]\,\mathrm{d}\xi\right] f^\perp(\bar{x}(\tau)) \cdot g(\bar{x}(\tau), \tau+t-p)\,\mathrm{d}\tau. \tag{3.80}$$

Similarly, points x_ε^s *on the stable flow barrier obey*

$$[x_\varepsilon^s(p,t) - \bar{x}(p)] \cdot \hat{f}^\perp(\bar{x}(p)) = \varepsilon \frac{M^s(p,t)}{|f(\bar{x}(p))|} + \mathcal{O}(\varepsilon^2) \tag{3.81}$$

for $p \in [P_s, \infty)$ *for any finite* P_s *and* $t \in \mathbb{R}$, *where*

$$M^s(p,t) := -\mathbb{1}_{[P_s,p_f]}(p) \int_{\max(p,p_i)}^{p_f} \exp\left[\int_\tau^p \mathrm{Tr}[Df(\bar{x}(\xi))]\,\mathrm{d}\xi\right] f^\perp(\bar{x}(\tau)) \cdot g(\bar{x}(\tau), \tau+t-p)\,\mathrm{d}\tau. \tag{3.82}$$

Proof. Fix $t \in \mathbb{R}$. The proof for the unstable flow barrier utilizes the fact that, under the construction of (3.77), the condition

$$x_\varepsilon^u(p,\tau) = \bar{x}(\tau - t + p) + \varepsilon x_1(p,\tau,\varepsilon)$$

for *bounded* x_1 is expected for all $\tau \in (-\infty, T]$ for some suitably large time. But this is (2.33) used in the Melnikov-type proof of Section 2.3.1. The development there goes through completely, with the exponential decay not being an issue since g is turned off outside $[p_i, p_f]$. This enables the reduction of the limits $(-\infty, p)$ in (2.26) to what is shown in (3.80), since $g(\bar{x}(p), \cdot) = 0$ for $p \notin [p_i, p_f]$. A similar argument works for the stable flow barrier. □

Remark 3.53. Theorem 3.52 can be extended to express the tangential corrections as in Theorems 2.12 and 2.23, but this shall be avoided here since the focus is on determining *flux* in this situation, for which the normal displacement is sufficient.

The unsteady flow barriers in a time-slice t are shown in Figure 3.15(b). To think of the flux "across" these, bearing in mind that the situation is that there were *two different fluids* on the two sides of the barrier when $\varepsilon = 0$, can be quantified exactly as in the heteroclinic instance described in Section 3.2. At some point $\bar{x}(p)$, where $p \in [P_s, P_u]$ chosen on the unperturbed flow barrier, consider the direction $\hat{f}^\perp(\bar{x}(p))$, and suppose a gate surface $\mathcal{G}$ is drawn in this direction. Now, $U_\varepsilon(t)$ is a flow barrier in the sense that its two sides has different fluids; the same is true for $S_\varepsilon(t)$. Transport between these can occur because of the intermingling of these two flow barriers occurring in the central region between $\bar{x}(p_i)$ and $\bar{x}(p_f)$. By "clipping off" these flow barriers at $\mathcal{G}$, a pseudoseparatrix can be defined in the usual way. Therefore, fluid transport from one region to the other is once again only achieved through crossing $\mathcal{G}$. In the situation pictured in Figure 3.15(b), the transport is instantaneously from the upper to the lower fluid, and this will change in the opposite direction at the time instance at

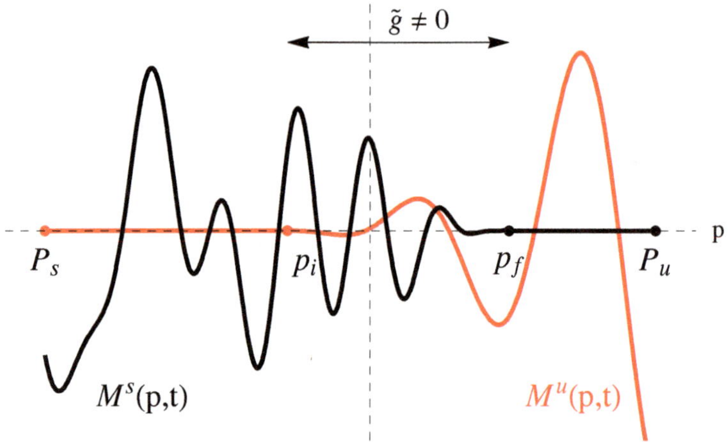

Figure 3.16. *Typical Melnikov functions in a time-slice t for the nonheteroclinic flow barriers; these can be thought of as proxies for the forward-time (red) and backward-time (black) streaklines.*

which the one visible intersection between the flow barriers crosses $\mathscr{G}$. Comparing with Figure 3.6 and Theorem 3.18, it is clear that the flux is given by subtracting x^s from x^u, which to leading-order is equivalent to $M^u - M^s$.

A typical situation illustrating $M^{u,s}$ is pictured in Figure 3.16, at a fixed time t. This might even be thought of as the streakline structure if the flow barrier happened to be a straight line; indeed, the Melnikov development offers exactly this intuition. The perturbation $\tilde{g}$ only applies for $p \in [p_i, p_f]$, and yet the perturbed flow barriers are influenced beyond this. For example, the red unstable barrier—the streakline going through the point $\bar{x}(p_i)$—is represented in this picture as the straight line coming into p_i (indicating that $M^u = 0$ for $p < p_i$), which then gets pushed as a result of the unsteady flow that it experiences, and this change continues beyond p_f. Thus, if a gate surface is drawn at a p value beyond p_f (but less than P_u, which is a finite value below which the perturbation expansion (3.79) can be deemed reliable), there will certainly still be a flux. The fact that gate surfaces can be drawn at p-values in each of the regions $[P_s, p_i]$, $[p_i, p_f]$, and $[p_f, P_u]$ has to be taken into account when computing the leading-order flux $M(p,t) = M^u(p,t) - M^s(p,t)$. When doing so, the cumbersome limits in Theorem 3.52 actually simplify:

Theorem 3.54 (instantaneous flux for nonheteroclinic barriers [40]). *Suppose Hypotheses 2.2, 2.3, and 3.48 are satisfied for the system (3.75). The instantaneous flux $\phi(p,t)$, parametrized by gate position $p \in [P_s, P_u]$ and time $t \in \mathbb{R}$, across the pseudoseparatrix in the direction of $\hat{f}^{\perp}$, is given by*

$$\phi(p,t) = \varepsilon M(p,t) + \mathcal{O}\left(\varepsilon^2\right), \tag{3.83}$$

in which the Melnikov function $M(p,t) := M^u(p,t) - M^s(p,t)$ can be represented as

$$M(p,t) := \int_{p_i}^{p_f} \exp\left[\int_{\tau}^{p} \mathrm{Tr}[Df(\bar{x}(\xi))]\,\mathrm{d}\xi\right] f^{\perp}(\bar{x}(\tau)) \cdot g(\bar{x}(\tau), \tau + t - p)\,\mathrm{d}\tau. \tag{3.84}$$

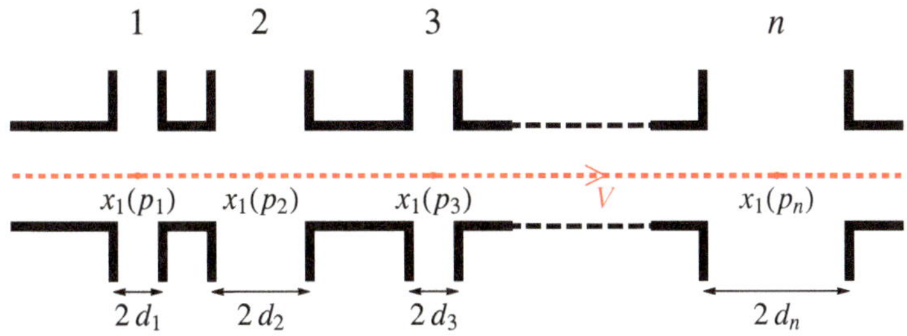

Figure 3.17. *Cross-channel micromixer with n cross-channels: the jth cross-channel is centered at* $x_1 = x_1(p_j)$ *and has width* $2d_j$.

Remark 3.55 (forward-time flux). The instantaneous flux has been characterized in Theorem 3.54 using a gate surface between forward- and backward-time streaklines. The backward-time streakline emanating from $\bar{x}(p_f)$ might be construed by some to violate the "arrow of time" concept, in which case constructing the pseudoseparatrix by involving this streakline would be problematic. If this is so, an alternative to constructing the pseudoseparatrix is to involve not this streakline but the *unperturbed flow barrier* instead. If viewing from the perspective of Figure 3.15(b), this would entail the pseudoseparatrix being formed by the red curve until it reaches the gate $\mathcal{G}$, the gate (extended to meet the dashed unperturbed flow barrier), and then the unperturbed flow barrier, which is the dashed curve which eventually gets covered by the black curve. If adopting this attitude, the flux computation requires only the relative displacement of the forward-time streakline emanating from $\bar{x}(p_i)$ from the unperturbed flow barrier. This is exactly what is given by the Melnikov function $M^u(p,t)$ in (3.80) and is represented by the red curve's relative position to the p-axis in Figure 3.16. Therefore, the *forward-time instantaneous flux* can be defined as

$$\phi_+(p,t) := \varepsilon M^u(p,t) + \mathcal{O}\left(\varepsilon^2\right), \quad p \in (-\infty, P_u], t \in \mathbb{R}. \tag{3.85}$$

Example 3.56 (cross-channel micromixer with constant velocity cont.). A standard device in microfluidics is the cross-channel micromixer [80, 79, 288, 390, 237]. Suppose two incompressible miscible fluids are flowing in a channel, parallel to one another, and tend not to mix across their mutual interface (along the center of the channel). Suppose we are in a regime where it is legitimate to consider that the flow along this is at a constant speed, V. This interface is shown by the red dotted curve in Figure 3.17; this is the x_1-axis where $x = (x_1, x_2)$ is the spatial coordinate. In many microfluidic applications the idea is to try to get the two fluids to mix across the interface by sloshing fluid back and forth in cross-channels. To account for the many possibilities which are available in the literature, a *general* geometry consisting of n cross-channels shall be assumed. The jth cross-channel is centered at the x_1 location $x_1(p_j)$ and is assumed to have width $2d_j$, where for consistency it is necessary that $x_1(p_j) + d_j < x_1(p_{j+1}) - d_{j+1}$. Since flow is at a constant speed along the main channel, the trajectory along the interface can be parametrized by time τ by $\bar{x}(\tau) = (x_1(\tau), 0) = (V\tau, 0)$, and thus $x_1(p_j) = V p_j$.

The standard approach, motivated by experimental evidence [390, 288, 237, 31], is to assume that the flow in the cross-channels takes on a parabolic profile. To achieve

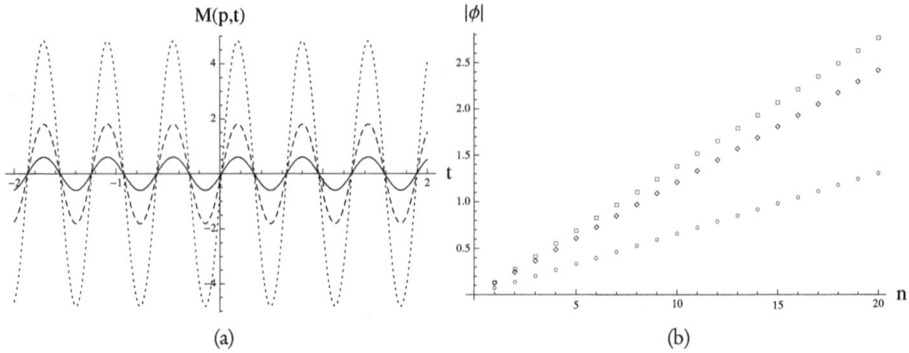

Figure 3.18. *Analysis of the flux (3.89) in a cross-channel micromixer with $V=1$, $v=1$, and $\omega=10$: (a) the flux with gate surface at $p=3$ and with $d=0.1$ with $n=5$ (solid), $n=15$ (dashed), and $n=40$ (dotted), and (b) the amplitude of the flux as a function of the number n of cross-channels, with $d=0.05$ (circles), $d=0.1$ (diamonds), and $d=0.3$ (squares).*

this, set the velocity in the jth cross-channel to be

$$g_j(x_1,x_2,t):=J_j\frac{v_j}{d_j^2}\left[\left(x_1-Vp_j\right)^2-d_j^2\right]\hat{x}_2\cos(\omega t)\,,\ Vp_j-d_j\le x_1\le Vp_j+d_j\,, \tag{3.86}$$

where $\hat{x}_2$ is the unit vector in the x_2-direction, $v_j>0$ is a velocity scale representing the speed at the center of the cross-channel, $\omega>0$ is the frequency of fluid sloshing, and $J_j\in\{-1,1\}$ is a factor which enables the specification of how the cross-channels are operating in relation to one another. For example, if adjacent cross-channels are sloshing fluid exactly out of phase with each other, $J_j=(-1)^j$ can be used; if all the cross-channels are exactly in phase, $J_j=1$ for all j. Therefore, the geometry and velocity specification can account for very general cross-channel configurations.

Since the cross-flow will only occur between the beginning of the first cross-channel and the ending of the nth one, Theorem 3.54 is applicable with $p_i=p_1-d_1/V$ and $p_f=p_n+d_n/V$. The additional simplification that $\mathrm{Tr}\,Df=0$ occurs since the flow is incompressible. Therefore,

$$\begin{aligned}
M(p,t)=&\int_{p_1-d_1/V}^{p_n+d_n/V}V\sum_{j=1}^{n}\mathbb{1}_{[Vp_j-d_j/V,p_j+d_j/V]}(\tau)\frac{J_jv_j}{d_j^2}\left[\left(V\tau-Vp_j\right)^2-d_j^2\right]\\
&\cdot\cos[\omega(\tau+t-p)]d\tau\\
=&\,V\sum_{j=1}^{n}\frac{J_jv_j}{d_j^2}\int_{p_j-d_j/V}^{p_j+d_j/V}\left[\left(V\tau-Vp_j\right)^2-d_j^2\right]\cos[\omega(\tau+t-p)]d\tau\\
=&\,\frac{4V^2}{\omega^3}\sum_{j=1}^{n}\frac{J_jv_j}{d_j^2}\left[\omega d_j\cos\frac{\omega d_j}{V}-V\sin\frac{\omega d_j}{V}\right]\cos\left[\omega\left(p_j-p+t\right)\right],
\end{aligned} \tag{3.87}$$

an *explicit* expression for the flux associated with the forward- and backward-time streaklines for general cross-channels. As is standard when the perturbed velocity is harmonic, the flux expression above depends on the combination $(t-p)$ and not on p and t separately and indeed can be written as a harmonic function of $(t-p)$ which has frequency ω.

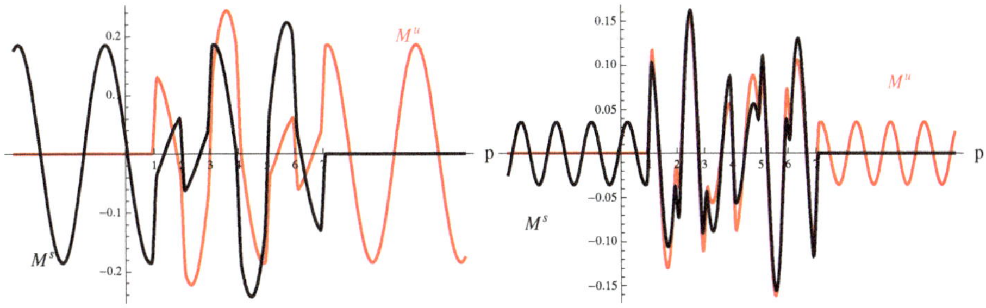

(a) $\omega = 2$ (b) $\omega = 5$

Figure 3.19. *The function M^u (red) and M^s (black) associated with the forward- and backward-time streaklines of the cross-channel micromixer with $V = 1$, $v = 1$, $d = 0.1$, $n = 7$ (cross-channels centered at $p = 1, 2, \ldots, 7$) and $t = 0$.*

The leading-order term of the forward-time flux ϕ_+ (considering only the unstable streakline) is M^u as given in (3.80), which becomes

$$M^u(p,t) = \int_{p_1 - d_1/V}^{\min(p, p_n + d_n/V)} V \sum_{j=1}^{n} \mathbb{1}_{[p_j - d_j/V, p_j + d_j/V]}(\tau) \frac{J_j v_j}{d_j^2} \times \left[\left(V\tau - V p_j \right)^2 - d_j^2 \right] \cos[\omega(\tau + t - p)] \, d\tau \quad (3.88)$$

for $p \geq p_1 - d_1/V$, and $M^u(p,t) = 0$ for p values less than this. Indeed, (3.88) when multiplied by ε/V gives a parametric representation for the streakline at each time t, as expressed in Theorem 3.52, and can be used directly to plot such streaklines. If the choice $p \geq p_i$ is made, the difference between the formulae (3.88) and (3.87) is the fact that the upper limit is the variable p as opposed to the constant p_f. Thus the integration needs to be done only over cross-channel intervals, or interval segments, which are contained in $[p_i, p]$. While an explicit expression can be given in piecewise form, it is not particularly illuminating. On the other hand, if $p > p_f$, the integration is done over all cross-channel intervals, and indeed then $M^u(p,t)$ is given by exactly the expression (3.87). This equivalence is because $M^s(p,t) = 0$ for $p > p_f$, and thus $M(p,t) = M^u(p,t) - M^s(p,t)$ in this domain.

Next, some calculations emerging from the flux and streakline expressions will be performed for the particular geometry of regularly spaced, identical cross-channels with the same velocity amplitude, but with alternating flow in adjacent ones. That is, set $p_j = j$, $d_j = d$, $v_j = v$ and $J_j = (-1)^j$. Under these conditions, (3.87) becomes

$$M(p,t) = \frac{4vd}{\alpha^3} (\alpha \cos\alpha - \sin\alpha) \sum_{j=1}^{n} (-1)^j \cos[\omega(j - p + t)], \quad (3.89)$$

where $\alpha := \omega d/V$ is nondimensional. It is possible to write the above as a harmonic function in the form $A\cos[\omega(t-p)+\beta]$ (left as an exercise), indicating that the flux across the pseudoseparatrix formed by taking portions of the forward-time and backward-time streaklines exhibits periodic behavior in time. In other words, these streaklines intersect infinitely many times. The flux (3.89) is shown in Figure 3.18(a)

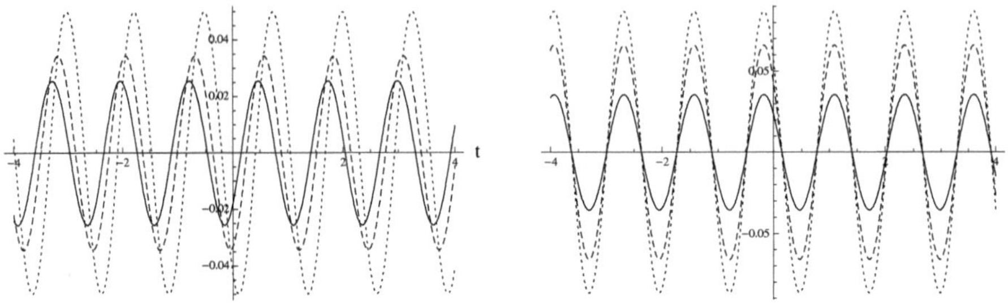

(a) $p=4$ (b) $p=12$

Figure 3.20. *The temporal variation of the forward-time flux M^u of the cross-channel micromixer with $V=1$, $v=1$, $\omega=5$, and $n=7$ (cross-channels centered at $p=1,2,\ldots,7$) at the width values $d=0.1$ (solid), $d=0.2$ (dashed), and $d=0.3$ (dotted), at two different gate locations.*

which illustrates this behavior at different values of n, the number of cross-channels. The amplitude of the flux function increases if the number of cross-channels is increased, which is as expected. A further analysis of how this amplitude depends on both n and the width d of the cross-channels is shown in Figure 3.18(b), indicating that the advantage of increasing the number of cross-channels is almost linear.

The forward-time streakline is associated with M^u in (3.88) which, after simplification based on the same assumptions, is shown by the red curves in Figure 3.19. Here, $n=7$, with the cross-channels centered at $p=1,2,\ldots,7$, each of width $d=0.1$, and the pictures are at time $t=0$. The black curve shows the backward-time streakline (i.e., M^s), and the figures are for $\omega=2$ and $\omega=5$. The intersection patterns between the red and black curves indicates the broken-up flow barrier; the flux associated with these intersections is what has been computed in Figure 3.18. The higher-frequency value has the streaklines aligning together more within the velocity agitation region, implying smaller transport within this region. Moreover, it is clear that once the streaklines have exited the agitation region, they gradually approach periodicity in p; the curved streaklines generated from the agitation region is now simply carried along with the constant flow. Indeed, since $M=M^u-M^s$, it is true that the difference between the red and the black curves in Figure 3.19 is periodic.

If considering only the forward-time flux at one instance in time, then one would think of M^u as a function of t at fixed p. For $p<p_i$ (which is at 0.9 in Figure 3.19), M^u will always be zero. In Figure 3.20, the temporal variation of M^u for the $n=7$ configuration is shown at the middle of the central channel location (at $p=4$) in (a), and at a far-field location ($p=12$) in (b), for several choices of the cross-channel width d. There is no qualitative difference between the time-variation at the two locations, except for the fact that in the far-field streakline is experiencing larger amplitude oscillations owing to the accumulated effect of all the cross-channels. Moreover, the out-of-phase behavior between the forward-time fluxes at different widths which is present in the agitation region due to the impacts of nearby cross-channels has diminished in the far-field. ■

In addition to the calculations similar to the above for different configurations of micromixers, the result of Theorem 3.54, or alternatively Remark 3.55, can be used to good effect in microfluidic situations in which the goal is to *optimize* transport; many examples shall be given in Chapter 4.

3.8 ▪ Transport due to fluid viscosity

This chapter has focused on the influence of a perturbation in the velocity field on the transport across a flow barrier. The perturbation has been prescribed in a generic fashion (and indeed permitted to be discontinuous in time), without necessarily specifying what particular physical phenomenon *causes* this perturbation. In this section, perturbations which can be thought of as the effect of viscosity are considered. That is, the unperturbed flow will be considered as resulting from an inviscid (nonviscous) fluid well-described by the Euler equations, and the perturbation shall be the change in this velocity due to using the Navier-Stokes equations under the condition of small viscosity.

The limit of small viscosity ν is of course a challenging one for many reasons. One is that this is the parameter regime usually associated with large Reynolds numbers, i.e., turbulent flows. However, the Reynolds number UL/ν also depends on velocity and lengthscale parameters U and L, and so small ν by itself does not necessarily imply turbulence. Moreover, in regions far away from boundaries, situations with small ν are often well-approximated by the Euler equations, in which the viscosity ν is specifically set to zero. A prime example of this is the surface ocean well away from continents, which behaves very much like an inviscid fluid. That is, small ν behavior is close to $\nu = 0$ behavior. This is by no means an obvious expectation since the ν in the Navier-Stokes equations can be thought of as a singular perturbation, operating as it does on the highest-order derivative and making the equations parabolic. In general—particularly in boundary layers—the solutions of the Navier-Stokes equations in the limit $\nu \downarrow 0$ do *not* converge to solutions of the Euler equations (where ν is explicitly zero). For situations such as the open ocean, however, the fluid velocity in the limit $\nu \downarrow 0$ does indeed converge to the velocity corresponding to $\nu = 0$. Such results are termed *vanishing viscosity* results, the obtaining of which depends very much on the particular framework of the governing equations, the spatial dimensionality, the initial conditions, the presence/absence of boundaries, any relevant boundary conditions, and the particular function spaces in which the convergence is being investigated [232, 138, 270, 173, 221, 22, 110, 112, 111, 261, 108, 24, 175, e.g.]. In these situations, *viscosity can be thought of as a perturbation.* How would it affect flow barriers?

To set the stage: Suppose a two-dimensional steady constant-density inviscid flow subject to conservative body forces engenders a velocity field $f(x)$ for which Hypothesis 3.5 applies. That is, there exists a one-dimensional heteroclinic manifold Γ, which is a branch of the unstable manifold of the saddle point a and also a branch of the stable manifold of the saddle point b (which may be the same as a), as shown in Figure 3.21. Since the fluid has constant density, it is incompressible. One consequence of this is that $\operatorname{Tr} Df = 0$, and thus it is possible to represent the velocity field f in terms of a streamfunction $\psi^0(x_1, x_2)$ such that the flow is given by

$$\dot{x}_1 = -\frac{\partial \psi^0}{\partial x_2}, \quad \dot{x}_2 = \frac{\partial \psi^0}{\partial x_1},$$

as argued in Remark 2.16, where $f(x)$ is the right-hand side of the above. (Note that the choice of ψ^0 in this section is *negative* that of the choice given in Remark 2.16, in keeping with the convention which is more prevalent in oceanography.) In terms of the perpendicular vector notation given in (2.13), the above unperturbed inviscid flow can be written as

$$\dot{x} = f(x) := \nabla^{\perp} \psi^0(x), \tag{3.90}$$

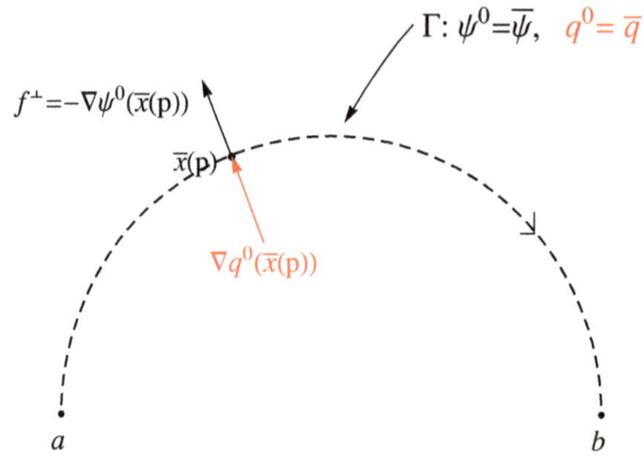

Figure 3.21. *The streamfunction ψ^0 and vorticity q^0 (in red) behavior along the inviscid heteroclinic manifold Γ, which is a contour of each of the ψ^0 or q^0 fields.*

where $x \in \Omega$, a two-dimensional open set. Another consequence of incompressibility is that $\lambda_s = -\lambda_u$ (because $\mathrm{Tr}\, Df = 0$), and thus the stable and unstable manifolds at a saddle fixed point are associated with identical decay rates.

Now, in this section the fact that the velocity field obeys a dynamical equation of interest—in this case the steady Euler equations with conservative body force—shall be assumed. If so, taking the curl of the Euler momentum equation gives the dynamical constraint [3, e.g.]

$$\nabla^{\perp}\psi^0(x)\cdot\nabla\left(q^0(x)\right)=0, \tag{3.91}$$

which simply states that the vorticity

$$q^0(x) := \nabla^2\psi^0(x)$$

is conserved following the flow (3.90). This is since for solutions $x(t)$ of (3.90), the rate of change of vorticity following the flow is

$$\frac{d}{dt}\left[q^0(x(t))\right] = \nabla q^0(x(t))\cdot\dot{x} = \nabla q^0(x(t))\cdot\nabla^{\perp}\psi^0(x(t)),$$

which by (3.91) is zero. Thus, (3.91) is the steady vorticity equation [3, e.g.] for this situation. The conservation means that the $q^0 =$ constant on streamlines of the steady flow (3.90). Since $\psi^0 =$ constant are *also* streamlines, the implication is that q^0 and ψ^0 are functionally dependent. What is important to note is that ∇q^0 will be normal to trajectories and in particular will be normal to the heteroclinic manifold Γ as shown in Figure 3.21. The idea is to determine flux in the direction across Γ given by $f^{\perp}$, and in this case $f^{\perp} = \left[\nabla^{\perp}\psi^0\right]^{\perp} = -\nabla\psi^0$. While it is not clear whether ∇q^0 is in the same direction of $f^{\perp}$, ∇q^0, just like $\nabla\psi^0$, is normal to Γ.

Suppose that the viscous term is now included in the Euler equations; that is, the Navier-Stokes equations are being considered. In this case, the flow becomes unsteady, since the viscosity dissipates energy. If the nondimensional viscous parameter is now designated by ε, the result of taking the curl of the Navier-Stokes momentum equation is

$$\frac{\partial q(x,t)}{\partial t}+\nabla^{\perp}\psi(x,t)\cdot\nabla(q(x,t))=\varepsilon\nabla^2(q(x,t)), \tag{3.92}$$

where the new unsteady streamfunction $\psi(x,t)$ is related to the new vorticity $q(x,t)$ by

$$q(x,t)=\nabla^2\psi(x,t). \tag{3.93}$$

Thus, (3.92) is the vorticity equation, whose left-hand side is the total derivative of the vorticity q following the flow

$$\dot{x}=\nabla^{\perp}\psi(x,t) \tag{3.94}$$

and whose right-hand side quantifies the viscous dissipation of vorticity [3, e.g.]. The viscous flow is *also* required to have a constant density, which enables the perturbed velocity field to also be represented in terms of a streamfunction ψ, which in this case is unsteady. The extra t-derivative on the left-hand side of (3.92) in comparison to (3.91) is because the vorticity is now explicitly time-dependent.

Remark 3.57 (potential vorticity on the barotropic β-plane). While this section has been presented with the understanding that q is the vorticity, the original development in [22, 48, 355] dealt with a more general oceanographic situation in which q is the *potential vorticity*, defined in this case by [310, 354, 314]

$$q(x,t)=\nabla^2\psi(x,t)+f_0+\beta x_2 \tag{3.95}$$

(and consequently $q^0(x)=\nabla^2\psi^0(x)+\beta x_2$), in which (x_1,x_2) represent the local eastward and northward variables at a location on the earth, f_0 is a constant representing the Coriolis effect at the latitude of interest, and β is a constant encapsulating the rate of change of the Coriolis effect with respect to the northward coordinate x_2 due to the earth's curvature. This is the so-called *β-plane approximation*, in which the effect of the earth's rotation and curvature are approximated on a tangent patch of the earth and in which additionally the *barotropic* approximation in which flow is confined to constant pressure surfaces is used to disregard the depth coordinate [310]. Conservation of potential vorticity in the form (3.91), or its near-conservation (in, for example, the form (3.92)) is an important ingredient in many mesoscale ocean modeling approaches [310, 314, 323, 82, 125, 277, 352, 48, 47, 24, 354]. The results to be stated work for the more general choice (3.95) but will be phrased in terms of (standard) vorticity for convenience.

The crucial issue is that the now nonautonomous velocity field as given in (3.94) would need to be thought of as a perturbation of (3.90) to proceed in analyzing the impact of viscosity on the flow barrier Γ. For this to be *genuinely* the case, two factors need to be satisfied, *neither of which is trivial*:

1. ***ε-closeness at any fixed time:*** If t is any fixed time, one would like think of ψ as a perturbation of ψ^0, to enable writing

$$\nabla^{\perp}\psi(x,t)=\nabla^{\perp}\psi^0(x)+\varepsilon\nabla^{\perp}\psi^1(x,t,\varepsilon) \tag{3.96}$$

for the velocity field, where ψ^1 needs to be appropriately bounded and smooth. This can be accomplished, for example, by proving the existence of C_t such that

$$\left|\nabla^{\perp}(\psi(\cdot,t)-\psi^0(\cdot))\right| \leq \varepsilon C_t \tag{3.97}$$

in some high-order norm on Ω, to enable the representation of $\nabla^{\perp}\psi = \nabla^{\perp}\psi^0 + \mathcal{O}(\varepsilon)$, and for this representation to be differentiable sufficiently many times to enable the sorts of computations that have been done in previous sections. *If* Ω *has no boundary*, and if the initial condition is sufficiently smooth, then (3.97) has been established in [232] for the L^2-norm on Ω, and in [22] for the much stronger C^3-norm even in the presence of $\beta \neq 0$ in (3.95). These are special cases of vanishing viscosity estimates [138, 270, 173, 221, 112, 110, 111, 261, 108, 24, 175], but in this case one specifically needs to know that this convergence occurs pointwise in an $\mathcal{O}(\varepsilon)$ way. In any case, the proof of (3.97) in the C^3-norm [22] justifies using an expansion in the form for suitably smooth and bounded ψ^1 *at any finite time.*

2. ***Extension to infinite times:*** In the Melnikov calculations that have been done, it has been necessary to integrate over time from $-\infty$ to ∞, as can be seen, for example, in (3.9). For this to be permitted for expansions of the form (3.96), the constants C_t in (3.97) would need to be uniformly bounded for $t \in \mathbb{R}$. *This is patently impossible*, since the viscous velocity field will decay to zero as $t \to \infty$, whereas there is no such requirement for the inviscid velocity. On the other hand, because of the exponential decays of the integrand in (3.9) as $t \to \pm\infty$, the contributions from infinite t might be considered small under some circumstances, and the Melnikov approach might be legitimized. In this spirit, Sandstede et al. [355] show that the Melnikov approach of writing (3.96) and utilizing a Melnikov integral will continue to be rigorous under the fact that (3.97) is only valid for
$$t \in \left[-\frac{2}{\lambda_u}\left|\ln\varepsilon\right|, \frac{2}{\lambda_u}\left|\ln\varepsilon\right|\right] =: [-L_\varepsilon, L_\varepsilon] =: \mathcal{T}_\varepsilon \tag{3.98}$$
rather than the physically unrealistic $t \in \mathbb{R}$. Indeed, a slightly more general result is established [355]: the Melnikov approach works if the closeness is $\mathcal{O}(\varepsilon^\mu)$, where $\mu \in (1/2, 1]$ and $t \in [-2\mu\left|\ln\varepsilon\right|/\lambda_u, 2\mu\left|\ln\varepsilon\right|/\lambda_u]$. In this case, the distance/flux would have leading-order term ε^μ and error $\mathcal{O}\left(\varepsilon^{2\mu}\right)$. It was later proven by Grenier et al. [175] that, subject to certain stability assumptions on the inviscid streamfunction ψ^0 in (3.91), the closeness of (3.97) was indeed correct in $\mathcal{T}_\varepsilon$, thereby justifying this hypothesis. Under (3.98), it is shown in [355] that the Melnikov function (3.9) and (3.21) remains legitimate with the integration domain restricted to $\tau \in \mathcal{T}_\varepsilon$. However, the development in [355] is necessarily functional-analytic, following the Lyapunov-Schmidt reduction idea of finding the perturbing velocity in the null space of a certain operator (building on [106, 305, 61]). Developing distances/fluxes in a physically and geometrically intuitive way may be better suited for assessing transport implications.

In view of the above two issues, it will be necessary to think of the velocities in a specific way. The ε-closeness of the velocities for the inviscid and viscous situations, as arising from the governing partial differential equations (3.91) and (3.92), can be proven in relation to assuming that the velocities of the two problems satisfy the *same* initial conditions [22]. Given the parabolic term on the right, the viscous equation (3.92) can moreover only be thought of in *forwards* time. Thus, it will be assumed that (3.91) and (3.92) are to have the same initial conditions at the time $-L_\varepsilon$ as given in (3.98). Thus,

$$\nabla^\perp \psi(x,-L_\varepsilon) = \nabla^\perp \psi^0(x), \tag{3.99}$$

with equality occurring also at higher derivatives (as was necessary in the proof presented in [22]). Then, the implications from the above two issues is that the expansion (3.96) is valid in the time domain $\mathcal{T}_\varepsilon = [-L_\varepsilon, L_\varepsilon]$. In other words, the quantity $\nabla^\perp \psi^1(x,t)$ which is the difference between the velocity fields divided by ε, remains bounded in $\mathcal{T}_\varepsilon$. Now, Theorem 1 in [355] shows that if this ψ^1 were extended to $t \in \mathbb{R}$ smoothly, then the difference in the Melnikov function (now defined for $t \in \mathbb{R}$) between any two such extensions is of higher-order. Thus, it is possible to proceed with a formal expansion, bearing in mind that the rigorous results in [355] justify the process in the sense of absorbing the error terms into higher-order terms in ε.

For the moment, let us ignore this finite-time issue and concentrate on obtaining an alternative expression for the flux in terms of vorticity rather than in terms of velocity. Indeed, this approach works for any conserved quantity q^0 in the unperturbed flow; for the argument leading to (3.104), the fact that q^0 is the vorticity is inconsequential. First consider the unstable Melnikov function (2.26) with $f = \nabla^\perp \psi$ and $g = \nabla^\perp \psi^1$. The first simplification is that $\operatorname{Tr} Df = 0$, and therefore

$$\begin{aligned} M^u(p,t) &= \int_{-\infty}^{p} \left[\nabla^\perp \psi^0(\bar{x}(\tau))\right]^\perp \cdot \left[\nabla^\perp \psi^1(\bar{x}(\tau), \tau + t - p)\right] d\tau \\ &= -\int_{-\infty}^{p} \nabla \psi^0(\bar{x}(\tau)) \cdot \nabla^\perp \psi^1(\bar{x}(\tau), \tau + t - p)\, d\tau . \end{aligned} \tag{3.100}$$

The trajectory $\bar{x}^u$ on the unperturbed unstable manifold has been replaced by $\bar{x}$ since Γ is a heteroclinic manifold in which the stable and unstable manifolds coincide. Now define a Melnikov-like function *based on the vorticity* instead by

$$Q^u(p,t) := \int_{-\infty}^{p} \nabla q^0(\bar{x}(\tau)) \cdot \nabla^\perp \psi^1(\bar{x}(\tau), \tau + t - p)\, d\tau . \tag{3.101}$$

The reason for this definition stems from the fact that ∇q^0, just like $\nabla \psi^0$, is a solution to the adjoint of the equation of variations of the unperturbed flow (3.90) [48, 355]. It is such a function which is needed in the integrand in the functional analytical approaches to discover persistent heteroclinics [106, 305, 61, 48, 355]. However, Q^u, unlike M^u, does not directly relate to a normal distance as expressed in Theorem 2.12. On the other hand, it is strongly related, as can be see by examining Figure 3.21. For the sake of argument, suppose that $\psi^0 = \bar{\psi}$ on Γ. Now, q^0 and ψ^0 are functionally dependent because of (3.91), and thus their contours coincide. Let $q^0 = \bar{q}$ on Γ, and let the functional relationship be given by

$$q^0 = F\left(\psi^0\right) \tag{3.102}$$

locally near $\psi^0 = \bar{\psi}$. Thus, $\psi^0 < \bar{\psi}$ on one side of Γ, and $\psi^0 > \bar{\psi}$ on the other, locally. A similar statement occurs with q replacing ψ in the previous sentence. Taking the gradient of (3.102), it is clear that $\nabla q^0 = F'(\psi^0)\nabla \psi^0$, and by evaluating on Γ, one sees that ∇q^0 and $\nabla \psi^0$ as illustrated in Figure 3.21 will have the same directionality if $F'(\bar{\psi}) > 0$, and the opposite if $F'(\bar{\psi}) < 0$. It is *not* possible for $F'(\bar{\psi})$ to be zero or infinite, since both ∇q^0 and $\nabla \psi^0$ are well defined on Γ and only approach zero in a limiting sense at a and b. Now, from (3.101),

$$
\begin{aligned}
Q^u(p,t) &= \int_{-\infty}^{p} \nabla\left[F\left(\psi^0(\bar{x}(\tau))\right)\right]\cdot\nabla^\perp\psi^1(\bar{x}(\tau),\tau+t-p)\,d\tau \\
&= \int_{-\infty}^{p} F'\left(\psi^0(\bar{x}(\tau))\right)\left[\nabla\psi^0(\bar{x}(\tau))\right]\cdot\nabla^\perp\psi^1(\bar{x}(\tau),\tau+t-p)\,d\tau \\
&= F'\left(\bar{\psi}\right)\int_{-\infty}^{p}\left[\nabla\psi^0(\bar{x}(\tau))\right]\cdot\nabla^\perp\psi^1(\bar{x}(\tau),\tau+t-p)\,d\tau \\
&= -F'\left(\bar{\psi}\right)M^u(p,t),
\end{aligned}
\tag{3.103}
$$

since the streamfunction $\psi^0(\bar{x}(\tau))$ takes the same value $\bar{\psi}$ at all points on $\bar{x}(\tau)$. A similar analysis (not shown) can be used to prove that $Q^s(p,t) = -F'(\bar{\psi})M^s(p,t)$, where Q^s is the obvious modification to M^s as given in (2.62), in which ∇q^0 is used instead of $\left[\nabla^\perp\psi^0\right]^\perp$. Combining these statements enables the representation

$$
\begin{aligned}
M(p,t) &= M^u(p,t) - M^s(p,t) \\
&= -F'(\bar{\psi})\left[Q^u(p,t) - Q^s(p,t)\right] \\
&= -F'(\bar{\psi})\int_{-\infty}^{\infty}\nabla q^0(\bar{x}(\tau))\cdot\nabla^\perp\psi^1(\bar{x}(\tau),\tau+t-p)\,d\tau \\
&=: -F'(\bar{\psi})Q(p,t),
\end{aligned}
\tag{3.104}
$$

where (3.104) serves as a definition of $Q(p,t)$, a Melnikov-like function which uses ∇q^0 rather than $f^\perp$ in its integrand. Thus, it is possible to determine a Melnikov function $Q(p,t)$ which is expressed in terms of a conserved quantity q^0 rather than the streamfunction ψ^0.

Remark 3.58 (extension to three dimensions). The expression (3.104) provides for a Melnikov function $Q(p,t)$ derived in relation to an unperturbed conserved quantity q^0 as opposed to an unperturbed velocity field. A similar procedure is possible in three dimensions, for example, in the presense of a Bernoulli function [3] which is the conserved quantity, and when a two-dimensional heteroclinic manifold Γ is given by a contour of this Bernoulli function. For this development, the reader is referred to [50], with the understanding that a full flux interpretation is yet to be developed in three dimensions.

So far, the q^0 in the definition of $Q(p,t)$ was arbitrary. Getting back to reality, it is now thought of as the vorticity of the inviscid flow, and $\nabla^\perp\psi^1$ the velocity arising through the inclusion of viscosity. Since $\nabla^\perp\psi^1$ only remains bounded on $\mathcal{T}_\varepsilon$, it will be necessary to use the definition

$$
Q(p,t) := \int_{-L_\varepsilon}^{L_\varepsilon}\nabla q^0(\bar{x}(\tau-t+p))\cdot\nabla^\perp\psi^1(\bar{x}(\tau-t+p),\tau)\,d\tau \tag{3.105}
$$

instead. Rather than the infinite limits, which represent an extension of $\nabla^\perp\psi^1$ in an artificial way [355], the above representation retains the true $\nabla^\perp\psi^1$. The essential equivalence to infinite time insofar as the Melnikov-type functions are concerned is given rigorously in [355], but the simple intuition is that in the infinite case, the increase of $\nabla^\perp\psi^1$ as $\tau\to\pm\infty$ does not matter since (3.105) keeps the increase controlled sufficiently to ensure that the exponential decay in the term ∇q^0 (at rate $\lambda_u = -\lambda_s$) can deal with it. In essence, [355] establishes that the difference between using the infinite and finite time-domains is of higher-order. The *viscous flux* is given by

$$\phi(p,t) = \varepsilon M(p,t) + \mathcal{O}\left(\varepsilon^2\right) = -\varepsilon F'(\bar{\psi}) Q(p,t) + \mathcal{O}\left(\varepsilon^2\right) \tag{3.106}$$

in terms of this function. Since $F'(\bar{\psi}) \neq 0$, (3.106) shows that $Q(p,t)$ can be used as a quantifier of transport flux across a gate drawn at $\bar{x}(p)$, at time t, in the usual way.

This long discussion has argued that (3.105) can be used for the leading-order flux, modulo a nonzero multiplier $-F'(\bar{\psi})$ which affects the directionality, when the perturbing velocity is allowed to increase in a certain sense. The actual details of the perturbing velocity have yet to be taken into account, and this is the next focus. The perturbation results directly from the inclusion of viscosity in the vorticity equation, as given by (3.92). This dynamical condition shall now be used to try to glean information on the viscosity-induced flux. A formal approach, highlighting intuition, shall be followed; these methods can be made rigorous by the arguments in [48, 355]. Since ψ which satisfies the viscous vorticity dissipating equation (3.92) is assumed to satisfy (3.96), similarly consider expanding the vorticity by

$$q(x,t) = q^0(x) + \varepsilon q^1(x,t,\varepsilon) \tag{3.107}$$

for $t \in \mathcal{T}_\varepsilon$, and note that

$$q(x,t) = \nabla^2 \psi(x,t) = \nabla^2 \psi^0(x) + \varepsilon \nabla^2 \psi^1(x,t,\varepsilon)$$

and hence

$$q^1(x,t,\varepsilon) = \nabla^2 \psi^1(x,t,\varepsilon).$$

Utilizing these expansions in (3.92),

$$\varepsilon \frac{\partial q^1}{\partial t} + \nabla^\perp \psi^0 \cdot \nabla q^0 + \varepsilon \nabla^\perp \psi^0 \cdot \nabla q^1 + \varepsilon \nabla^\perp \psi^1 \cdot \nabla q^0 + \mathcal{O}\left(\varepsilon^2\right) = \varepsilon \nabla^2 q^0.$$

The ε^0 term disappears because of (3.91), and thus

$$\nabla q^0 \cdot \nabla^\perp \psi^1 = \nabla^2 q^0 - \left[\frac{\partial q^1}{\partial \tau} + \nabla^\perp \psi^0 \cdot \nabla q^1 \right] + \mathcal{O}(\varepsilon), \tag{3.108}$$

where now τ is being used as the time variable to enable thinking of (p,t) as being fixed, as is the standard approach in this chapter. The left-hand side contains the integrand of $Q(p,t)$ in (3.105), and thus the above will need be evaluated along trajectories $\bar{x}$ which are the solution to the unperturbed flow (3.90). For such solutions, note that

$$\begin{aligned} \frac{d}{d\tau}\left[q^1(\bar{x}(\tau-t+p),\tau)\right] &= \frac{\partial q^1(\bar{x}(\tau-t+p),\tau)}{\partial \tau} + \nabla q^1(\bar{x}(\tau-t+p),\tau) \cdot \frac{d\bar{x}(\tau-t+p)}{d\tau} \\ &= \frac{\partial q^1(\bar{x}(\tau-t+p),\tau)}{\partial \tau} + \nabla q^1(\bar{x}(\tau-t+p),\tau) \cdot \nabla^\perp \psi^0(\bar{x}(\tau-t+p)), \end{aligned}$$

since $\bar{x}$ is a solution to (3.90). Thus, the term in the square brackets of (3.108) is the *derivative following the flow* of (3.90), also called the total derivative or material derivative. Tossing out the higher-order term as befitting this formal approach, (3.108) therefore gives the dynamical constraint

$$\nabla q^0 \cdot \nabla^\perp \psi^1 = \nabla^2 q^0 - \frac{dq^1}{d\tau} \tag{3.109}$$

for $\tau \in \mathscr{T}_\varepsilon$. If (3.109) is evaluated at (a,τ)—which is a trivial solution to (3.90)—the left-hand side is zero because $\nabla q^0(a)=0$; a is a point at which the q^0 contours cross and is therefore a saddle point of the scalar field q^0. Since the same argument works at b,

$$\nabla^2 q^0(a) - \frac{dq^1(a,\tau)}{d\tau} = 0 \quad \text{and} \quad \nabla^2 q^0(b) - \frac{dq^1(b,\tau)}{d\tau} = 0 \tag{3.110}$$

for all times τ. Evaluating (3.109) at $(\bar{x}(\tau-t+p),\tau)$ gives

$$\begin{aligned}\nabla q^0(\bar{x}(\tau-t+p))\cdot\nabla^\perp\psi^1(\bar{x}(\tau-t+p),\tau) &= \nabla^2 q^0(\bar{x}(\tau-t+p)) - \frac{dq^1(\bar{x}(\tau-t+p),\tau)}{d\tau} \\ &= \nabla^2 q^0(\bar{x}(\tau-t+p)) - \nabla^2 q^0(a) - \frac{d}{d\tau}\left[q^1(\bar{x}(\tau-t+p),\tau) - q^1(a,\tau)\right],\end{aligned}$$

where the second line is using (3.110). If $Q_-(p,t)$ is the integral given in (3.105) but with integration domain $\tau\in(-L_\varepsilon,t]$, integrating the above from $\tau=-L_\varepsilon$ to t yields

$$\begin{aligned}Q_-(p,t) &= \int_{-L_\varepsilon}^{t}\left[\nabla^2 q^0(\bar{x}(\tau-t+p)) - \nabla^2 q^0(a)\right]d\tau \\ &\quad - \int_{-L_\varepsilon}^{t}\frac{d}{d\tau}\left[q^1(\bar{x}(\tau-t+p),\tau) - q^1(a,\tau)\right]d\tau \\ &= \int_{-L_\varepsilon}^{t}\left[\nabla^2 q^0(\bar{x}(\tau-t+p)) - \nabla^2 q^0(a)\right]d\tau - q^1(\bar{x}(p),t) + q^1(a,t)\end{aligned}$$

since at the lower limit $q^1(\bar{x}(\tau-t+p),\tau)\to q^1(a,\tau)$. If similarly $Q_+(p,t)$ is the integral in (3.105) restricted to the domain $[t,L_\varepsilon]$, then

$$Q_+(p,t) = \int_{t}^{L_\varepsilon}\left[\nabla^2 q^0(\bar{x}(\tau-t+p)) - \nabla^2 q^0(b)\right]d\tau + q^1(\bar{x}(p),t) - q^1(b,t).$$

Since $Q(p,t)=Q_-(p,t)+Q_+(p,t)$, it is clear that

$$\begin{aligned}Q(p,t) &= q^1(a,t) - q^1(b,t) \\ &\quad + \int_{-L_\varepsilon}^{t}\left[\nabla^2 q^0(\bar{x}(\tau-t+p)) - \nabla^2 q^0(a)\right]d\tau \\ &\quad + \int_{t}^{L_\varepsilon}\left[\nabla^2 q^0(\bar{x}(\tau-t+p)) - \nabla^2 q^0(b)\right]d\tau.\end{aligned} \tag{3.111}$$

To further represent (3.111) in terms of the inviscid quantities, note that by integrating the first expression in (3.110) from time $-L_\varepsilon$ to a general time τ one gets

$$q^1(a,\tau) = q^1(a,-L_\varepsilon) + \nabla^2 q^0(a)\left[\tau+L_\varepsilon\right].$$

A similar treatment for the second expression yields

$$q^1(b,\tau) = q^1(b,-L_\varepsilon) + \nabla^2 q^0(b)\left[\tau+L_\varepsilon\right].$$

Now, the expansion (3.96) was legitimate in the time-domain $\mathscr{T}_\varepsilon$ under the condition (3.97); i.e., the velocities associated with the inviscid and viscous situations were $\mathcal{O}(\varepsilon)$-close. This closeness condition was established in [22] under the assumption that the inviscid and viscous equations (3.91) and (3.92) had identical initial conditions at time $-L_\varepsilon$ as expressed in (3.99), but also assuming equality of higher-order derivatives of this initial velocity. In other words, assuming equality in a higher Sobolev norm [22]. Under this assumption, $q^1(a,-L_\varepsilon)=q^1(b,-L_\varepsilon)$ in the above two expressions, leading to

$$q^1(a,t)-q^1(b,t)=\left[\nabla^2 q^0(a)-\nabla^2 q^0(b)\right][t+L_\varepsilon].$$

Substituting into (3.111) gives the expression, valid for $t\in[-L_\varepsilon,L_\varepsilon]$,

$$\begin{aligned}Q(p,t)=&\left[\nabla^2 q^0(a)-\nabla^2 q^0(b)\right][t+L_\varepsilon]\\&+\int_{p-t-L_\varepsilon}^{p}\left[\nabla^2 q^0(\bar{x}(\tau))-\nabla^2 q^0(a)\right]\mathrm{d}\tau\\&+\int_{p}^{p-t+L_\varepsilon}\left[\nabla^2 q^0(\bar{x}(\tau))-\nabla^2 q^0(b)\right]\mathrm{d}\tau,\end{aligned}\tag{3.112}$$

where the integral terms have also been modified by the change of variable $\tau-t+p\to\tau$. The above is therefore modulo a factor of $-F'(\bar{\psi})$, the leading-order viscous flux across Γ in relation to a gate at $\bar{x}(p)$, and in a time-slice t, which is positive in the direction $-\nabla\psi^0(\bar{x}(p))$.

The crucial quantity in the leading-order flux is the Laplacian of the inviscid vorticity, i.e.,

$$\nabla^2 q^0=\nabla^2\nabla^2\psi^0.$$

If this is constant along Γ, then $Q(p,t)=0$ above, and the analysis fails. This means that higher-order terms are important in the flux. If this changes along Γ, then there will be a contribution.

Unlike in the purely ordinary differential equations situation in which time-reversal is no issue, the fact that the velocity field is generated from a partial differential equation which was well-posed in forwards time only has led to some interesting consequences. Consider the flux at $t=-L_\varepsilon$, the smallest time for which (3.112) is defined. Then

$$Q(p,-L_\varepsilon)=\int_{p}^{p+2L_\varepsilon}\left[\nabla^2 q^0(\bar{x}(\tau))-\nabla^2 q^0(b)\right]\mathrm{d}\tau,$$

which is generically nonzero. Therefore, although the velocities have been set to be equal to one another at this time according to (3.99), the stable and unstable manifolds do not coincide. The stable manifold at this point in time is identical for both the viscous and the inviscid flow, since one can think of the viscous velocity being extended for $t\in(-\infty,L_\varepsilon]$ to be identical to the inviscid flow, and thus the stable manifold is exactly Γ. However, the unstable manifold is not Γ, since it is generated from fluid trajectories for times greater than $-L_\varepsilon$, which experience the viscous velocity field. The time-variation of Q at a fixed gate location is

$$\begin{aligned}\frac{\partial Q(p,t)}{\partial t}&=\left[\nabla^2 q^0(a)-\nabla^2 q^0(b)\right]+\left[\nabla^2 q^0(\bar{x}(p-t-L_\varepsilon))-\nabla^2 q^0(a)\right]\\&\quad-\left[\nabla^2 q^0(\bar{x}(p-t+L_\varepsilon))-\nabla^2 q^0(b)\right]\\&=\nabla^2 q^0(\bar{x}(p-t-L_\varepsilon))-\nabla^2 q^0(\bar{x}(p-t+L_\varepsilon)),\end{aligned}\tag{3.113}$$

which might in some ways be thought of as a more natural representation than (3.112), in which the impact of integrals and boundary terms appears together.

3.8.1 ▪ Viscous flux for homoclinic flow barriers

The most remarkable reduction happens when $\nabla^2 q^0(a) = \nabla^2 q^0(b)$. This will automatically occur in the homoclinic situation in which $a = b$, which was the more restrictive situation addressed in [48, 47]. An example of this is the eddy illustrated in Figure 1.1(b). The homoclinic assumption could also be applicable when, for example, there are spatially periodic boundary conditions the imposition of which renders two different saddle points identical. Thus, the cat's-eyes flow of Figure 1.1(c), if subject to spatial periodicity in the x_1-direction (eastward direction) for the perturbed flow as well, could be represented in a cylindrical phase space in which the left and right saddle points of the upper cat's-eye eddy are the same. Indeed, the saddle point flanking the lower cat's-eyes will connect back to itself in the cylindrical phase space as well, and thus *all* flow barriers visible in Figure 1.1(c) could be thought of as homoclinic if in the cylindrical phase space. This sort of spatial periodicity is common in many toy oceanographic models [314, 125, 417, 225, 323]. For homoclinic Γ, (3.112) collapses to

$$Q(p,t) = \int_{p-t-L_\varepsilon}^{p-t+L_\varepsilon} \left[\nabla^2 q^0(\bar{x}(\tau)) - \nabla^2 q^0(a) \right] \mathrm{d}\tau \tag{3.114}$$

by combining the integrals and using $\tau - t + p \to \tau$ in the integration variable. Thus, the viscous flux obeys

$$\phi(p,t) = -\varepsilon F'(\bar{\psi}) \int_{p-t-L_\varepsilon}^{p-t+L_\varepsilon} \left[\nabla^2 q^0(\bar{x}(\tau)) - \nabla^2 q^0(a) \right] \mathrm{d}\tau + \mathcal{O}\left(\varepsilon^2\right). \tag{3.115}$$

If L_ε is large, the (p,t)-dependence above is diminished, and indeed goes away completely if $L_\varepsilon = \infty$ is formally inserted. If so, the flux is a constant, which is generically nonzero. This means that the stable and unstable manifold primary segments split apart so that they do *not* intersect. Flux occurs, not in a back-and-forth way due to intersections crossing the gate as time progresses, but unidirectionally. That is, a channel opens up, along which transport occurs. The result would appear as a filament spreading along the channel lying between the stable and unstable manifold, when an Eulerian snapshot of a relevant quantity (say, the vorticity) is considered. This is quite consistent with physical intuition on viscous fingering, or filamentation. Balasuriya and Jones [47] examined the dependence of the size of the viscous flux ϕ on the geometry of a homoclinic eddy (as shown in Figure 1.1(b)) in an attempt to analyze which eddies are more likely to survive in the ocean. If including the finiteness of L_ε but still thinking of it as large because of ε being small, while $Q(p,t)$ would not necessarily be a constant with respect to (p,t), it will be *approximately* a constant, and the generic expectation of viscous filamentation would still persist. This explanation will now be verified for a specific example.

Example 3.59 (Kelvin-Stuart cat's-eyes flow). A classical exact solution to the steady Euler flow equation (3.91) is the streamfunction

$$\psi^0(x_1,x_2) = -\ln\left[c\cosh x_2 + \sqrt{c^2-1}\cos x_1 \right], \tag{3.116}$$

in which $c \geq 1$ is a parameter [384, 365, 343, 69, 337]. The corresponding velocity field is given by

$$\dot{x}_1 = \frac{c\sinh x_2}{c\cosh x_2 + \sqrt{c^2-1}\cos x_1}, \quad \dot{x}_2 = \frac{\sqrt{c^2-1}\sin x_1}{c\cosh x_2 + \sqrt{c^2-1}\cos x_1}. \tag{3.117}$$

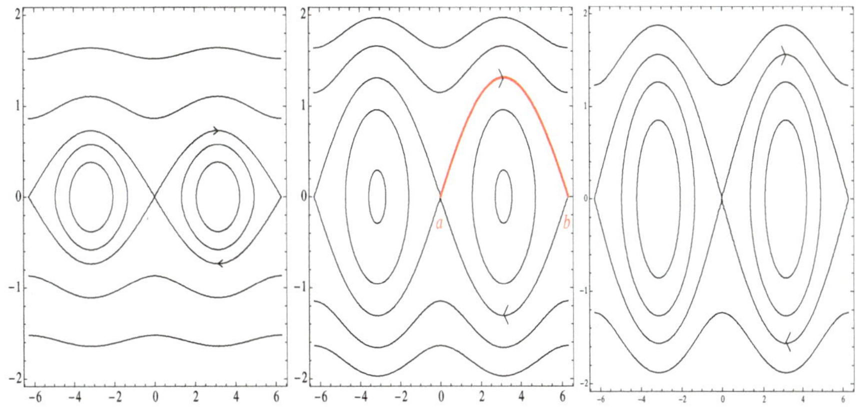

(a) $c = 1.01$ (b) $c = 1.15$ (c) $c = 1.5$

Figure 3.22. *The Kelvin-Stuart cat's-eyes flow (3.117) at different values of the parameter c.*

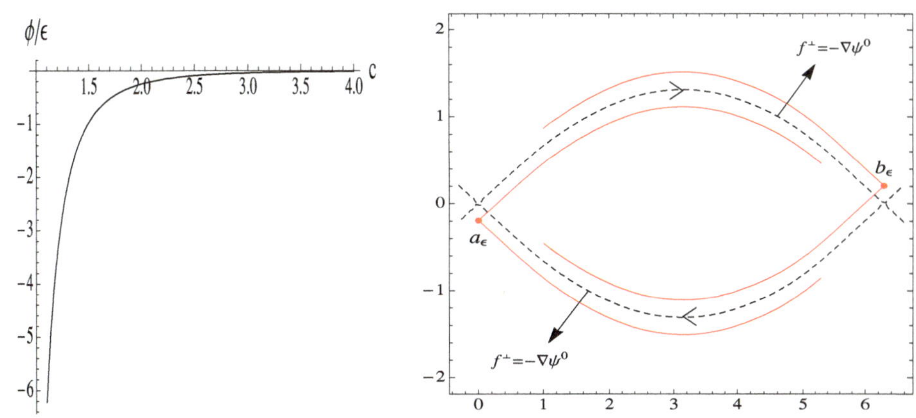

(a) (b)

Figure 3.23. *Viscous dissipation of a Kelvin-Stuart cat's-eye eddy due to viscosity-induced transport flux: (a) its quantification by (3.119), and (b) its consequence.*

When $c > 1$, this flow is associated with a row of cat's-eye eddies; one might think of Figure 1.1(c) as consisting of two rows of such Kelvin-Stuart cat's-eyes eddies. The parameter c is associated with the width of the cat's-eyes, which is $2\cosh^{-1}(1 + 2\sqrt{c^2-1}/c)$. As $c \downarrow 1$, these collapse to the x_1-axis, and then the flow is simply that of a horizontal shear flow. As $c \to \infty$, the width approaches $2\cosh^{-1}3$, and indeed does so very rapidly. Figure 3.22 shows the phase portrait for different values of c on the same scale. Now, the vorticity in this case satisfies [384, 69]

$$q^0 = \nabla^2\psi^0 = -\exp\left(2\psi^0\right) =: F(\psi^0).$$

Let Γ be the upper heteroclinic shown as a red curve in Figure 3.22(b), or the equivalent curve in any of the other figures. This connects the saddle point $a \equiv (0,0)$ to $b \equiv (2\pi, 0)$, and the 2π-periodicity of ψ^0 in (3.117) ensures that all flow quantities are the same when evaluated at a or b. Thus, Γ is effectively homoclinic, and (3.115) can be used to quantify the transport flux occurring as the result of including viscosity via the vorticity equation (3.92). Along Γ, the perpendicular to the flow velocity, $f^\perp = \left[\nabla^\perp \psi^0\right]^\perp = -\nabla \psi^0$, points upwards, i.e., out of the Kelvin-Stuart eddy, and thus flux out of this eddy across Γ will be quantified as positive in (3.115). Since $\psi^0(a) = \psi^0(b) = -\ln\left[c + \sqrt{c^2-1}\right] = \bar{\psi}$, the (x_1, x_2) coordinates on Γ obey $\psi^0(x_1, x_2) = \psi^0(a)$, which gives the condition

$$\cosh x_2 = 1 + \frac{\sqrt{c^2-1}}{c}(1 - \cos x_1)\,. \tag{3.118}$$

Computing $\nabla^2 q^0$ on Γ and utilizing (3.118) to simplify the expression yields

$$\nabla^2 q^0\Big|_\Gamma - \nabla^2 q^0(a) = \frac{8\left[-1 + c\left(c + \sqrt{c^2-1}\right)\right]\left[\cos x_1 - 1\right]}{\left[1 - 2c\left(c + \sqrt{c^2-1}\right)\right]^2}$$

in terms of the x_1 variable. Moreover, on Γ,

$$\dot{x}_1\Big|_\Gamma = \frac{c \sinh x_2}{c + \sqrt{c^2-1}} = \frac{c\sqrt{\left[\frac{\sqrt{c^2-1}(\cos x_1 - 1)}{c} - 1\right]^2 - 1}}{c + \sqrt{c^2-1}}\,.$$

Now using (3.115), approximating the integral with the infinite one, and transferring the variable of integration from τ to x_1, gives the leading-order transport flux as

$$\begin{aligned}
\frac{\phi(p,t)}{\varepsilon} &= -F'(\bar{\psi}) \int_{-\infty}^{\infty} \left[\nabla^2 q^0(\bar{x}(\tau)) - \nabla^2 q^0(a)\right] \mathrm{d}\tau \\
&= -\left(-2e^{2\bar{\psi}}\right) \int_0^{2\pi} \frac{\left[\nabla^2 q^0(x_1, x_2(x_1)) - \nabla^2 q^0(a)\right]}{\dot{x}_1(x_1)} \mathrm{d}x_1 \\
&= \frac{2}{\left(c + \sqrt{c^2-1}\right)^2} \int_0^{2\pi} \frac{8\left(c + \sqrt{c^2-1}\right)\left(c\left(c + \sqrt{c^2-1}\right) - 1\right)(\cos x_1 - 1)\,\mathrm{d}x_1}{c\left[1 - 2c\left(c + \sqrt{c^2-1}\right)\right]^2 \sqrt{\left(\frac{\sqrt{c^2-1}(\cos x_1 - 1)}{c} - 1\right)^2 - 1}}\,,
\end{aligned} \tag{3.119}$$

where the intermediate calculations and the higher-order error term in the viscous parameter ε are omitted for brevity. At this approximation, the flux depends on neither the gate position p nor time t. The variation of ϕ/ε is shown in Figure 3.23(a), which shows that it is always negative, with larger absolute values for smaller c (Kelvin-Stuart cat's-eyes with smaller widths), and that it approaches zero for large cat's-eyes. Larger flux for smaller eddies is entirely consistent with the understanding that smaller coherent structures are influenced in a stronger way than larger ones; the viscous flux destroys them more quickly. The negativity means that the flux is in the *opposite* direction of $f^\perp = -\nabla \psi^0$ (see Theorem 3.18), and thus fluid flows *into* the cat's-eye eddy from above. If viewing an instantaneous quantity such as the vorticity, this would be visible as a tendril or filament emanating along the channel which has opened up between the stable and unstable manifolds; see Figure 3.23(b). This is because the flow along this channel causes the vorticity to equalize somewhat along the channel and in the part of the eddy to which it flows. A similar calculation (not shown) for the lower heteroclinic which formed the inviscid flow barrier for this eddy shows that the flux across it is given by the identical expression (3.119), which then again is negative. Thus, here too a channel opens up, pushing fluid into the cat's-eye, and once again potentially visible as a tendril. The generic expectation of viscosity to create viscous fingering around eddies is therefore well illustrated (in Figure 3.23(b)) when the theory is applied to the Kelvin-Stuart eddies.

As remarked in [48], pictures such as Figure 3.23(b) might appear to contradict fluid incompressibility since fluid is entering the eddy from two directions. However, incompressibility need not be compromised since the eddy will expand with time due to the inflow. That is, the stable and unstable manifolds, while maintaining the relative positioning as shown in Figure 3.23(b), will gradually push outwards. The eddy becoming "larger" is in fact a consequence of its slow dissipation into the surrounding waters. Due to the influx of fluid of smaller vorticity from outside, the vorticity in the eddy becomes diluted, and so while the eddy is becoming larger, the vorticity gradient at its flanges is diminishing. The eddy experiences this viscous dissipation until time L_ε, after which the theory becomes inapplicable (as $t \to \infty$, of course, the entire flow decays to zero velocity, and there are no eddies whatsoever). ∎

Chapter 4
Optimizing transport across flow barriers

If you optimize everything, you will always be unhappy.
—Donald Knuth

4.1 ▪ Flux measures for time-harmonic incompressible flows

In this chapter the focus is on utilizing previous results towards optimizing transport. The application at the back of our minds is that of optimizing transport in microfluidic devices, concerning which the background was discussed in Section 1.3. A frequently considered situation in such devices, both experimental and conceptual, is causing a velocity agitation which is *time-sinusoidal* with the hope of improving mixing within the device [374, 407, 79, 291, 394, 333, 244, 161, 336, 367, 430, 414, 242, 172, 413, 376, 288, 287, 284, 387, 385, 235]. This is natural in particular when using alternating currents in generating the velocity agitation through, for example, attached electromagnets. This is what was defined to be *time-harmonic* in Section 3.4. That is, the flow is governed by (3.33), which is repeated here for convenience:

$$\dot{x} = f(x) + \varepsilon \tilde{g}(x) \cos(\omega t) .$$

Under the extra assumption that the unperturbed flow is incompressible, it is possible to represent f in terms of a streamfunction as explained in Remark 2.16. Utilizing the perpendicular vector notation, $f = -[\nabla \psi_0]^{\perp}$, where ψ_0 is the streamfunction associated with f, as specified in (2.30). That is, the flow is now represented by

$$\dot{x} = -[\nabla \psi_0(x)]^{\perp} + \varepsilon \tilde{g}(x) \cos(\omega t) , \tag{4.1}$$

where it shall be assumed that $\omega > 0$. This chapter addresses the question of how best to choose $\tilde{g}$ and ω to optimize transport across a flow barrier.

Much of the analysis is confined to the situation of *heteroclinic* flow barriers (Section 4.4 deals with nonheteroclinic barriers). Therefore, when $\varepsilon = 0$, suppose the system (4.1) possesses a heteroclinic manifold Γ as illustrated in Figure 4.1. It proves convenient to think of Γ not only as being parametrized by τ in the representation $\bar{x}(\tau)$ of the heteroclinic trajectory but also in terms of arclength. Unlike in (3.19), let

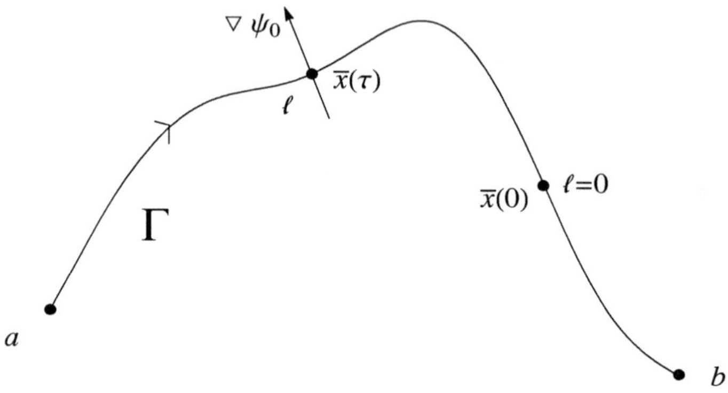

Figure 4.1. *The flow barrier (the heteroclinic manifold Γ) parametrized in terms of signed arclength ℓ as given by (4.2) and in terms of τ.*

ℓ represent not the arclength, but a *signed arclength* along Γ, in the sense that

$$\ell(\tau) = \int_0^\tau |\nabla\psi_0(\bar{x}(\xi))|\, d\xi. \tag{4.2}$$

In (4.2), ℓ is *signed* in that it is permitted to be negative. Basically, the choice of setting $\ell = 0$ at the point $\bar{x}(0)$ has been made, which proves to be convenient in the optimization methods to follow. Points on Γ given by $\bar{x}(\tau)$ for $\tau < 0$ will therefore have negative ℓ values.

The fact that the flux in this case can be represented to leading-order by the Melnikov function $M(p,t)$ has already been established in Theorem 3.18. Moreover, the Melnikov function has been expressed in terms of a Fourier transform in Theorem 3.28 for harmonic perturbations. Theorem 3.28 with f represented by the streamfunction (see also Corollary 3.33) enables the leading-order flux to be written as

$$M(p,t) = |\Lambda(\omega)| \cos\left[\omega(t-p) - \operatorname{Arg}\left[\Lambda(\omega)\right]\right], \tag{4.3}$$

in which

$$\Lambda(\omega) := \mathscr{F}\{\lambda(t)\}(\omega) \quad \text{where} \quad \lambda(t) := \nabla\psi_0(\bar{x}(t)) \cdot \tilde{g}(\bar{x}(t)). \tag{4.4}$$

Observe that the Melnikov function is itself harmonic in $(t-p)$. This harmonic function has an amplitude and phase given, respectively, by the amplitude and argument of the complex number $\Lambda(\omega)$.

One consequence of (4.3) is the presence of infinitely many zeros with respect to p, a signature usually associated with chaotic transport. In particular, since the Melnikov function as a function of p (at fixed t) shows the topology of intersections between the stable and unstable manifolds in the time-slice t, this represents exactly the classical situation of Hypothesis 3.2. When thinking in relation to the Poincaré map of period $T = 2\pi/\omega$ (the period of the perturbation) being applied to (4.1), notice that the replacement $t \to t + 2\pi/\omega$ in (4.3) leaves it invariant. An intersection point which was at q in the time-slice t would progress to $P_t(q)$ in the time-slice $t + 2\pi/\omega$, representing the fact illustrated in Figure 3.2 that it "jumps" across one intersection point to reach the next under each iteration of the Poincaré map. The jump across another intersection point occurs since there is also an intermediate zero; this represents the other heteroclinic solution (with respect to the Poincaré map) as stated in Hypothesis 3.2(b).

Theorem 4.1 (lobe areas and average flux in incompressible harmonic situations [341, 26, 27, 35]). *Consider the flow (4.1), and suppose the intersection pattern of the stable and unstable manifolds Γ splits into is drawn in the time-slice t. If $\Lambda(\omega) \neq 0$, then this results in a picture qualitatively similar to Figure 3.2 with the following features:*

(a) *All lobes have equal areas to $\mathcal{O}(\varepsilon)$.*

(b) *This lobe area is given by*

$$\textit{Lobe Area} = \varepsilon \frac{2|\Lambda(\omega)|}{\omega} + \mathcal{O}(\varepsilon^2). \tag{4.5}$$

(c) *The average flux is given by*

$$\textit{Average Flux} := \frac{\textit{Lobe Area}}{\textit{Period}} = \varepsilon \frac{|\Lambda(\omega)|}{\pi} + \mathcal{O}(\varepsilon^2). \tag{4.6}$$

Proof. The fact that the lobe area of a lobe flanked by the two zeros p_1 and p_2 of the Melnikov function are given by (3.31) has already been discussed. Suppose p_1 and $p_2 = p_1 + \frac{\pi}{\omega}$ are two adjacent zeros of (4.3) at fixed t. Then,

$$\begin{aligned}\text{Lobe Area} &= \varepsilon\,|\Lambda(\omega)| \int_{p_1}^{p_1+\pi/\omega} |\cos[\omega(t-p) - \operatorname{Arg}(\Lambda(\omega))]|\,\mathrm{d}p + \mathcal{O}(\varepsilon^2) \\ &= \varepsilon\,|\Lambda(\omega)| \int_{-\pi/(2\omega)}^{\pi/(2\omega)} \cos[-\omega p]\,\mathrm{d}p + \mathcal{O}(\varepsilon^2),\end{aligned}$$

where this equality arises since the absolute value of the cosine function, when integrated between adjacent zeros, is the same wherever the zeros are chosen. One may as well shift the integral to a convenient location. This of course can be easily integrated to give (4.5). Since the same result would be obtained whatever the value of p_1, all lobe areas are equal to $\mathcal{O}(\varepsilon^2)$. The average flux expression is obtained by dividing the lobe area by the time taken in each iteration of the Poincaré map, *when there is a well-defined lobe area which is transported* [342, 341] (see also Hypothesis 3.2). In this case, the lobe area is well-defined to leading-order, and (4.6) results. □

Remark 4.2. The discussion of the classical situation in Section 3.1 had the requirement that the Poincaré map be area-preserving. This gave the fact that the lobes which got mapped to one another had areas that were not just equal to $\mathcal{O}(\varepsilon)$, but were equal to one another. The assumptions of Theorem 4.1 are slightly weaker: while the dominant velocity $-[\nabla\psi_0]^\perp$ is associated with area-preservation, it was not assumed that $\tilde{g}$ had to also be area-preserving. If this assumption, as befitting realistic fluid flows, was also included, then the equality of lobe areas—and not just to $\mathcal{O}(\varepsilon^2)$—would result.

Example 4.3 (double-gyre cont.). The double gyre was examined in Example 3.27 to determine lobe areas, and previously in Example 3.25 to find the flux function, when the standard time-periodic form $q(t) = \sin \omega t$ was used. These calculations can both be recovered by using the Fourier transform representation $\Lambda(\omega)$. The function whose Fourier transform needs to be computed in this case turns out to have been calculated in (2.67) in Example 2.25. Examining (2.67) and the next displayed equation tells us that

$$\lambda(t) = -\frac{\pi^3 A^2}{2} \sin\left[4\cot^{-1} e^{\pi^2 At}\right] = -\pi^3 A^2 \tanh\left(\pi^2 At\right) \operatorname{sech}\left(\pi^2 At\right).$$

The Fourier transform of this is explicitly computable as

$$\Lambda(\omega) = i\omega \operatorname{sech} \frac{\omega}{2\pi A} \quad \text{and so} \quad |\Lambda(\omega)| = \omega \operatorname{sech} \frac{\omega}{2\pi A}.$$

Utilizing the phase and the amplitude of the above expression in the Fourier transform flux formula (4.3) gives

$$M(p,t) = \omega \operatorname{sech} \frac{\omega}{2\pi A} \cos\left[\omega(t-p) - \frac{\pi}{2}\right] = \omega \operatorname{sech} \frac{\omega}{2\pi A} \sin\left[\omega(t-p)\right],$$

exactly the answer obtained in (3.29). Applying the lobe formula (4.5) also recovers the previously obtained lobes measure (3.32). If optimizing the flux was the goal, one needs to maximize $|\Lambda(\omega)|$; as also discussed in Example 3.27, this occurs by choosing frequencies ω satisfying a transcendental equation from which $\omega \approx 7.53781A$. ∎

An alternative representation of the Melnikov function in terms of arclength ℓ along the unperturbed flow barrier Γ was given in Corollary 3.13. Under the simplifications of incompressibility and time-harmonicity, and with the *signed* arclength definition (4.2) instead, (3.18) reduces to

$$M(p,t) = \int_\Gamma \tilde{g}^n\left(\bar{x}(\tau(\ell))\right) \cos\left[\omega\left(\tau(\ell) + t - p\right)\right] d\ell, \tag{4.7}$$

where $\tau(\ell)$ is the inverse of the mapping (4.2). The quantity $\tilde{g}^n$ is the component of $\tilde{g}$ in the direction of $f^\perp = \left(\left[-\nabla\psi^0\right]^\perp\right)^\perp = \nabla\psi_0$, that is, normal to Γ. The expression (4.7) provides nice intuition on how such a perpendicular velocity influences the flux. However, this is a genuinely Lagrangian flux which somehow quantifies the fact that the stable and unstable manifolds are moving. A purely Eulerian flux definition would simply have a $\cos(\omega t)$ as opposed to the more complicated cosine expression, since the local perpendicular velocity component at ℓ at the instance in time t is $\tilde{g}^\perp(\ell)\cos\omega t$. The "correct" expression in (4.7) encapsulates—in a not very obvious fashion—the fact that points on the manifolds are moving, and a Lagrangian definition is in use for the instantaneous flux. Now, the expression (4.7), since it is equivalent to (4.3), must provide a different way of defining $\Lambda(\omega)$ rather than through a temporal Fourier transform. It shall be left as an exercise to check that

$$\Lambda(\omega) = \int_\Gamma \tilde{g}^n(\ell) e^{-i\omega\tau(\ell)}\, d\ell \tag{4.8}$$

works as a *direct* definition of $\Lambda(\omega)$ to be used in (4.3). Here the abuse of notation $\tilde{g}^n(\ell) = \tilde{g}^n\left(\bar{x}(\tau(\ell))\right)$ is used to indicate that g^n needs to be evaluated at a point parametrized by ℓ. The formulation (4.8) highlights the spatial velocity agitation in terms of a *location*, which is more intuitive than the representation in terms of time τ

via $\bar{x}(\tau)$. Under the choice (4.2), moreover, it is is necessary that $\tau(\ell)$ in (4.8) has been chosen such that $\tau(0)=0$.

For the situation of introducing a velocity agitation to optimize transport across Γ, it seems that there are three possible methods for quantifying transport:

1. The instantaneous flux function whose leading-order term is (4.3).
2. The lobe area given by (4.5).
3. The average flux given by (4.6).

What is pleasing to note is that when considering the leading-order terms in *all* these three methods, it is clear that the term $|\Lambda(\omega)|$ is the crucial quantity. This is particularly so when considering fixed ω and choosing $\tilde{g}$ to optimize flux, as in Section 4.3. If considering the reverse question of fixed $\tilde{g}$ but choosing the optimum frequency ω, more care is needed in this comparison since the lobe area equation (4.5) has an ω in the denominator, indicating that the lobes might be made larger by choosing ω smaller. This is indeed the case (subject to the complication of the numerator's dependence on ω). However, in this situation considering lobe areas as a flux measure is not reasonable, since by the turnstile argument a lobe area only gets transported over the time period $2\pi/\omega$, which will itself diminish as ω gets smaller. The average flux (4.6) continues to be a good measure. Therefore, in either of these optimization regimes, one can choose $|\Lambda(\omega)|$ as the entity to optimize to maximize transport across a fluid interface in a microfluidic device.

4.2 ▪ Optimal frequency

Time-harmonic velocity agitations are the norm in microfluidic mixing attempts [374, 407, 79, 291, 394, 333, 244, 161, 336, 367, 430, 414, 242, 172, 413, 377, 288, 287, 284, 387, 385, 235]. The question of whether there is an *optimum* frequency for achieving maximum mixing has intrigued many researchers, and there is strong experimental [244, 161, 336, 367, 414, 242, 413, 376, 387, 272] and numerical [288, 244, 161, 336, 414] evidence for its presence. These studies involve a laborious investigation of many frequency (or Strouhal number) values, after which some mechanism for quantifying the *global* mixing is used for comparing values. In contrast, there are some studies which seem to indicate that the mixing increases [333, 79, 284] or decreases [288, 385]. This is not inconsistent with the presence of an optimal mixing frequency, since it is possible that the values obtained in [333, 79, 284, 288, 385] were taken over regions which did not include a maximum, that is, by sampling a nonmonotonic function in a monotonic regime. Rom-Kedar and Poje [342] argue the necessity of there being an optimal frequency, since mixing is zero when the frequency is zero, and must diminish towards zero at large frequencies because the lobes created have much smaller sizes.

In the context of the flow (4.1), the previous section has shown that $|\Lambda(\omega)|$ provides an excellent quantifier of transport, consistent with both the instantaneous and average flux definitions. The transport is guaranteed to be chaotic if $a=b$, but will usually also be so if $a \neq b$ but the lobes are confined to a region, which would make them wrap around and re-enter the heteroclinic tangle region, effectively resulting in horseshoe type dynamics and hence chaos in the sense of the Smale-Birkhoff Theorem [69]. The decay as $\omega \to \infty$ argued by [342] is recovered by the fact that the amplitude of the Fourier transform decays to zero by the Riemann-Lebesgue Lemma [382, e.g.]. In the limit $\omega \to 0$, however, $|\Lambda(\omega)|$ is *not* guaranteed to decay to zero. This is since when $\omega=0$, the perturbation in (4.1) has no time-dependence, and hence a nonautonomous viewpoint is not necessary. This precludes transverse intersections between perturbed stable and unstable manifolds since each of these manifolds would represent a trajectory in $\mathbb{R}^2$, and trajectories are not allowed to cross. The conclusion—implicit in the arguments in [342]—is that the manifolds will continue to coincide, thereby preventing any transport across them. If so, of course $|\Lambda(0)|$ would be zero. However, there is another possibility: the stable and unstable manifolds split apart such that they do *not* intersect at all. This occurs, for example, when the Duffing oscillator is considered with no forcing and no damping, and then damping is turned on as a small parameter (see Example 2.21). So when considering the modification to Figure 2.7, the unstable manifolds emanating from the origin will wrap around, initially being close to the unperturbed heteroclinic as shown by the red curve, but then will slowly spiral into the nontrivial equilibria located at $(\pm 1, 0)$. The initial part of this behavior is shown in Figure 2.8(a). The corresponding stable manifold will go outside this curve. Consider creating a pseudoseparatrix with a gate located along the x_1-axis, which creates a boundary between a closed region (interior to the pseudoseparatrix) and outside it. Since the unstable manifold will lie inside the stable manifold, there will be a flux into this region. Thus, even though there is no time-variation, there is still a flux. This flux is sign-definite (i.e., unidirectional) with no time-variation since the system is autonomous. This will be reflected in the theory by the fact that $\Lambda(0)$ is *not* zero. It must be accepted, though, that the flux associated with $\omega = 0$ does *not* cause chaotic transport and is a purely autonomous, unidirectional flux. In this sense, the statement by Rom-Kedar and Poje [342] for there being zero *chaotic* flux when $\omega = 0$ is not contradicted.

Given the possibility of $\Lambda(0)$ not being zero, can the presence of an optimal frequency for maximum transport still be argued? If $|\Lambda(\omega)|$ in the domain $\omega \in [0, \infty)$ does indeed take its maximum at $\omega = 0$, then the flux behavior for $\omega \in (0, \infty)$ (i.e., when the perturbed flow in (4.1) is genuinely nonautonomous) does not possess a global maximum. On the other hand, one can reach the conclusion that flux can be enhanced by choosing smaller and smaller frequencies, which is still valuable information. In the most generic situation in which the maximum of $|\Lambda(\omega)|$ does not occur at $\omega = 0$, however, there will be a genuine value $\omega^* \in (0, \infty)$ corresponding to a maximum; this is the frequency one must choose if aiming for maximum cross-interface transport in the microfluidic device. Both of these situations will be illustrated in the examples to follow.

The idea of using $|\Lambda(\omega)|$ to determine optimal frequency was introduced in [31] and is further discussed in [36]. Its usage shall be illustrated via several examples.

Example 4.4 (Cross-channel micromixer with T-mixer velocity). In Example 3.21, the situation in which two fluids came into a channel and then flowed along it with a fluid interface in between was considered. The particular strategy of shaking the device, in the direction transverse to the main flow, was considered as a mechanism for generating transport across the flow interface shown in Figure 3.7. To phrase already known facts from Example 3.21 in current terminology, $\tilde{g} = \hat{x_2}$, and $\nabla \psi_0(\bar{x}(t)) = -\dot{x}_1(t)\hat{x_2} = -L^2 e^{-Lt}/(1+e^{-Lt})^2 \hat{x_2}$, where $\hat{x_2}$ is the unit vector in the x_2-direction. Thus, $\lambda(t) = -L/\left(1+e^{-Lt}\right)$, whose Fourier transform is

$$\Lambda(\omega) = -\pi\omega \operatorname{cosech} \frac{\pi\omega}{L}.$$

(This can be used in conjuction with (4.3) as an alternative method for deriving the flux function (3.24).) For $\omega \geq 0$, $|\Lambda(\omega)|$ monotonically decreases, implying that for greater cross-interface transport, one needs to shake the apparatus as slowly as possible. This

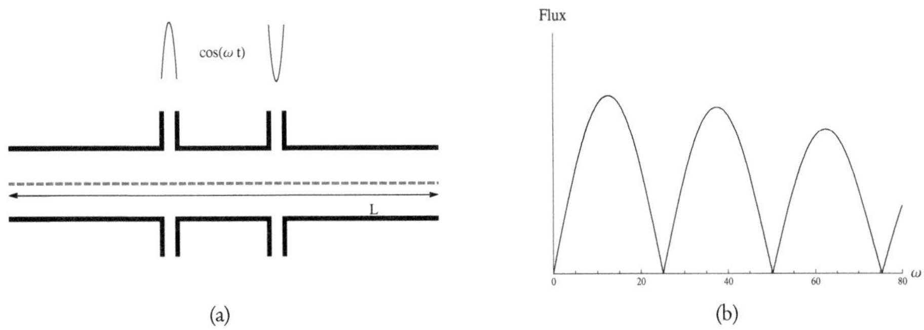

Figure 4.2. *(a) Schematic of cross-channel configuration by Bottausci et al. [79], and (b) the frequency-dependence of its flux, as addressed in Example 4.5.*

might be slighly counterintuitive, but the reason for this is that if the shaking is slow, that permits lobes which are elongated in the x_1-direction to occur, as it takes a longer time to "close off" the lobe. These are therefore long filaments, and indeed diffusion can help in breaking these up since they are filamented. Quicker shaking makes these lobes have small extent in the x_1-direction; although more lobes are created, they are very tiny, and the transported area per unit time across the flow interface is small. In this situation, to truly determine the best optimal frequency, one would require a more sophisticated analysis which is able to quantify the role of diffusion in a more concrete sense. The purely advective mechanism indicates that vanishingly small frequencies are best. ∎

Example 4.5 (Bottausci et al. [79] cross-channel micromixer). Consider fluid injected into a main channel, such as the T-mixer as shown in Figure 3.7. As argued there, the two fluids tend not to mix across the separating interface shown by the red dashed line, with flow velocity v_m along the central interface. Suppose flow across this interface is to be enhanced by introducing cross-channels and sloshing fluid back and forth; this is a *cross-channel micromixer* [80, 79, 288, 390, 237]. Consider the particular experimental set up of Bottausci et al. [79], with two syringes pumping fluid into two cross-channels as shown in Figure 4.2(a), modulated by a term $\cos \omega t$. Figure 4.2 is drawn to scale based on the dimensions of the device of Bottausci et al. [79]: the main channel has length about 1400 μm, and the cross-channels' width is about 1/28th of this. Thus, Figure 4.2 supposes a main channel going from $x_1 = 0$ to L, with the side channels centered at $x_1 = 21L/56$ and $35L/56$, each with width $L/28$. Moreover, following experimental [288, 390, 80] and standard theoretical fluid mechanics considerations, suppose the flow in these cross-channels is modeled by a parabolic profile with maximum value $v_c \ll v_m$, and following [79], suppose the flow in these channels is exactly out of phase with each other, as shown in Figure 4.2(a).

To apply the theory, one needs to know the function $\lambda(t)$ in (4.4) along the flow interface. The parabolic profile gives the $\tilde{g}$, which is zero except for $x_1 \in [10L/28, 11L/28] \cup [17L/28, 18L/28]$. The value of $\nabla\psi_0$ is the velocity along the centerline of the main channel in the absence of the cross-channel agitation. One can use a particular model such as that for the T-mixer (Example 3.21) to estimate this, but since the cross-channels are sufficiently far away from the main channels ends, a reasonable assumption is that the speed is approximately constant, at a value v_m, in this entire region of interest (this approximation will be addressed in more generality in Section 4.4). It is not necessary to specify the velocity in the main channel outside of the central interface; since the theory is perturbative in nature, it does not care whether one uses a parabolic flow profile in the main channel or not. In contrast, the flow profile in the cross-channels *does* affect $\lambda(t)$. If time 0 is considered at the exact center of the main channel (at $x_1 = L/2$), then

$$\lambda(t) = v_m v_c \left\{ \left[1 - \left(\frac{56 v_m t}{L} - 21 \right)^2 \right] \mathbb{1}_{\left[-\frac{4L}{28 v_m}, -\frac{3L}{28 v_m}\right]}(t) + \left[\left(\frac{56 v_m t}{L} - 35 \right)^2 - 1 \right] \mathbb{1}_{\left[\frac{3L}{28 v_m}, \frac{4L}{28 v_m}\right]}(t) \right\}.$$

While $\Lambda(\omega)$ can be determined explicitly, it is not particularly illuminating. Instead, $|\Lambda(\omega)|$ is shown in Figure 4.2(b) for the choice of parameters $v_m = 1$, $L = 1$, and $v_c = 0.1$. It appears that there is an optimal frequency at around $\omega = 12.5$ in this particular situation. (Dimensionally, the optimum frequency is around $12.5 v_m / L$.) Now, this optimum frequency was obtained by using the assumption that the flow along the interface has approximately constant speed in the region of influence of the cross-channels. This same configuration has been analyzed [31] using a *time-varying* velocity along the flow interface, specified by the T-mixer velocity. The frequency-dependence of the flux obtained there (Figure 5 in [31]) is not very different from Figure 4.2(b). Again, an optimum frequency of about 12.5, with additional peaks similarly placed to those in Figure 4.2(b), was obtained. This illustrates the robustness of the method, which is particularly important since the idealizations used for its development can only hoped to be approximately met in reality. ■

An interesting feature in Figure 4.2(b) is that it appears that there are frequency values at which the cross-interface flux is zero. This is intriguing in that it displays an "anti-resonance" type phenomenon between the imposed velocity perturbation and the flow along the channel. The frequency of flushing back and forth of fluid conspires with the time evolution along the main channel to create lobes with zero size to leading-order. In realistic situations, of course, this does not mean that the cross-interface transport will be exactly zero, for two reasons. First, the theory is perturbative in nature, and higher-order terms will change this. Second, in realistic situations diffusion will play a role on top of this advective mechanism, resulting in mild transport.

This issue is well-represented by the example given in [36], which compares the output of the technique desribed above with *experimentally* obtained mixing in an actual microfluidic device fabricated by Lee et al. [237]. Remarkably good accuracy is obtained from the usage of this theoretical method [36], *despite* the fact that Lee et al. [237] compute their mixing not based on cross-interface transport but by using a concentration variance norm in a cross-section of the channel well beyond the region of cross-channel velocity agitation. In contrast, the optimum frequency determination in [36] is entirely based on the arguments of this chapter and obtained in terms of the Fourier transform $\Lambda(\omega)$. Similarly, values of frequency for which *zero* flux are predicted according to the theory turn out to have *small* experimentally measured flux [36]. Thus, this provides *empirical* evidence that by optimizing transport across a flow barrier, one does indeed obtain optimal *global* mixing. Therefore, the approach detailed in this section has promise which is yet to be exploited in genuine microfluidic design and fabrication.

Further examples of the usage of the $|\Lambda(\omega)|$ diagnostic for determining optimal frequencies are given in [31] and [36]. Some additional simplifications—based on non-heteroclinic flow barriers associated with constant speed—will be addressed in Section 4.4.

4.3 ▪ Optimal energy-constrained flow protocol

While the previous section discussed the issue of whether one can find an optimum frequency for cross-interface transport for a velocity agitation whose spatial form is given, this section addresses the question of whether one can determine the optimal *spatial form* of the agitation. This is in the spirit of several other studies [267, 246, 117, 408] seeking the best velocity agitations for mixing. In contrast with these studies, the focus is on optimizing cross-interface transport, with the picture of the cross-channel micromixer (Example 3.56) in mind. This methodology explained in this section is based on the idea developed in [42], with some additional modifications.

The setting shall be as in Section 4.1, with a two-dimensional flow given by (4.1), which is incompressible when $\varepsilon = 0$, and all functions are assumed to be smooth. The $\varepsilon = 0$ flow is assumed to possess a heteroclinic connection $\bar{x}(t)$ between two hyperbolic saddle-like fixed points a and b (which might be the same point). This heteroclinic barrier, Γ, is assumed to have length L, and suppose that the arclength parametrization ℓ is used on Γ, as shown in Figure 4.1. This can be connected to the τ-parametrization $\bar{x}(\tau)$ by

$$\frac{d\ell}{d\tau} = \left|\dot{\bar{x}}(\tau)\right| = \left|-[\nabla\psi_0(\bar{x}(\tau))]^{\perp}\right| = |\nabla\psi_0(\bar{x}(\tau))|, \tag{4.9}$$

where $\ell = 0$ corresponds to the point a and hence $\tau = -\infty$, whereas $\ell = L$ is b, where $\tau = \infty$. This enables the definition of an invertible function $\tau(\ell)$ for $\ell \in (0, L)$ such that $\tau(0^+) = -\infty$ and $\tau(L^-) = \infty$. At any ℓ value on Γ, $\nabla\psi_0$ is normal to Γ as shown in Figure 4.1. If thinking of (4.1) in terms of the more general form $\dot{x} = f(x) + \varepsilon g(x, t)$, here $f^{\perp}$ is exactly in the direction of $\nabla\psi_0$ (see Figure 3.5).

Now, the fact that the leading-order transport across Γ is given by $|\Lambda(\omega)|$ as given in (4.4) has already been established. By choosing $\tilde{g}$ larger and larger, the transport can be made greater. Therefore, determining an optimal $\tilde{g}$ is an ill-posed question. To resolve this, and to simultaneously incorporate a physically relevant scenario, suppose the condition

$$\int_0^L |\tilde{g}(\ell)|^2 \, d\ell = G^2 L, \tag{4.10}$$

where G is a given constant with dimensions of velocity, is imposed. In (4.10), the cumbersome notation $\tilde{g}(\bar{x}(t(\ell)))$ has been replaced by $\tilde{g}(\ell)$—a simplifying abuse of notation which shall be persisted with throughout this section, since there is no scope for confusion. The reason for using $|\tilde{g}|^2$ is that this represents the kinetic energy, and thus (4.10) effectively limits the amount of kinetic energy which can be supplied to the flow interface.

Theorem 4.6 (spatial velocity for energy-constrained flux optimization [42]). *From among the $\tilde{g}$s defined for $\ell \in (0, L)$ which satisfy the energy contraint (4.10) and the additional condition that*

$$\mathscr{F}\{\nabla\psi_0(\bar{x}(t)) \cdot \tilde{g}(\bar{x}(t))\}(\omega) \in \mathbb{R}, \tag{4.11}$$

the functions

$$\tilde{g}_{\pm}(\ell) = \pm \frac{G\sqrt{L}}{\left[\int_0^L \cos^2[\omega\tau(\ell)]\, d\ell\right]^{1/2}} \cos[\omega\tau(\ell)] \frac{\nabla\psi_0(\ell)}{|\nabla\psi_0(\ell)|} \tag{4.12}$$

achieve the maximum of the flux measure $|\Lambda(\omega)|$.

Proof. Given the fact that the relevant Fourier transform is confined to be real, it is explicity given by

$$\Lambda(\omega) = \int_{-\infty}^{\infty} \nabla\psi_0(\bar{x}(\tau)) \cdot \tilde{g}(\bar{x}(\tau)) \cos[\omega\tau]\, d\tau .$$

In maximizing the size of the above, it is clearly counterproductive to utilize any components of $\tilde{g}$ in the direction orthogonal to $\nabla\psi_0$. Therefore, the first observation is that we must choose $\tilde{g}$ to be in the direction of $\nabla\psi_0$. Under the definition

$$\tilde{g}^n(\ell) := \tilde{g}(\ell) \cdot \frac{\nabla\psi_0(\ell)}{|\nabla\psi_0(\ell)|},$$

for the *scalar* function $\tilde{g}^n$, the constraint becomes

$$\int_0^L (\tilde{g}^n(\ell))^2\, d\ell = G^2 L \tag{4.13}$$

and moreover

$$\Lambda(\omega) = \int_0^L \nabla\psi_0(\ell) \cdot \tilde{g}(\ell) \cos[\omega\tau(\ell)] \frac{d\tau}{d\ell}\, d\ell = \int_0^L \tilde{g}^n(\ell) \cos[\omega\tau(\ell)]\, d\ell \tag{4.14}$$

by virtue of (4.9). Now since $\Lambda(\omega)$ is a scalar, the maximum of $|\Lambda(\omega)|$ occurs when $\Lambda(\omega)$ has an extremum (maximum or minimum). Thus, (4.14) needs to be optimized subject to the constraint (4.13). A direct application of the Euler-Lagrange equation with Lagrange multiplier λ yields the equation

$$\frac{\partial}{\partial \tilde{g}^n} \left(\tilde{g}^n \cos\omega\tau(\ell) - \lambda(\tilde{g}^n)^2 \right) = 0 .$$

The fortunate absence of a $(\tilde{g}^n)'$-dependence in the integrand enabled avoidance of having to perform integration by parts and specifying boundary conditions at $\ell = 0$ and L. Therefore, the only potential extremal functions satisfy

$$\tilde{g}^n(\ell) = \frac{1}{2\lambda} \cos\omega\tau(\ell).$$

Substituting into the constraint (4.13) gives

$$2\lambda = \pm \frac{1}{G\sqrt{L}} \left(\int_0^L \cos^2 \omega\tau(\ell)\, d\ell \right)^{1/2},$$

and therefore (4.12) is derived. There are two solutions arising from this process because of the ± signs. It now remains to verify whether either of these corresponds to a maximum of $|\Lambda(\omega)|$. Substitution of *either* of these values into (4.14) gives the fact that

$$|\Lambda(\omega)| = G\sqrt{L}\left(\int_0^L \cos^2 \omega\tau(\ell)\,\mathrm{d}\ell\right)^{1/2}. \qquad (4.15)$$

On the other hand, applying the Cauchy-Schwarz inequality to (4.14) yields

$$|\Lambda(\omega)| \le \left(\int_0^L (\tilde{g}^n)^2 \mathrm{d}\ell\right)^{1/2} \left(\int_0^L \cos^2[\omega\tau(\ell)]\,\mathrm{d}\ell\right)^{1/2} = G\sqrt{L}\left(\int_0^L \cos^2[\omega\tau(\ell)]\,\mathrm{d}\ell\right)^{1/2},$$

proving therefore that for either of the choices $\tilde{g}_{\pm}$, a maximum of the flux measure is achieved. □

Remark 4.7. It is possible to use elementary arguments such as those in [27] to show that (4.15) decays monotonically from GL at $\omega = 0^+$ to $GL/\sqrt{2}$ as $\omega \to \infty$.

While Theorem 4.6 provided a suprisingly clean answer to the maximum flux question by straightforward applications of variational calculus and the Cauchy-Schwarz inequality, the astute reader may question the *additional* constraint (4.11) slipped into Theorem 4.6. This turns out to *not* be a serious issue, and shall be addressed in a moment. First, let us consider the implications of the optimal velocity agitations $\tilde{g}_{\pm}$ predicted by this process. The first concern is that there are *two* protocols given by the two signs. Since $\tilde{g}$ will be modulated by the cosine function in (4.1) which itself changes sign continually, the sign difference in $\tilde{g}$ is seen to be no more than a phase-shift in time of size $\pi/(2\omega)$ in (4.1). Thus, the same intersection pattern between stable and unstable manifolds will occur in time-slices which are $\pi/(2\omega)$ apart. The *average flux* will be exactly the same for the two signs. Therefore, any sign of convenience can be chosen.

The next consideration is to understand the form of the velocity agitation predicted by Theorem 4.6. The cosine function switches sign at values where $\omega\tau(\ell) = (2m+1)\pi/2$ for $m \in \mathbb{Z}$. The corresponding ℓ values can be indexed by $\{\ell_m\}_{m\in\mathbb{Z}}$, where this bi-infinite sequence converges to 0 and L in the limits $m \to \pm\infty$. Thus, the perpendicular component of velocity on Γ has to switch signs at these values. The necessity for the velocity agitation to switch directions across Γ was first observed in [27] based on which a micromixer design was suggested in [25]; however, as acknowledged in [27], the optimization problem was ill-posed, and the switching suggested was abrupt. The methodology leading to Theorem 4.6 was suggested afterwards in [42], which (by setting up the problem as an energy-constrained optimization) has a well-posed solution which is moreover *continuous* along Γ. In any event, the optimized cross-interface transport strategy needs to switch sign infinitely often across Γ (since this infinitude occurs in approaching the accummulation points $\ell \to 0$ and $\ell \to L$, continuity is compromised at a and b). Having these infinite switchings turns out to not be an issue in a practical sense, since it turns out the contributions from the switchings near a and b are exponentially small by the fact that $|\nabla\psi_0(\bar{x}(\tau))|$ has exponential decay in approaching these limiting values. Therefore, in practice (as shall be shown in Example 4.9), it is possible to limit this switching to a finite number of values.

The assumption of the Fourier transform being real appears to unduly restrict the types of velocity agitations being used, and therefore is Theorem 4.6 meaningless in an applied context? It turns out that whatever the form of $\tilde{g}$ that is imposed across

Γ, it is *always* possible, by a suitable choice of parametrization along Γ, to ensure that this condition is met. To explain how this can be done, consider any parametrization $\bar{x}(\tau)$. This obeys (4.1) when $\varepsilon = 0$ and moreover satisfies $\lim_{t\to-\infty} \bar{x}(t) = a$ and $\lim_{t\to\infty} \bar{x}(\tau) = b$. *Any* alternative reparametrization of Γ in the form $\bar{x}(\tau - \tau_0)$ for any real τ_0, *also* satisfies all these conditions. This is simply the statement that different points on Γ may be chosen as initial conditions for (4.1) with $\varepsilon = 0$, such that the trajectory generated in forwards and backwards time traces out Γ. From a physical perspective the solution found in Theorem 4.6 should *not* depend on the choice of parametrization $\bar{x}(\tau)$. The parametrization is an intermediate calculation to enable *points* on Γ to be identified, and (4.12) tells us unequivocally the value of the velocity agitation required at each such point. Therefore, we should have the freedom to choose any parametrization when selecting $\bar{x}(\tau)$. This fact can be used to deal with the condition of $\Lambda(\omega)$ being real, as required by Theorem 4.6, in order to generate an algorithm for finding the velocity:

Remark 4.8 (algorithm for determining optimal spatial velocity agitation). Theorem 4.6 can be used to find the optimal velocity by the following procedure.

1. For some choice of $\bar{x}(\tau)$, compute $\tau(\ell)$ by inverting (4.9);
2. Compute $\tilde{g}^n$ at ℓ values on Γ using the positive sign in (4.12);
3. Determine $\Lambda(\omega)$ as defined in (4.4);
4. If Λ is real, no reparametrization is necessary, and the solution has been found;
5. If Λ is not real, let $\beta = \operatorname{Arg}\Lambda$, and then $e^{-i\beta}\Lambda$ (which rotates the complex number Λ by an angle $-\beta$) *must* be real;
6. Since $e^{-i\beta}\Lambda = \mathscr{F}\left\{\nabla\psi_0(\bar{x}(\tau - \beta/\omega)) \cdot \tilde{g}\left(\bar{x}(\tau - \beta/\omega)\right)\right\}(\omega)$ by standard Fourier transform properties, this provides a (nonunique) reparametrization that can be used;
7. Employ the new parametrization $\bar{x}(\tau - \beta/\omega)$ on Γ which ensures that the new Λ is real;
8. Recalculate $\tau(\ell)$ from (4.9) based on the new parametrization;
9. Recompute $\tilde{g}^n$ in (4.12) using the new $\tau(\ell)$; this is the optimal cross-barrier velocity.

This method is now applied to optimizing transport across a flow interface, much as in the previous section. However, here the focus is on determining the spatial structure of the velocity agitation using the above algorithm. An important observation must be made: numerical experiments indicate that the precise form of the flow model used along the interface does not seem to have a major impact on flux-optimization results in either this or the previous section. This is why examples using a T-mixer flow and a constant velocity were examined in Section 4.2. To explore this further, the following example uses for variety the Taylor-Green flow as the model for the flow near the flow interface.

Example 4.9 (cross-channel micromixer with Taylor-Green velocity). As in [42], the optimal spatial velocity agitation is determined when the velocity in the flow interface is given by the Taylor-Green flow [330, 100, 25, 26, 368, 5, 373, 1] already examined in Example 3.41, but with a slightly different parametrization. The unperturbed flow is assumed to be kinematically equivalent to Figure 3.7, with the point a being $(x_1, x_2) = (0, 0)$ and b being $(L, 0)$; the straight line between these points forms the flow barrier Γ. This structure is achievable by utilizing a dimensional shift of the Taylor-Green flow given in (3.62). So the Taylor-Green flow in use for this example satisfies

$$\dot{x}_1 = V \sin\frac{\pi x_1}{L} \cos\frac{\pi x_2}{L}, \quad \dot{x}_2 = -V \cos\frac{\pi x_1}{L} \sin\frac{\pi x_2}{L}, \tag{4.16}$$

corresponding to a streamfunction

$$\psi_0(x_1, x_2) = \frac{LV}{\pi} \sin\frac{\pi x_1}{L} \sin\frac{\pi x_2}{L}.$$

The positive parameters L and V have dimensions of length and velocity, and V has been chosen to ensure that the velocity at the center of the flow barrier has size V. The heteroclinic trajectory along Γ, spanning the line segment $0 \le x_1 \le L$, $x_2 = 0$, is given by

$$\bar{x}(\tau) = \left(\frac{2L}{\pi} \tan^{-1}\left(e^{\pi V \tau / L}\right), 0\right) \tag{4.17}$$

for the symmetrically chosen time-parametrization. Since the arclength $\ell = \bar{x}_1$ (the first coordinate above), this enables the representation

$$\tau(\ell) = \frac{L}{\pi V} \ln\left(\tan\frac{\pi \ell}{2L}\right).$$

Since $\nabla \psi_0$ is in the x_2-direction, using (4.12) gives the optimal cross-barrier velocity agitation as

$$\tilde{g}(x_1, 0) = \frac{G\sqrt{L}}{\left(\int_0^L \cos^2\left[\frac{\omega L}{\pi V} \ln\left(\tan\frac{\pi x_1}{2L}\right)\right] dx_1\right)^{1/2}} \cos\left[\frac{\omega L}{\pi V} \ln\left(\tan\frac{\pi x_1}{2L}\right)\right] \hat{x}_2. \tag{4.18}$$

Here, $\Lambda(\omega)$ is real since both $\tilde{g}$ and $\nabla \psi_0$ are even in τ, and so no readjustment of parametrization is necessary. Indeed, this is typical if the speed is symmetric about the point chosen as $\tau = 0$: an observation that can be used to advantage in other calculations as well. The x_2-component of (4.18) is pictured in Figure 4.3 for several values of ω, showing how $\tilde{g}$ must be chosen to flip signs at the "turning point" values x_1^m given by

$$\frac{\omega L}{\pi V} \ln\left(\tan\frac{\pi x_1^m}{2L}\right) = \frac{2m+1}{2} \pi, \; m \in \mathbb{Z}. \tag{4.19}$$

Since these accumulate towards 0 and L, for practical purposes one can limit to only the central changes, and set $\tilde{g}^n = 0$ outside of this. Figure 4.3(a) shows the behavior required of $\tilde{g}^n$ in the domain $x_1 \in (L/100, 99L/100)$, and it is seen that in this region, which for all practical purposes encapsulates the entire flow barrier, there is only one unidirection segment observable for $\omega = 1$ (i.e., two turning points), while there are 4 and 14 for $\omega = 4$ and 15, respectively. The velocity can simply be set to zero outside of some chosen turning points near to 0 and L. The error in this procedure is small since exponential decay in $|\nabla \psi_0(\bar{x}(t))|$ as the endpoints are approached means that any velocity agitation contributions in these regions to $\lambda(t)$ in (4.4) get substantially squashed anyway. A more detailed assessment of this is presented in Figure 4.3(b), where the effect of including only a finite number of turning points, and setting $\tilde{g}$ to zero outside of this region, on lobe areas is examined. The lobes are computed by evolving streaklines on the x_1-axis near $x_1 = 0$ according to the perturbed velocity

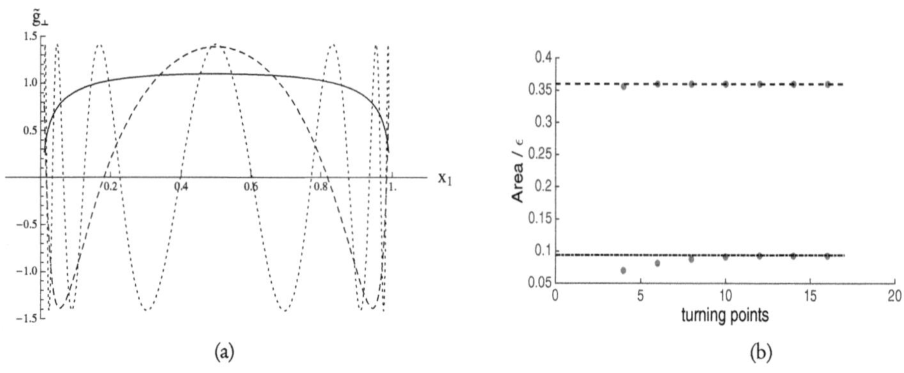

Figure 4.3. *(a) Optimal $\tilde{g}^n$ in Example 4.9 as expressed in (4.18) with $L = 1$, $V = 1$, and $G = 1$: $\omega = 1$ (solid), $\omega = 4$ (dashes), and $\omega = 15$ (dots and dashes), and (b) Rapid convergence in using a finite number of turning points.*

field (4.1). In doing so, the velocity $\tilde{g}$ in (4.12), which is only defined on Γ, has been extended uniformly in x_2—a reasonable approach which preserves incompressibility. Lobes are generated as these streaklines pass through the region and pile up near $x_1 = L$. There areas are computed numerically and reported by the red dots in Figure 4.3(b). As the number of turning points increases, there is rapid convergence to the dashed lines, which are the theoretically predicted lobe areas (divided by ε) as predicted in (4.5). Therefore, for all intents and purposes, having a finite number of turning points (of the order of 10) appears sufficient. Figure 4.3(b) shows this calculation for $\omega = 4$ and $\omega = 15$; as expected from examining Figure 4.3(a), the larger ω value requires a smaller number of turning points for good accuracy than the smaller ω.

A test for optimality is shown in Figure 4.4. The top left panel shows in black the optimal $\tilde{g}^n$ chosen for $\omega = 6$, $V = 1$, and $L = 1$, with the choice to include agitations only between x_1-values indicated by $m = -2$ and $m = 1$ in (4.19) (this is symmetric about $L/2$). An exaggerated scale in the x_2-direction has been used to visualize structures; this scale is consistent across all the panels in Figure 4.4. While $\tilde{g}$'s value along the flow barrier is the only information required for the theory, it must be extended in a consistent way. Here, the choice of extending it uniformly in the x_2-direction has been taken; this automatically complies with incompressibility. The streaklines which were evolved by the full flow (4.1) are shown in red after the process has been going on for some time. Three other forms of $\tilde{g}^n$ (black dashed curves) which conform to the kinetic energy constraint are shown in the other panels in Figure 4.4. The flux measures computed in these instances, as a percentage of the flux of the optimum $\tilde{g}^n$, are indicated above each panel. The closest approach is by the velocity form on the top right, which is in many ways quite close to the optimal one. The panels at the bottom show that concentrating too much of the energy in the central region, or putting too much energy near the endpoints, leads to a smaller flux. The optimal solution shows how best to distribute energy along the flow barrier. ∎

One question arising from the above example is how one might be able to generate the required velocity agitations. As can be seen, these velocity perturbations correspond to flow being directed across the main channel. Thus, approximations to these velocities can be obtained in a natural way using cross-channel micromixers [409, 438, 334, 123, 411, 80, 79, 390, 237, 279, 424, 198], with the cross-channels

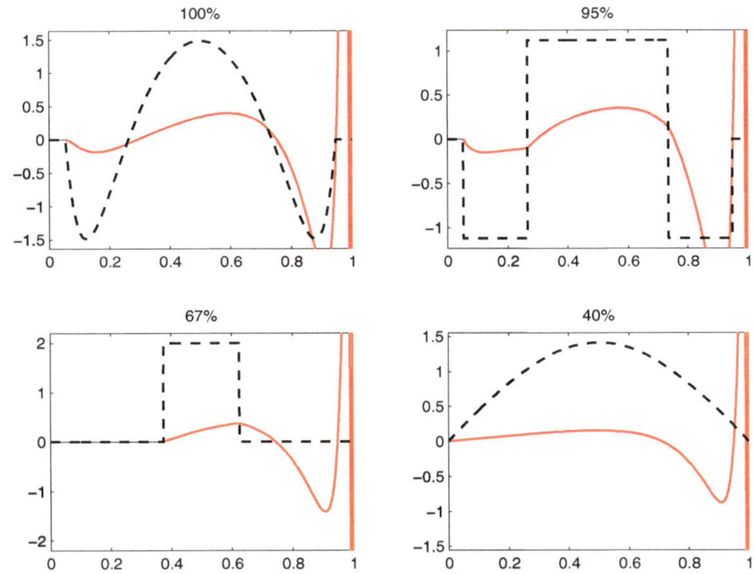

Figure 4.4. *Comparison of performance of optimal agitation (4.18) (top left panel) with several other options (black dashed curves); the heading of each figure shows the percentage of optimal flux obtained.*

positioned at the relevant values. An interesting insight based on the structure of the cross-velocity is that the cross-channels, unlike in standard multi-cross-channel designs [409, 438, 334, 123, 411, 80, 79, 279, 424, 198], need to be placed right against each other for optimal effect. An alternative method for generating these velocities is to have the upper and lower boundaries in the device (as shown in Figure 3.7, for example) to be flexible membranes which can be vibrated according to our specifications [123, 411]. If this approach is followed, the locations x_1^m as specified in (4.19) give positions which do not move under the $\cos \omega t$ time-modulation and therefore are locations at which these flexible membranes can be anchored. A third possibility is to achieve fluid velocities within the device by judiciously positioned electromagnets which influence charged particles in the flow [242, 376, 288, 289, 437, 62, 4, 206]. The optimization theory described here provides direct input into how any of these velocity agitation strategies can be designed, as, for example, explained further in [25, 42, 36].

4.4 ▪ Optimization for nonheteroclinic flow barriers

The fact that the model for flow barriers in microfluidic devices considered so far—such as the T-mixer or the Taylor-Green models—have extra assumptions in them, notably, the presence of saddle-fixed points at the beginning ($\ell = 0$) and the end ($\ell = L$) of the flow interface, is of slight concern. Such saddle points at the ends are not always visible in most applied descriptions in the microfluidic literature. The fixed point at a is sometimes visible when a "Y-mixer" geometry [237, 36] is used, but there are hardly any instances in which the fixed point b is obvious at the outlet. This brings to light the fact that in microfluidic applications, there are often situations in which the

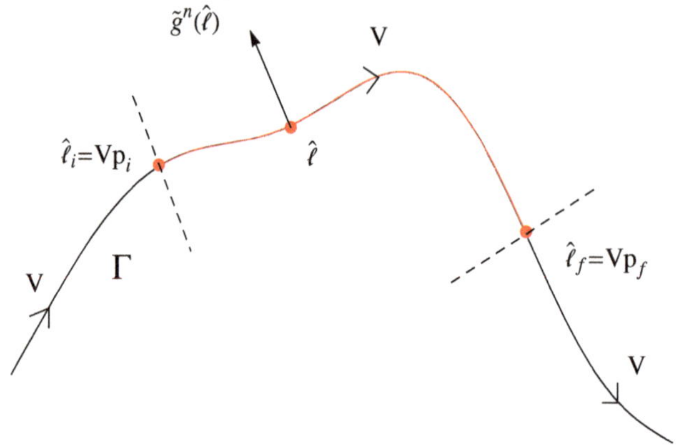

Figure 4.5. *Nonheteroclinic* Γ *along which the unperturbed speed is a constant* V*; velocity agitation is nonzero only in the red part of* Γ *between the signed arclength values* ℓ_1 *and* ℓ_2.

flow interface is not a heteroclinic trajectory at all. Fortunately, the development of Section 3.7 applies to exactly this situation. In this section, modifications of previous flux optimization ideas are adapted for nonheteroclinic flow barriers. More details on the material in this section are available in ongoing work [40, e.g.].

4.4.1 ▪ Quantifying time-harmonic flux

To establish the setup: The equation considered is (4.1), where an unperturbed steady flow with streamfunction ψ_0 is perturbed by a harmonic velocity agitation to cause flux across a flow interface Γ which is not necessarily a heteroclinic trajectory of (4.1) when $\varepsilon = 0$. Nevertheless, Γ is a trajectory $\bar{x}(t)$ of (4.1) when $\varepsilon = 0$. The fluids on the two sides of Γ are *different* miscible fluids, and the idea is to impose a velocity agitation *over a defined finite range of* Γ in order to maximize transport. Some simplifying assumptions—consistent with Section 3.7 and previous experiences in optimizing transport in the heteroclinic context—will be made.

Hypothesis 4.10 (velocity agitation conditions).

1. *The velocity agitation* $\tilde{g}$ *is assumed to be nonzero on* $\bar{x}(p)$ *only for* $p \in [p_i, p_f]$*;*
2. $\tilde{g}$ *shall be chosen to be purely in the direction* $\pm\nabla\psi_0$ *at all points, i.e.,*

$$\tilde{g}(x) = \tilde{g}^n(x)\frac{\nabla\psi_0(x)}{|\nabla\psi_0(x)|}; \quad \textit{and}$$

3. *the unperturbed speed* $|\nabla\psi_0(\bar{x}(p))|$ *at all points* $p \in [p_i, p_f]$ *is approximately a constant value* V.

This constitutes a velocity agitation region across Γ, which is confined to be between $\bar{x}(p_i)$ and $\bar{x}(p_f)$. To take advantage of the fact that velocities tangential to Γ have no influence on the leading-order flux, $\tilde{g}$ has been set to be purely in the normal direction. The final assumption is based on the fact that flows in microchannels, before any agitation is imposed, tend to be well-developed in applications. The success of using such a "constant speed" assumption on the flow interface in [36] on the experimental device of Lee et al. [237] motivates this further.

The "double-streakline" situation, in which flux across the interface is assessed in terms of a pseudoseparatrix connecting the unstable and stable streaklines, is the main focus in flux assessment. The relevant Melnikov function, which encodes the leading-order flux, has already been computed in (3.84). In this instance,

$$f^{\perp}\tilde{g} = \left(-\nabla^{\perp}\psi_0\right)^{\perp} \cdot \tilde{g} = \nabla\psi_0 \cdot g = V\tilde{g},$$

and moreover $\mathrm{Tr}\, Df = 0$. Thus, the expression (3.84) simplifies to

$$M(p,t) = \int_{p_i}^{p_f} V\tilde{g}^n\left(\bar{x}(\tau)\right)\cos\left[\omega\left(\tau+t-p\right)\right]\, \mathrm{d}\tau .$$

So in this case the definition

$$\lambda(t) = \begin{cases} V\tilde{g}^n\left(\bar{x}(t)\right) & \text{if } t \in [p_i, p_f], \\ 0 & \text{if } t \notin [p_i, p_f] \end{cases} \tag{4.20}$$

is called for, with $\Lambda(\omega)$ being its Fourier transform. Once again, the representation (4.3) is possible, and the relationships to lobe areas and average flux given in Theorem 4.1 hold.

As a useful alternative, one can use a signed arclength version of $\Lambda(\omega)$ instead of taking the Fourier transform of the appropriate $\lambda(t)$. That is, $\Lambda(\omega)$ can be represented in the form (4.8) directly, but care must be taken to put in the correct ℓ-limits associated with the various p-limits given above. This is accomplished by using (4.2) with τ being equivalent to p. Under the constant speed assumption, it is possible to simply set

$$\hat{\ell}(p) = Vp \tag{4.21}$$

for $p \in [p_i, p_f]$, with the understanding that $\hat{\ell}$ is a signed arclength parametrization of the agitation region of Γ, as shown in Figure 4.5. (The hat notation is used since this shall be adjusted to something else shortly.) Let $\hat{\ell}_i = Vp_i$ and $\hat{\ell}_f = Vp_f$ be the endpoints of this region. Now, using the arclength representation of the function $\Lambda(\omega)$ (whose modulus is the flux measure) as given in (4.8) under these simplifications leads to

$$\hat{\Lambda}(\omega) = \int_{\hat{\ell}_i}^{\hat{\ell}_f} \tilde{g}^n\left(\hat{\ell}\right) e^{-i\omega\hat{\ell}/V}\, \mathrm{d}\hat{\ell} . \tag{4.22}$$

The fact that $\hat{\Lambda}$ is in general complex leads to some annoyance in determining optimization strategies. This can be removed by a reparametrization of the signed arclength. To do this, let $\beta = \mathrm{Arg}\left(\hat{\Lambda}(\omega)\right)$. Then define

$$\ell(p) = \hat{\ell}(p) + \frac{V\beta}{\omega} = V\left(p + \frac{\beta}{\omega}\right), \tag{4.23}$$

where now $\ell_i = V(p_i + \beta/\omega)$ and $\ell_f = V(p_f + \beta/\omega)$ are the endpoints of the agitation region. Now if $\Lambda(\omega)$ is the flux measure written in this new parametrization,

$$\Lambda(\omega) = \int_{\ell_i}^{\ell_f} \tilde{g}^n(\ell) e^{-i\omega\ell/V} \,\mathrm{d}\ell = \int_{\hat{\ell}_i}^{\hat{\ell}_f} \hat{\tilde{g}}^n\left(\hat{\ell}\right) e^{-i\omega(\hat{\ell}+V\beta/\omega)/V} \,\mathrm{d}\hat{\ell}$$

$$= e^{-i\beta} \int_{\hat{\ell}_i}^{\hat{\ell}_f} \hat{\tilde{g}}^n\left(\hat{\ell}\right) e^{-i\omega\hat{\ell}/V} \,\mathrm{d}\hat{\ell} = e^{-i\beta}\hat{\Lambda}(\omega), \tag{4.24}$$

where $\hat{\tilde{g}}$ is the same as $\tilde{g}$ but written in terms of the $\hat{\ell}$ parametrization. Now since $\beta = \mathrm{Arg}\left(\hat{\Lambda}(\omega)\right)$, the final answer above is real. The above argument serves to show that it is always possible to find a signed arclength parametrization such that $\Lambda(\omega)$ is real. In doing so, note that

1. The reparametrization simply shifts which point is called $\ell = 0$. Whichever parametrization is used, the value of $\tilde{g}$ is unambiguously defined at each point.
2. Since from (4.24), $|\Lambda(\omega)| = \left|\hat{\Lambda}(\omega)\right|$, the flux measure is *not* affected by the reparametrization.

Henceforth in this section, it shall be assumed that a signed arclength parametrization which renders $\Lambda(\omega)$ real has been judiciously chosen, and so

$$\Lambda(\omega) = \int_{\ell_i}^{\ell_f} \tilde{g}^n(\ell) \cos\frac{\omega\ell}{V} \,\mathrm{d}\ell\,. \tag{4.25}$$

This quantifies the flux across the nonheteroclinic flow barrier. It will turn out that in applications this parametrization can be done quite easily using symmetry, and a simple check for ensuring this is whether

$$\int_{\ell_i}^{\ell_f} \tilde{g}^n(\ell) \sin\frac{\omega\ell}{V} \,\mathrm{d}\ell \stackrel{?}{=} 0. \tag{4.26}$$

If (4.26) is not met, then one can perform the shift defined by (4.23) before proceeding. Now with (4.25) in hand, the two optimization questions posed for the heteroclinic flow barrier situation can be addressed in exactly the same manner.

If one chooses the "single-streakline" representation as expressed in Remark 3.55, the flux is associated with the displacement of the unstable streakline from its unperturbed location and therefore is quantified by M^u in (3.80) rather than M in (3.84). Unlike in the double-streakline situation, the amplitude of the harmonic flux in this case turns out to depend on the location at which the gate is chosen. If this is at $\bar{x}(p)$, corresponding to a signed arclength location $\hat{\ell}$, then $\hat{\ell} = Vp$. Imposing Hypotheses 4.10 on the expression (3.80) under the current assumptions leads to

$$M^u(p,t) = \mathbb{1}_{[p_1,P_u)}(p) \int_{p_1}^{\min(p,p_f)} V\tilde{g}^n(\tau)\cos[\omega(t+\tau-p)] \,\mathrm{d}\tau$$

$$= \left|\Lambda_p(\omega)\right| \cos\left[\omega(t-p) - \mathrm{Arg}\left(\Lambda_p(\omega)\right)\right]$$

by the standard procedure, where now the relevant $\Lambda(\omega)$ is p-dependent, taking the form

$$\Lambda_p(\omega) = \mathbb{1}_{[p_1,P_u]}(p) \int_{p_1}^{\min(p,p_f)} V\tilde{g}^n(\tau).e^{-i\omega\tau} \,\mathrm{d}\tau\,.$$

Converting these expressions to the arclength representation, with the gate at $\hat{\ell}$, yields

$$M^u\left(\hat{\ell},t\right)=\left|\Lambda_{\hat{\ell}}(\omega)\right|\cos\left[\omega\left(t-\frac{\hat{\ell}}{V}\right)-\operatorname{Arg}\left(\Lambda_{\hat{\ell}}(\omega)\right)\right], \tag{4.27}$$

where

$$\Lambda_{\hat{\ell}}(\omega)=\mathbb{1}_{\left[\hat{\ell}_i,VP_u\right)}(\hat{\ell})\int_{\hat{\ell}_i}^{\min\left(\hat{\ell},\hat{\ell}_f\right)}\tilde{g}^n(\hat{u})e^{-i\omega\hat{u}/V}\,d\hat{u}. \tag{4.28}$$

The modulus of (4.28) gives the flux measure for the single-streakline attitude. As before, suppose $\beta=\operatorname{Arg}\left(\Lambda_{\hat{\ell}}\right)$, and suppose an alternative arclength parametrization $\ell=\hat{\ell}+V\beta/\omega$ is used. The argument leading to (4.24) goes through exactly as above, indicating that there is an arclength parametrization for which Λ_ℓ can be made real, and moreover, this parametrization does not affect the value of the leading-order flux. In this parametrization, the flux measure for the single-streakline displacement is

$$\Lambda_\ell(\omega)=\mathbb{1}_{[\ell_i,V(P_u+\beta/\omega))}(\ell)\int_{\ell_i}^{\min(\ell,\ell_f)}\tilde{g}^n(u)\cos\frac{\omega u}{V}\,du. \tag{4.29}$$

The test for ensuring that the correct signed arclength parametrization ℓ has been used is to check whether

$$\int_{\ell_i}^{\min(\ell,\ell_f)}\tilde{g}^n(u)\sin\frac{\omega u}{V}\,du\stackrel{?}{=}0$$

and, if not, perform a shift in ℓ as specified in (4.23) before continuing an analysis for flux optimization.

4.4.2 ▪ Optimal frequency

Henceforth, the double-streakline form, in which the flux is quantified in terms of the relative displacement between the perturbed stable and unstable streaklines, will be used. In this subsection, it is assumed that the normal velocity's spatial form $\tilde{g}^n$ is given, and the idea is to determine ω which maximizes $|\Lambda(\omega)|$ for $\Lambda(\omega)$ given in (4.25). By taking the modulus on both sides of (4.24) it is clear that this value *does not depend on the parametrization used.* Therefore, one may as well *directly* calculate the modulus of (4.25), *where or whether not* a shift has been employed. That is, for *any* choice of signed arclength parametrization, (4.25) can be computed, and its absolute value plotted, as a function of ω. Visually, the optimal value(s) of ω can then be determined from this graph.

Example 4.11 (cross-channel micromixer cont.). The cross-channel micromixer of Example 3.56, and illustrated in Figure 3.17, is reexamined. There are n cross-channels, located at the x_1-locations $x_i(p_j)$, $j=1,2,\ldots,n$, each with width $2d_j$. The flow in these is parabolic with speed v_j in the center of the jth channel, and $J_j\in\{1,-1\}$ quantifies the relationship between adjacent channels ($J_j=1$ means flow in the cross-channels is all in the same direction; $J_j=(-1)^j$ implies that adjacent channels are exactly out of phase with one another). Now, setting $\ell=x_1$ as the signed arclength, the relationship to the p-variable is simply $\ell=Vp$, and so the cross-channels are centered

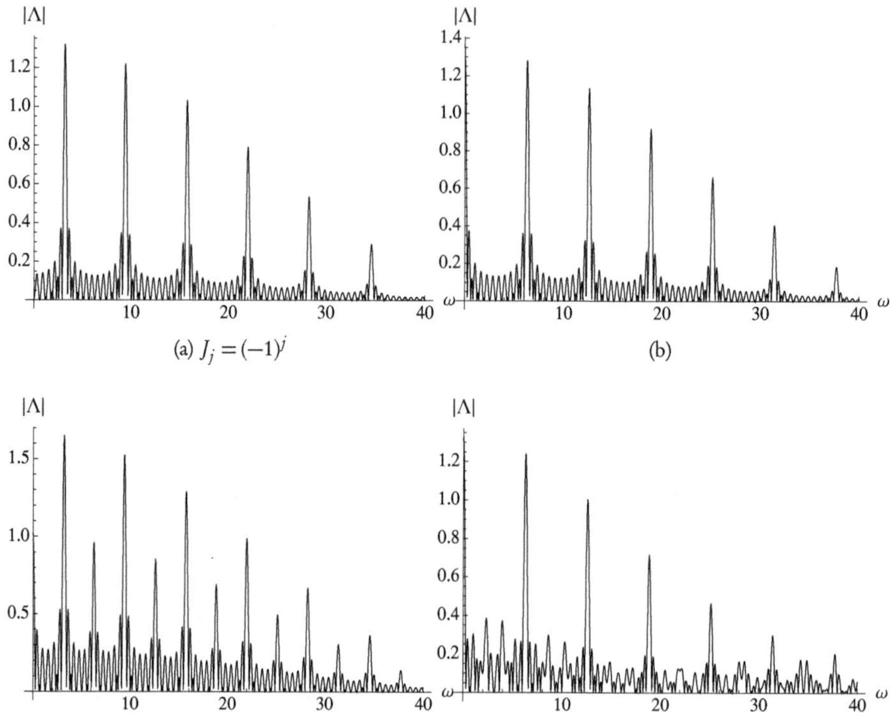

(a) $J_j = (-1)^j$

(b)

(c) $v = \{0.5, 2, 0.5, 2, 0.5, 2, 0.5, 2, 0.5, 2\}, J_j = (-1)^j$

(d) $d = \{0.1, 0.15, 0.05, 0.2, 0.1, 0.05, 0.1, 0.05, 0.15, 0.05\}$

Figure 4.6. *Flux measure (4.30) for several cross-channel micromixer designs with unequal configurations, with* $V = 1$, $n = 10$, $\ell_j = j$, *and, unless otherwise stated,* $J_j = 1$, $v_j = 1$, *and* $d_j = 0.1$.

at $\ell_j = V p_j$. Examining the second line of (3.87) and formulating the corresponding flux measure (4.25) gives the formula

$$\begin{aligned} |\Lambda(\omega)| &= \left| \sum_{j=1}^{n} \frac{J_j v_j}{d_j^2} \int_{\ell_j - d_j}^{\ell_j + d_j} \left[\left(\ell - \ell_j \right)^2 - d_j^2 \right] \cos \frac{\omega \ell}{V} \mathrm{d}\ell \right| \\ &= \left| \frac{4V^2}{\omega^3} \sum_{j=1}^{n} \frac{J_j v_j}{d_j^2} \cos \frac{\omega \ell_j}{V} \left(\omega d_j \cos \frac{\omega d_j}{V} - V \sin \frac{\omega d_j}{V} \right) \right|, \end{aligned} \tag{4.30}$$

which is therefore the leading-order flux measure in a very general cross-channel micromixer. Note that the ℓ_js are the signed arclength locations of the center of the cross-channels arranged in order, and (to avoid overlap between cross-channels) the condition $\ell_j + d_j < \ell_{j+1} - d_{j+1}$ is needed for all relevant j.

The variation of (4.30) with the frequency ω is shown in Figure 4.6 for a variety of configurations in which $V = 1$ and 10 cross-channels are chosen, to be centered at the locations $\ell_j = j$. In (a), adjacent channels are exactly out of phase, whereas in (b) they are in phase. In (c) they are out of phase, but the cross-channel maximum speeds v_j have been chosen to alternate between 0.5 and 2, whereas in (d) the channels are in phase with the same value of this maximum speed, but the channel widths d_j have been chosen to have a fairly random variation which nonetheless has a total width identical to each of the other cases. For each particular configuration, the flux measure goes to zero as the frequency increases, according to standard expectations [342, 31]. Moreover, the flux exhibits significant variation with ω, with well-defined local peaks, some of which are much more prominent than others. Using calculations such as these based on (4.30) for any particular design of cross-channel micromixer, it is easy to produce the relevant form of Figure 4.6, and by zooming in the optimal frequency is easily obtained. For example, for the configurations in Figure 4.6, the best frequency values are, respectively, $\omega \approx 3.14, 6.28, 3.14$, and 6.28; these are clearly associated with π and 2π, and it is easy to show that these values represent local maxima of (4.30). The *global* maximum nature of these values is harder to establish theoretically, since there are infinitely many turning points in each situation. These "nice" values for the maximal ω occur because the configurations chosen are somewhat regular, and for a very irregular choice of the v_j J_j, ℓ_j, and d_j, less obvious optimal frequencies would occur. Such are easily obtainable from a practical perspective by plotting the explicit formula (4.30) and zooming in on the graphs. ∎

4.4.3 ▪ Optimal spatial form

Akin to Section 4.3, the next issue is to determine the optimal spatial form of the perturbation to maximize transport across a nonheteroclinic flow barrier. The flow is given by (4.1), but the assumptions of Hypothesis 4.10 apply. Thus, the perturbation only applies between the p-values p_i and p_f. The goal is to determine the optimal form of the perturbation $\tilde{g}^n$ (normal to Γ), subject to an enegy constraint, which maximizes the flux measure.

It proves convenient to choose the arclength parametrization in a symmetric way which takes into account physical intuition. Let $L = V(p_f - p_i)$ be the length of the nonheteroclinic flow barrier across which the perturbation applies. Define the signed arclength parametrization by

$$\ell(p) = V\left(p - \frac{p_i + p_f}{2}\right), \tag{4.31}$$

which ensures that $\ell_i = \ell(p_i) = -L/2$ and $\ell_f = L/2$. Thus, it is necessary to determine $\tilde{g}^n(\ell)$, which satisfies the kinetic energy constraint

$$\int_{-L/2}^{L/2} [\tilde{g}^n(\ell)]^2 \, d\ell = G^2 L \tag{4.32}$$

for a constant $G > 0$ with the dimensions of velocity, which moreover maximizes the quantity

$$F := |\Lambda| = \left| \int_{-L/2}^{L/2} \tilde{g}^n(\ell) e^{-i\omega\ell/V} \, d\ell \right|^2$$

as given in (4.22). Since the signed arclength parametrization has been specified, at this point it is not clear whether Λ would be real. However, the problem remains invariant under the transformation $\ell \to -\ell$, and thus restricting to a symmetric (i.e., even) $\tilde{g}^n$ is appropriate. In this case Λ would be real, and therefore it suffices to consider extremizing

$$\tilde{F} = \int_{-L/2}^{L/2} \tilde{g}^n(\ell) \cos\frac{\omega\ell}{V} \, d\ell \, . \tag{4.33}$$

Of course, the requirement is that $|\tilde{F}|$ be maximized; seeking extrema for $\tilde{F}$ will encapsulate such solutions. The problem of extremizing (4.33) subject to (4.32) is essentially similar to what has been done in Section 4.3, but in this case, in which an explicit and constant form for the velocity along the flow barrier is used, an explicit formula sans integrals is possible. As in Section 4.3, applying the Euler-Lagrange formula with a Lagrange multiplier λ gives

$$\tilde{g}^n(\ell) = \frac{1}{2\lambda} \cos \frac{\omega \ell}{V},$$

which upon insertion into (4.32) gives

$$\frac{1}{2\lambda} = \pm \frac{G\sqrt{L}}{\left| \int_{-L/2}^{L/2} \cos^2 \frac{\omega \ell}{V} \, d\ell \right|}.$$

Evaluating the integral above and replacing in the $\tilde{g}^n$ expression leads to the explicit formula

$$\tilde{g}^n(\ell) = G\sqrt{\frac{2\alpha}{\alpha + \sin\alpha}} \cos \frac{\alpha \ell}{L}, \quad -L/2 \le \ell \le L/2, \tag{4.34}$$

in which a nondimensional Strouhal number $\alpha := \omega L/V$ has been used. As explained in Section 4.3, the negative sign has been ignored since this merely represents an exactly out-of-phase spatial form. Unlike in the heteroclinic situation, there is no difficulty here of the solution exhibiting infinitely many oscillations since for any ω there is a finite number of oscillations of the function (4.34) in the domain $[-L/2, L/2]$. Another feature is that since the zeros of $\cos(\alpha\ell/L)$ are equally spaced, the subintervals of ℓ in which $\tilde{g}^n$ is unidirectional are of equal widths, except possibly the subintervals nearest $\ell = \pm L/2$, which may get clipped at those values.

To check maximality, the application of the Cauchy-Schwarz inequality gives

$$\begin{aligned} \left| \int_{-L/2}^{L/2} \tilde{g}^n(\ell) \cos \frac{\omega \ell}{V} \, d\ell \right| &\le \left(\int_{-L/2}^{L/2} [\tilde{g}^n(\ell)]^2 \, d\ell \right)^{1/2} \left(\int_{-L/2}^{L/2} \cos^2 \frac{\omega \ell}{V} \, d\ell \right)^{1/2} \\ &= \left(G\sqrt{L} \right) \left[\frac{L}{2\alpha} (\alpha + \sin\alpha) \right]^{1/2} = GL \sqrt{\frac{\alpha + \sin\alpha}{2\alpha}}, \end{aligned} \tag{4.35}$$

and the right-hand side is exactly what one would obtain if inserting $\tilde{g}^n$ from (4.34) into the left-hand side. Thus, (4.34) is a maximal solution to the problem.

The right-hand side of (4.35) therefore gives the maximum flux attainable for a given frequency, or equivalently Strouhal number. The variation of this maximum flux with α (scaled by GL) is shown in Figure 4.7(a). The maximum of $|\Lambda|$ varies from GL as $\alpha \downarrow 0$, to $GL/\sqrt{2}$ as $\alpha \to \infty$, exhibiting a wiggling variation along the way. It is instructive to note that for any given α, the amplitude of the required spatial velocity perturbation $\tilde{g}^n$ is *reciprocal* to this and is shown in Figure 4.7(b). Indeed, $\tilde{g}^n$ varies between $\pm$ this value, modulated by the cosine variation in ℓ as given in (4.34).

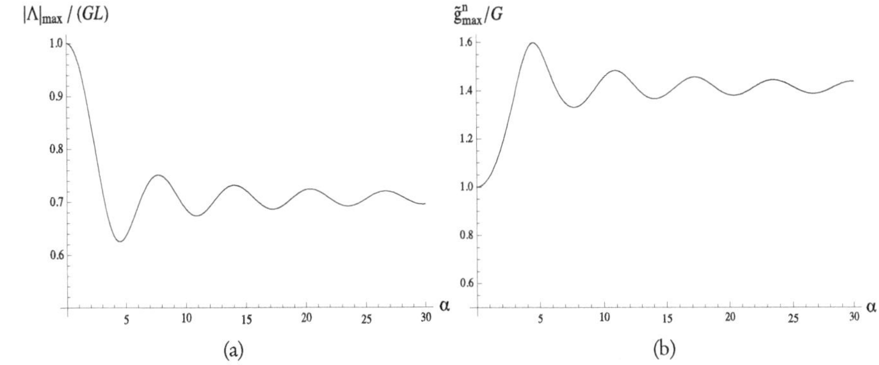

Figure 4.7. *(a) The maximum flux attainable across a nonheteroclinic flow barrier as a function of Strouhal number $\alpha = \omega L/V$, and (b) corresponding amplitude of velocity $\tilde{g}^n$.*

Example 4.12 (cross-channel micromixer cont.). A straight cross-channel micromixer with constant velocity was addressed in Examples 3.56 and 4.11, where the idea was to quantify and optimize with respect to frequency the flux across the nonheteroclinic flow interface for a *specified* device geometry. The above results can be used to help address the reverse question of where to position cross-channels in such a configuration. Under the assumptions above, the cross-channels are to be positioned between locations $-L/2$ and $L/2$. Subject to the energy constraint of constant kinetic energy across the nonheteroclinic flow barrier in this segment, the maximal solution of the cross-interface velocity is given by (4.34). Bear in mind, however, that this is modulated by a time-varying term; i.e., the actual flow velocity across the interface must be $\varepsilon \tilde{g}^n \cos(\omega t)$.

From a practical perspective, the solution (4.34) offers several difficulties. The first issue is that, as discussed in Example 3.56, flow in cross-channels is best modeled with a parabolic velocity variation. However, (4.34) tells us to impose a sinusoidal variation. Happily, these two scenarios are not that different; one may expect that if cross-channels were appropriately positioned, a velocity which is *approximately* that of (4.34) might be achieved. The second issue is that the zero-velocity condition at the endpoints of cross-channels which arises from the parabolic profile is not necessarily attainable by (4.34). For example, the value of (4.34) at the purported end of the agitation region, $\ell = L/2$, is not necessarily zero, whereas if there were a cross-channel ending at this point the resulting velocity would be zero. Basically, (4.34) indicates that ends of cross-channels must be positioned where $\cos(\alpha\ell/L) = 0$, i.e., at the finite locations

$$\ell_m = \frac{(2m+1)\pi}{2\alpha} L, \quad m \in \{0, \pm 1, \pm 2, \cdots\}, \tag{4.36}$$

chosen such that $\ell_m \in [-L/2, L/2]$. A reasonable option would be to enforce such an endpoint at each of $\pm L/2$, which would require $(\alpha - \pi)/(2\pi)$ to be a nonnegative integer. Thus, it is possible to achieve this if $\alpha = \pi, 3\pi, 5\pi, \cdots$, and the corresponding $\tilde{g}^n(\ell)$ variation is shown in Figure 4.8(a) for the first three of these cases. The fact that the "amplitude" of $\tilde{g}^n$ in these cases is $G\sqrt{2}$ is clear from (4.34) and is obvious in Figure 4.8(a). The design implications for $\alpha = \pi$ is that there would be one cross-channel spanning the length of the agitation region, whereas for $\alpha = 3\pi$ there would be three cross-channels stacked against each other, with the flow in adjacent ones being exactly out of phase. The possible design for $\alpha = 5\pi$ involves five cross-channels, as is illus-

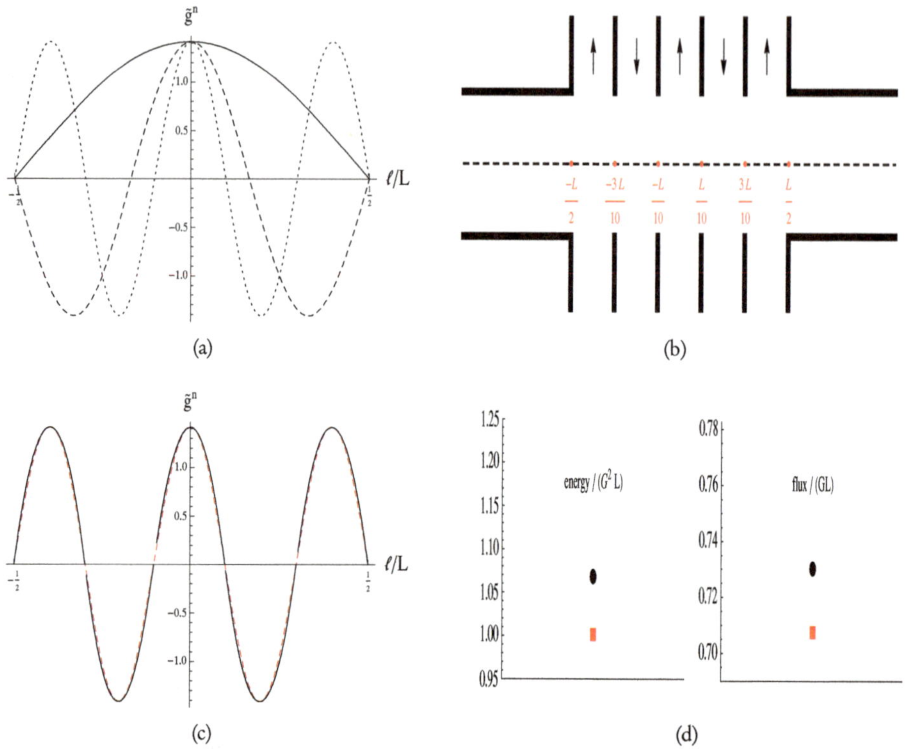

Figure 4.8. *(a) Optimal energy-constrained velocity perturbation in the agitation region for $\alpha = \pi$ (solid), 3π (dashed), and 5π (dotted), (b) designing cross-channels to approximately achieve optimality for $\alpha = 5\pi$, (c) parabolic fit (black solid) to the $\alpha = 5\pi$ optimal curve (red dashed), and (d) comparison between energy and flux values for parabolic (black circle) and true optimal agitation (red square) for this situation.*

trated in Figure 4.8(b). The cross-channel boundaries are at $\ell/L = \pm 0.1, \pm 0.3$, and ± 0.5 (i.e., the endpoints of the agitation region). Unlike in "standard" cross-channel mixer designs [438, 334, 123, 411, 79, 390, 237, 279, 424, 198], it is interesting to note that the cross-channels here are stacked right against each other. Fluid needs to be sloshed in these exactly out of phase with adjacent channels, as shown in Figure 4.8(b), with the intention of obtaining a maximum speed (in the center of each cross-channel) of size $G\sqrt{2}$. Figure 4.8(c) shows the parabolic flow profile with maximum velocity $G\sqrt{2}$ fitted within each cross-channel (solid black curve) in comparison with the optimal spatial form (red dashed curve) when $\alpha = 5\pi$; the curves are virtually indistinguishable. Thus, the required velocity agitation can be obtained with excellent accuracy by using these channels since the parabolic profile generated through fluid mechanics closely matches the required velocity agitation obtained via optimization theory. A further analysis of this is shown in Figure 4.8(d), in which the energy (4.32) and the flux measure (4.33) are compared for the parabolic (black) and actual optimal (red) velocity agitations. The parabolic agitation appears to achieve greater flux (by about 3.2%), but the reason for this is that it contains more energy (about 6.7% more) than the true optimal agitation. The lesson here is that, to an excellent approximation, one can in practical situations achieve the optimal agitation using cross-channels. ■

Chapter 5

Controlling unsteady flow barriers

You can act to change and control your life; and the procedure, the process is its own reward.
—Amelia Earhart

5.1 ▪ Leading-order hyperbolic trajectory control in $\mathbb{R}^2$

This chapter addresses the question of *control*: whether important flow entities in unsteady flows can be made to follow a user-specified time variation. This form of control is different from the standard control theory literature, in which (usually) the goal is to ensure that the the long-term behavior of the system follows a particular trend, such as following a periodic trajectory or approaching a fixed point. These long-term trends are usually *unstable* entities, hence the need to apply a control to achieve the desired outcome, such as in "controlling chaos" [301, 146, 243, 77]. A basic idea in this area is the pioneering OGY method due to Ott, Grebogi, and Yorke [301], in which an "unsteady saddle" (a hyperbolic trajectory according to the definitions used here) is stabilized by pushing trajectories onto the stable manifold (see also [428]). A related chaos control method is due to Pyragas [326], which employs a continuous feedback to once again stabilize unstable entities. The reader is referred to the review [147] for the considerable literature in the control of chaos.

Unlike in classical control theory, the focus here is not on stabilization, or long-term attractiveness, of some trajectory. Rather, it is in determining how to control entities associated with flow barriers, notably hyperbolic trajectories, and stable and unstable manifolds, in some manner. The typical question would be if a particular time-variation of such an entity is specified, what nonautonomous control velocity would be required in order to achieve this time-variation? In this section, the focus is on hyperbolic trajectories in $\mathbb{R}^2$, for which rigorous results can be presented. This will be extended to formal results in $\mathbb{R}^n$ in Section 5.2 and then to local (Section 5.3) and global (Section 5.4) control of stable and unstable manifolds in the subsequent sections.

Hyperbolic trajectories are of course the nonautonomous analogue of saddle stagnation points. The importance of saddle stagnation points in steady flow situations is well established in fluids applications ranging from groundwater modeling [253, 405, e.g.], macro- and micro-mixing devices [166, 13, 65, 427, 379, 207, 272, e.g.], and oceanographic flows [419, 23, 314, 125, 86, 349, e.g.]. In groundwater modeling, for example, these saddle points essentially establish the watershed, which consists of stable or unstable manifold curves coming into the saddle point. Thus, controlling the location of saddle-like entities [13, 207, 379, e.g.] is of interest in fluid flows. Since hyperbolic trajectories in nonautonomous flows also have attached to them time-varying stable and unstable manifolds which once again partition the phase space into time-varying regions of distinct behavior, being able to control them would impact fluid transport. Thus, the control of hyperbolic trajectories is connected to the the research area of controlling and optimizing fluid mixing, which has recently elicited much interest [267, 246, 117, 408, 379, 207, 25, 31, 42, 145, 44, 361, 151, 182, 255, e.g.].

In fluids applications, the physical space is the phase space of the dynamical system. However, controlling hyperbolic trajectories is of interest even when the phase space is not physical space. As an example think of a nonlinear oscillator subject to time-dependent forcing [183, 243, 239, 222, 404, 304]. Chaotic behavior of such entities is well-understood in terms of intersections of stable and unstable manifolds of hyperbolic trajectories, and thus the set of initial conditions which experiences chaotic motion is located in regions strongly tied to the hyperbolic trajectories. Indeed, the associated manifolds sometimes form boundaries for basins of attraction, and thus controlling hyperbolic trajectories can have implications in defining stability boundaries of biological and mechanical systems [7, 139, 344, e.g.]. Since the trajectory will in general have both a stable and an unstable manifold, both varying nonautonomously, it is not stable. An intriguing idea is whether classical control theory can be *combined* with the methods of this section to achieve a long-term nonautonomous behavior which is prescribed but which would be unachievable without control because of the instability of this behavior.

The flow to be considered in this section will take the form

$$\dot{x} = u(x) + c(x,t), \tag{5.1}$$

in which full information is assumed for the velocity u, and c is the control velocity which will be used for the desired outcome. Here, $x \in \Omega$, which is a *two*-dimensional connected open set, and the uncontrolled velocity u is *steady*. Both these restrictions will be relaxed in Section 5.2, but for this section, the hypotheses are given as follows.

Hypothesis 5.1. *The uncontrolled system (5.1) with $c \equiv 0$ satisfies the following:*

1. $u : \Omega \to \mathbb{R}^2$ *is such that* $u \in C^2(\Omega)$.
2. *There exists $a \in \Omega$ such that $u(a) = 0$, and $Du(a)$ has one positive and one negative eigenvalue, given, respectively, by $\lambda_u > 0$ and $\lambda_s < 0$, with corresponding normalized eigenvectors v_u and v_s.*

With the control velocity c switched off, (5.1) has a hyperbolic saddle point a with one-dimensional stable and unstable manifolds $\Gamma^{s,u}$ as pictured in Figure 2.3. The goal is to determine c such that a *specified* trajectory (sufficiently close to a) is a hyperbolic trajectory of (5.1).

Theorem 5.2 (control velocity in $\mathbb{R}^2$ [52]). *Assume Hypothesis 5.1 is satisfied. Let $\tilde{a}(t)$ be a desired hyperbolic trajectory of (5.1), for which there exists ε such that*

$$\left|\frac{\partial \tilde{a}(t)}{\partial t}\right| + |\tilde{a}(t) - a| \leq \varepsilon \quad \textit{for all } t \in \mathbb{R}.$$

Suppose the control velocity c is subject to the existence of C such that

$$|c(x,t)|+\|Dc(x,t)\|+\left|\frac{\partial c(x,t)}{\partial t}\right|\le C\varepsilon \quad \textit{for all } (x,t)\in\Omega\times\mathbb{R},$$

and the specification

$$c(a,t)=\gamma_u(t)v_u^{\perp}+\frac{\gamma_u(t)(v_u\cdot v_s)-\gamma_s(t)}{v_u^{\perp}\cdot v_s}v_u, \tag{5.2}$$

in which c's projections in the $v_u^{\perp}$ and $v_s^{\perp}$ directions are, respectively,

$$\gamma_u(t):=\left[\dot{\tilde{a}}(t)-\lambda_s(\tilde{a}(t)-a)\right]\cdot v_u^{\perp} \quad \textit{and} \quad \gamma_s(t):=\left[\dot{\tilde{a}}(t)-\lambda_u(\tilde{a}(t)-a)\right]\cdot v_s^{\perp}. \tag{5.3}$$

Then, there exist a hyperbolic trajectory $a(t)$ of (5.1) and a constant K such that

$$|\tilde{a}(t)-a(t)|\le\varepsilon^2K \quad \textit{for all } t\in\mathbb{R}.$$

Proof. The main observation here is that Theorem 5.2 is an "inversion" of Theorem 2.7, with the choice $f(x)\equiv u(x)$. Suppose c were chosen according to the given specification. By the fact that c is bounded by $C\varepsilon$, one can express c in the form and the choice $c(x,t)=\varepsilon g(x,t)$, where there exists a constant C such that $|g(x,t)|+\|Dg(x,t)\|\le C$ for all $(x,t)\in\Omega\times\mathbb{R}$. Thus, Theorem 2.7 applies, and there exists a hyperbolic trajectory $a(t)$ as shown in Figure 2.3. The projection of $a(t)-a$ in the $v_s^{\perp}$ and $v_u^{\perp}$ at each time t are given by (2.15). Thus, if

$$\beta_s(t)=-\int_0^{\infty}g(a,t+\tau)\cdot v_s^{\perp}e^{-\lambda_u\tau}\,\mathrm{d}\tau \quad \text{and} \tag{5.4}$$

$$\beta_u(t)=\int_0^{\infty}g(a,t-\tau)\cdot v_u^{\perp}e^{\lambda_s\tau}\,\mathrm{d}\tau, \tag{5.5}$$

then $[a(t)-a]\cdot v_{s,u}^{\perp}=\varepsilon\beta_{s,u}(t)+\varepsilon^2E_{s,u}(t)$, where there exists $K_{s,u}$ such that $|E_{s,u}(t)|+|\dot{E}_{s,u}(t)|\le K_{s,u}$ for all $t\in\mathbb{R}$. The ability to bound the t-derivative of the error term $E_{s,u}$ comes from the extra condition imposed on c: that its t-derivative is bounded. This was not required for Theorem 2.7, in which essentially the conclusion is that $E_{s,u}(t)$ was bounded, but not necessarily its derivative. Bounding the derivative only requires the understanding that a t-derivative needs to be taken after the expression for $a(t)-a$ is derived, and since the error terms include c within them, all that is needed is to have the t-derivative of c bounded. In any case, for the $v_s^{\perp}$ component, using (5.4) with $c=\varepsilon g$ gives

$$\begin{aligned}[a(t)-a]\cdot v_s^{\perp}&=-\int_0^{\infty}c(a,t+\tau)\cdot v_s^{\perp}e^{-\lambda_u\tau}\,\mathrm{d}\tau+\varepsilon^2E_s(t)\\&=-\int_t^{\infty}c(a,\mu)\cdot v_s^{\perp}e^{-\lambda_u(\mu-t)}\,\mathrm{d}\mu+\varepsilon^2E_s(t)\\&=e^{\lambda_u t}\int_{\infty}^{t}c(a,\mu)\cdot v_s^{\perp}e^{-\lambda_u\mu}\,\mathrm{d}\mu+\varepsilon^2E_s(t),\end{aligned}$$

and so

$$e^{-\lambda_u t}[a(t)-a]\cdot v_s^{\perp}=\int_{\infty}^{t}c(a,\mu)\cdot v_s^{\perp}e^{-\lambda_u\mu}\,\mathrm{d}\mu+\varepsilon^2E_s(t)e^{-\lambda_u t}.$$

Now, differentiating the above with respect to t yields

$$e^{-\lambda_u t}\left[\dot{a}(t)-\lambda_u(a_\varepsilon(t)-a)\right]\cdot v_s^\perp = c(a,t)\cdot v_s^\perp e^{-\lambda_u t}+\varepsilon^2\left[\dot{E}_s(t)-\lambda_u E_s(t)\right]e^{-\lambda_u t},$$

which gives

$$c(a,t)\cdot v_s^\perp=\left[\dot{a}(t)-\lambda_u(a(t)-a)\right]\cdot v_s^\perp+\varepsilon^2\tilde{K}_s(t),$$

in which $\tilde{K}_s(t)$ is bounded (call the bound $\tilde{K}_s$). However, c was chosen to satisfy

$$c(a,t)\cdot v_s^\perp=\gamma_s(t)=\left[\dot{\tilde{a}}(t)-\lambda_u(\tilde{a}(t)-a)\right]\cdot v_s^\perp,$$

and hence

$$\left[\dot{\tilde{a}}(t)-\lambda_u(\tilde{a}(t)-a)\right]\cdot v_s^\perp=\left[\dot{a}(t)-\lambda_u(a(t)-a)\right]\cdot v_s^\perp+\varepsilon^2\tilde{K}_s(t).$$

Defining $\eta(t):=\left[\tilde{a}(t)-a(t)\right]\cdot v_s^\perp$, this gives the equation

$$\dot{\eta}(t)-\lambda_u\eta(t)=\varepsilon^2\tilde{K}_s(t).$$

Multiplying by an integrating factor and integrating from a general t to ∞ yields

$$e^{-\lambda_u t}\eta(t)-\lim_{\tau\to\infty}e^{-\lambda_u\tau}\eta(\tau)=\varepsilon^2\int_t^\infty e^{-\lambda_u\tau}\tilde{K}_s(\tau)\,\mathrm{d}\tau.$$

Now, each of $a(t)$ and $\tilde{a}(t)$ remain $\mathcal{O}(\varepsilon)$-close to a for all t, and thus $\eta(\tau)$ remains bounded as $\tau\to\infty$. Therefore,

$$\left|e^{-\lambda_u t}\eta(t)\right|\le\tilde{K}_s\varepsilon^2\int_t^\infty e^{-\lambda_u\tau}\,\mathrm{d}\tau=\varepsilon^2\frac{\tilde{K}_s}{\lambda_u}e^{-\lambda_u t},$$

which leads to the bound $|\eta(t)|\le\varepsilon^2\tilde{K}_s/\lambda_u$. Working similarly with (5.4) enables bounding $[\tilde{a}(t)-a(t)]\cdot v_u^\perp$, and thus the difference between the desired hyperbolic trajectory $\tilde{a}(t)$ and the actual hyperbolic trajectory $a(t)$ remains $\mathcal{O}(\varepsilon^2)$. □

Example 5.3 (four-roll mill). The four-roll mill was initially conceptualized by Taylor [398] for manipulating deformations of droplets which are positioned between four rolling cylinders. By varying the roller speeds, it is possible to obtain a combination of straining and rotational motion at the origin, providing a mechanism for different forms of drop deformation [65, 262, 431, 423, 237, 210, 418, 208, 388]. A major difficulty in such four-roll mills is maintaining a particle precisely at the point of interest, particularly under unsteady rotation protocols. If the rollers are rotated at the same angular speed ω with directionality as shown in Figure 5.1, the origin would be a saddle point, with a droplet positioned there experiencing elongation along the unstable manifold direction (i.e., x_2-direction). This is a situation in which there is pure straining motion at the origin, and it will be assumed that the strain rate induced is $\mu>0$. That is, the equation of motion near the origin will take the form

$$\dot{x}_1=-\mu x_1,\quad \dot{x}_2=\mu x_2,$$

and so the quantities associated with Hypothesis 5.1 are $\lambda_u=\mu$, $v_u=(0,1)$, $\lambda_s=-\mu$, and $v_s=(-1,0)$. (The eigenvectors could be taken to be in the opposite direction, as

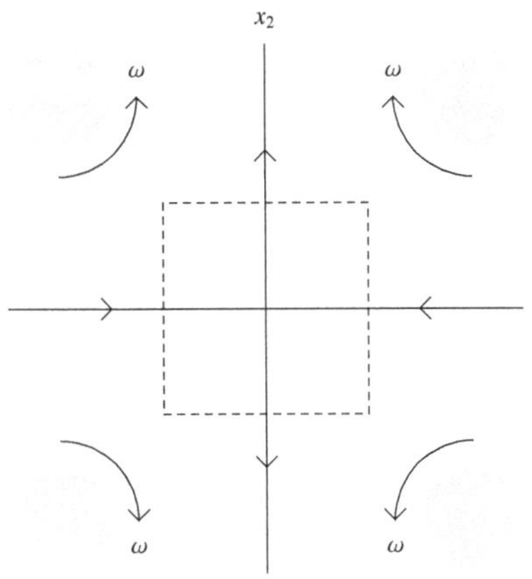

Figure 5.1. *The four-roll mill with rotations leading to a hyperbolic point at the origin.*

long as one is consistent in this; here, the choice of using the two branches of manifolds which border the first quadrant has been used for defining $v_{s,u}$.) Now, suppose that it is required to control the flow of the four-roll mill such that the hyperbolic trajectory moves with time, according to $\tilde{a}(t) = \delta\,(x_1(t), x_2(t))$ for specified $x_i(t)$ and $\delta > 0$. From the conditions in Theorem 5.2, this means that one can take ε according to

$$\varepsilon = \delta \sup_{t \in R} \left(\sqrt{x_1(t)^2 + x_2(t)^2} + \sqrt{\dot{x}_1(t)^2 + \dot{x}_2(t)^2} \right),$$

and the errors of the control procedure would be bounded by $\varepsilon^2 K$ for some K. Using (5.2), the control velocity must satisfy

$$c(0,t) = \delta \begin{pmatrix} \dot{x}_2(t) - \mu x_2(t) \\ -\dot{x}_1(t) - \mu x_1(t) \end{pmatrix}. \tag{5.6}$$

If, for example, it was required to have the hyperbolic trajectory rotating near the origin, one could take $x_1(t) = \cos\omega_0 t$ and $x_2(t) = \sin\omega_0 t$; a detailed evaluation of the efficacy of the control strategy is performed in [52], which demonstrates excellent performance even with $\delta = 0.1$. If the hyperbolic trajectory is required to remain on the curve $x_2 = x_1^3$, one could choose arbitrary time-dependence in the form $x_1(t) = h(t)$ and $x_2(t) = h(t)^3$, for which the control condition is

$$c(0,t) = \delta \begin{pmatrix} 3h(t)^2 \dot{h}(t) - \mu h(t)^3 \\ -\dot{h}(t) - \mu h(t) \end{pmatrix}.$$

If, for example, a steady location on this curve is chosen (with $h(t) = h$, a constant), the above would reduce to the steady control velocity $-\delta\mu h\left(h^2, 1\right)^\top$. For several other examples, with numerical verification and error analysis, the reader is referred to [52]. A practical factor here is how one might actually *achieve* the desired control velocity at the origin. The actual control that is available is the rotation of the rollers; a time-varying $\varepsilon\omega_i(t)$ perturbation can be added to the constant angular velocity ω for each roller. The exact relationship between this added nonautonomous rotation and the velocity it generates at the origin is somewhat unclear, but a simple kinematic argument would be that each $\delta\omega_i(t)$ generates a velocity $\delta u_i(t)$ in the direction perpendicular to the line connecting the orgin to the center of the roller, where the $u_i(t)$ is proportional to $\omega_i(t)$. If each of the $\delta\omega_i(t)$s were taken to be positive in the counterclockwise direction, the $x_{1,2}$-components of the combined effect could be set equal to the required control (5.6), and (with some assumptions on the proportionality constants), it will be possible to find the required $\omega_i(t)$s. There will be a nonunique way of doing this, since there will be two equations for the four unknowns. ■

Example 5.4 (Duffing oscillator cont.). The Duffing oscillator [183, 243, 239, 222, 404, 254], previously examined in Examples 2.11, 2.21, 2.32, 3.34, and 3.16, is now reconsidered. With damping $\gamma > 0$ and a time-varying forcing $\varepsilon\phi(x_1, x_2, t)$, the dynamics are given by

$$\frac{d}{dt}\begin{pmatrix} x_1 \\ x_2 \end{pmatrix} = \begin{pmatrix} x_2 \\ x_1 - x_1^3 - \gamma x_2 + \phi(x_1, x_2, t) \end{pmatrix}. \tag{5.7}$$

Note that this is slightly different from (2.19) in that the forcing is now permitted to depend on both x_1 and $x_2 = \dot{x}_1$ (i.e., the position and velocity of the oscillator), in addition to just time. This forcing can be thought of as a control whose amplitude is chosen depending on the current position, velocity, and time, in some way which is to be chosen in order to achieve a desired behavior of the oscillator. With $\phi = 0$ and $\gamma = 0$, the phase portrait of (5.7) is shown in Figure 2.7. When $\gamma > 0$ but is small, the branches of the unstable manifold of the origin extending into the first and third quadrants spiral into the fixed points at $(1,0)$ and $(-1,0)$, respectively; these are now attracting fixed points. The branches of the stable manifold coming in from the second and fourth quadrants, "spiral outwards," wrapping around the figure-eight structure visible in Figure 2.7 but getting further away each time a full circle is completed. The curve along this stable manifold forms the separator between the basins of attraction of $(1,0)$ and $(-1,0)$ (see, for example, Figure 2.2.7 in [183] and Figure 8 in [52]). Being the stable manifold of the hyperbolic point $(0,0)$, this curve which demarcates the eventual fate of trajectories is intimately tied to the location of the hyperbolic point. Now, by turning on the forcing, the question that is to be asked is whether the hyperbolic trajectory could be moved nonautonomously in some specified fashion, thereby adjusting the location of a point to which the time-varying basin boundaries are anchored. This would be a preliminary step in being able to control the basin boundaries in a nonautonomous fashion. Thus, the form of the control forcing $\phi(x_1, x_2, t)$ is sought, in order to achieve a *specified* hyperbolic trajectory location $\tilde{a}(t) = \delta\,(x_1(t), x_2(t))$. Note, however, that in this case $x_2(t) = \dot{x}_1(t)$ and cannot be specified independently, and thus the requirement is that

$$\varepsilon = \delta \sup_{t \in R}\left(\sqrt{x_1(t)^2 + \dot{x}_1(t)^2} + \sqrt{\dot{x}_1(t)^2 + \ddot{x}_1(t)^2}\right),$$

based on which the error will be $\mathcal{O}(\varepsilon^2)$. (One can indeed set $\varepsilon = K\delta$, with K arising from the supremum above; thus, $\mathcal{O}(\varepsilon^2)$ is no different from saying $\mathcal{O}(\delta^2)$.) The problem now boils down to determining the control forcing ϕ in terms of the quantity $x_1(t)$. The stable eigenvalue and its corresponding eigenvector associated with the origin of (5.7) when there is no forcing is

$$\lambda_s = \frac{-\gamma - \sqrt{4+\gamma^2}}{2}, \quad v_s = \frac{1}{\sqrt{4 + \left(\gamma - \sqrt{4+\gamma^2}\right)^2}}\begin{pmatrix} \gamma - \sqrt{4+\gamma^2} \\ 2 \end{pmatrix}.$$

The unstable counterparts are

$$\lambda_u = \frac{-\gamma + \sqrt{4+\gamma^2}}{2}, \quad v_u = \frac{1}{\sqrt{4+\left(\gamma+\sqrt{4+\gamma^2}\right)^2}} \begin{pmatrix} \gamma + \sqrt{4+\gamma^2} \\ 2 \end{pmatrix}.$$

The expressions (5.3) are complicated, but when finally evaluating (5.2) after some algebra, the expression

$$c(0,0,t) = \delta \begin{pmatrix} 0 \\ \ddot{x}_1(t) + \gamma \dot{x}_1(t) - x_1(t) \end{pmatrix}$$

results. However, this control must be consistent with the perturbing term in (5.7). Thus, the controlling force (with the abuse of notation of suppressing the second argument, which is the derivative of the first) needs to satisfy

$$\phi(x_1, t) = \delta \left[\ddot{x}_1(t) + \gamma \dot{x}_1(t) - x_1(t) \right] \tag{5.8}$$

in order to achieve a user-specified hyperbolic trajectory $\delta\,(x_1(t), \dot{x}_1(t))$. Numerical investigations of this result are available in [52]. ∎

The straightforward expression (5.8) which emerged from the usage of Theorem 5.2 is suggestive. The *formal* substitution of $a(t) = \delta\,(x_1(t), x_2(t))$ into (5.7) in order to ensure that it is a solution to the equation yields $x_2(t) = \dot{x}_1(t)$ and

$$\delta \ddot{x}_1(t) = \delta x_1(t) - \delta^3 x_1(t)^3 - \delta \gamma \dot{x}_1(t) + \phi(x_1, t).$$

Now, if higher-order terms in ε are discarded from the above expression, then (5.8) appears immediately! This formal process—unlike Theorem 5.2—does *not* ensure that higher-order terms do not contribute. In other words, there is no proof that the hyperbolic trajectory resulting from this choice is correct to $\mathcal{O}(\varepsilon)$, whereas this is an integral part of Theorem 5.2. Somewhat mysteriously, the *hyperbolicity* seems not to have been a part of this "derivation" at all. The reason why this works is somewhat hidden, and it turns out that the hyperbolicity is crucial to why it works. This formal approach will enable the control of hyperbolic trajectories to be enunciated for higher dimensions than 2 and also for the situation in which the unperturbed flow (associated with the vector field u in this case) is itself nonautonomous. This is the content of the next section.

5.2 ▪ Higher-order hyperbolic trajectory control in $\mathbb{R}^n$

The previous section provided a method for controlling a hyperbolic trajectory, with a rigorous justification (Theorem 5.2) that the method worked. In doing so, there were several assumptions: the flow was in two spatial dimensions, it was nearly autonomous (measured by a parameter ε), and the hyperbolic trajectory was saddle-like in that it possessed one-dimensional stable and unstable manifolds, respectively. Under these conditions, the control velocity that was needed to move a saddle point to a nearby time-varying hyperbolic trajectory was quantified, and the fact that the resulting error was higher-order was proven. It was noticed in the initial development of these results [52] that a *formal* perturbative approach led to identical expressions. The purpose of this section is to adapt this formal approach, with some intuitive justification for its applicability, in relaxing the assumptions used. That is, the flow will now be in $\mathbb{R}^n$

for arbitrary $n \geq 2$, it will not need to be nearly autononomous, and the concept of hyperbolicity will encompass a range of trajectories which include saddle-like, purely attracting or purely repelling fixed points *and also nonfixed points*. Examples of the latter include limit cycles or other trajectories to which there is exponential attraction, repulsion, or both; these are all encapsulated in the concept of exponential dichotomies (1.22). Moreover, rather than having the error be $\mathcal{O}(\varepsilon^2)$, a method for determining the control velocity in order to achieve *arbitrarily* high-order accuracy in ε will be presented. This section is based on the results of [53].

A particular example of such a trajectory for a nonautonomous flow in $\mathbb{R}^n$ would be a purely attracting trajectory. All initial conditions chosen in a neighborhood of this trajectory will get attracted towards it as time progresses. This is therefore an *attractor* in whose time-variation is encoded the eventual behavior of *all* sufficiently nearby trajectories. Being able to control such a trajectory has clear implications on controlling the *system* in the classical sense of control literature [301, 75, 326], since the eventual behavior of the system will follow the controlled trajectory. This will allow the long-term behavior to be made *time-varying* in a desired fashion; essentially, trajectories will approach a user-specified *nonautonomous attractor*. Such a method would extend more classical control strategies which, for example, attempt to have a chaotic system approach a *fixed* point [301, 18, 213, 435, 393], a periodic orbit [209, 326, 327, 313, 401, 320], or even a chaotic regime [371] by taking advantage of the relevant entity's stable manifold.

Rather than the more limited situation of (5.1), here the flow results from the system

$$\dot{x} = u(x,t) + c(x,t,\varepsilon), \tag{5.9}$$

for a *parameter-dependent* control velocity c, with Hypothesis 5.1 replaced by the following.

Hypothesis 5.5. *The system (5.9) satisfies the following:*

(a) $u : \Omega \times \mathbb{R} \to \mathbb{R}^n$, $n \geq 1$, *is such that* $u \in C^{r+1}(\Omega \times \mathbb{R})$, *where* $r \geq 1$, *and moreover there exists a constant* C_u *such that*

$$\sup_{t \in \mathbb{R}} \left(\|u(\cdot,t)\|_{r+1,\infty} + \left\| \frac{\partial u}{\partial t}(\cdot,t) \right\|_{r+1,\infty} \right) \leq C_u,$$

where $\|\cdot\|_{r+1,\infty}$ *is the Sobolev norm on* Ω *(i.e., all spatial derivatives up to the* $(r+1)$*st order are bounded).*

(b) *The control velocity* $c : \Omega \times \mathbb{R} \times [0, \varepsilon_0) \to \mathbb{R}^n$ *may be chosen subject to*

- $c(x,t,0) = 0$ *for all* $(x,t) \in \Omega \times \mathbb{R}$;
- $c(\cdot,t,\varepsilon) \in C^{r+1}(\Omega)$ *for all* $(t,\varepsilon) \in \mathbb{R} \times [0,\varepsilon_0)$;
- *there exists a constant* C_c *such that*

$$\sum_{i=0}^{r+1} \left\| \frac{\partial^i c(x,t,\varepsilon)}{\partial \varepsilon^i} \right\|_{r+1,\infty} \leq C_c \quad \textit{for } (x,t,\varepsilon) \in \Omega \times \mathbb{R} \times [0,\varepsilon_0). \tag{5.10}$$

(c) *When* $\varepsilon = 0$, $a^{(0)}(t)$ *is a hyperbolic trajectory of (5.9) in the sense of the exponential dichotomy conditions (1.22), where the projection matrix* P *may be either the identity or the zero matrix, or any other projection.*

Recall from the form in which the exponential dichotomy conditions were stated in (1.22) that the projection P was onto the unstable manifold. If P were the zero matrix, this means that only the second of the exponential dichotomy conditions (1.22) comes into operation (the first equation in (1.22) is automatically satisfied since the left-hand side is zero). This implies that $I-P=I$ projects onto the entire space $\mathbb{R}^n$, i.e., that the stable manifold of the hyperbolic trajectory is *full-dimensional*. In other words, the hyperbolic trajectory $a^{(0)}(t)$ is attracting, with the attraction occurring at an exponential rate. Thus, the behavior is *not* saddle-like in the sense of there being both a stable *and* an unstable manifold; there is only one of these present. Using the term "hyperbolic" for this situation is analogous to the classical characterization of a fixed point as hyperbolic if it has no eigenvalues on the imaginary axis, without necessarily insisting on the presence of eigenvalues in *both* the left- and right-half planes. If P is the identity matrix, that would mean that the second of the exponential dichotomy conditions (1.22) is redundant while the first applies. This would mean that the hyperbolic trajectory $a^{(0)}(t)$ possesses a full-dimensional unstable manifold, and is purely repelling. In general, $I-P$ would project onto an m-dimensional subspace of $\mathbb{R}^n$ with $0 \le m \le n$ corresponding to the local stable manifold. The equalities correspond to the two degenerate situations discussed so far, but if $0<m<n$, then $a(t)$ would have an m-dimensional stable manifold and an $(n-m)$-dimensional unstable manifold when pictured in any time-slice. In the augmented $(n+1)$-dimensional phase space (i.e., $\dot{x}=u(x,t)$, $\dot{t}=1$), the hyperbolic trajectory $\left(a^{(0)}(t),t\right)$ would possess an $(m+1)$-dimensional stable manifold and a $(n-m+1)$-dimensional unstable manifold. Therefore, unlike in previous sections, the situations of the hyperbolic trajectory being either purely attractive or purely repulsive *are permitted* in this section.

The implication of hyperbolicity (Hypothesis 5.5(c)) is that there is a tubular neighborhood $\mathcal{V} := \cup_{t\in\mathbb{R}} \mathcal{U}_t$ in which $\mathcal{U}_t \subset \Omega$ is an open neighborhood of $a^{(0)}(t)$ at each time t, such that the *only* trajectory of $\dot{x}=u(x,t), \dot{t}=1$ which remains within $\mathcal{V}$ in both backwards and forwards time is $\left(a^{(0)}(t),t\right)$. This is because if there is a stable manifold (if $m>0$), then in backwards time nearly all nearby trajectories will get pulled away from $a^{(0)}(t)$ due to its influence. The only exceptions to this are trajectories which lie precisely on the unstable manifold, which will get attracted towards the hyperbolic trajectory. Thus, if there is a stable manifold, then all trajectories which are not on the unstable manifold exit $\mathcal{V}$ as $t\to-\infty$. Conversely, if there is an unstable manifold (if $m<n$), then all trajectories which are not on the stable manifold exit $\mathcal{V}$ as $t\to\infty$. It is therefore only trajectories which are on *both* the stable and the unstable manifold which remain within $\mathcal{V}$ for all $t\in\mathbb{R}$. This is trivially the set $\left(a^{(0)}(t),t\right)$ itself. This statement is still true even if one or the other of the stable and unstable manifolds did not exist, that is, if $m=0$ or $m=n$.

The basic premise now is that a desired hyperbolic trajectory $a(t,\varepsilon)$ for (5.9) is *specified*, and the control velocity $c(x,t,\varepsilon)$ is to be found in order to achieve this. For this section only, the notation $a(t,\varepsilon)$ will be used for $a_\varepsilon(t)$ since ε-derivatives will need to be specified. An accuracy of ε^r is sought in the process; this is why the $(r+1)$st spatial derivatives are assumed bounded in Hypothesis 5.5. It must be emphasized that the approach to be discussed, unlike much of the previous development, is purely formal, and thus this section will not state any theorems but rather will explain the process to be followed.

Remark 5.6 (controlling the nonautonomous attractor). The instance in which $m=n$ (i.e., the stable manifold of $a^{(0)}(t)$ is full-dimensional and hence forms an attractor) is particularly interesting. In this case, a nearby attractor persists under perturbation (see, for example, Theorem 11.1 in [224]). This attractor is of course the perturbed hyperbolic trajectory. The control process to be outlined describes how this attractor's time-variation can be *prescribed.*

Hypothesis 5.7. *The desired hyperbolic trajectory $a(t,\varepsilon)$ of (5.9) satisfies the following:*

(a) $a(t,0)=a^{(0)}(t)$.

(b) *There exists a constant A such that*

$$\sum_{i=0}^{r+1}\left(\left|\frac{\partial^i a(t,\varepsilon)}{\partial \varepsilon^i}\right|+\left|\frac{\partial^i}{\partial \varepsilon^i}\frac{\partial a(t,\varepsilon)}{\partial t}\right|\right)\leq A \quad \text{for } t\in\mathbb{R}. \tag{5.11}$$

Since $a(t,\varepsilon)$ is required to be a hyperbolic trajectory of (5.9) and is close to $a^{(0)}(t)$ for sufficiently small ε because of Hypothesis 5.7, it too inherits a tubular neighborhood $\mathscr{V}(\varepsilon)$ to which the only trajectory confined for $t\in\mathbb{R}$ is $(a(t,\varepsilon),t)$. Within $\mathscr{V}(\varepsilon)$, each component $a_i(t,\varepsilon)$ $(i=1,2,\ldots,n)$ of $a(t,\varepsilon)$ possesses a Taylor expansion

$$a_i(t,\varepsilon)=a_i^{(0)}(t)+\varepsilon a_i^{(1)}(t)+\varepsilon^2 a_i^{(2)}(t)+\cdots+\varepsilon^r a_i^{(r)}(t)+\frac{\varepsilon^{r+1}}{(r+1)!}\frac{\partial^{r+1}a_i(t,\tilde{\varepsilon})}{\partial\varepsilon^{r+1}}, \tag{5.12}$$

where $\tilde{\varepsilon}$ is between 0 and ε, and the Taylor coefficients are

$$a_i^{(j)}(t):=\frac{1}{j!}\frac{\partial^j a_i}{\partial\varepsilon^j}(t,0),$$

with the superscript $j\in\{0,1,\ldots,r\}$ identifying the order of the ε-derivative. Now since $a(t,\varepsilon)$ is the *only* trajectory which remains within $\mathscr{V}(\varepsilon)$ in both backwards and forwards time, if this Taylor expansion is considered valid for all times, it *must* represent exactly a hyperbolic trajectory. Thus, (5.12) unambiguously represents the perturbed hyperbolic trajectory, and not just any other trajectory in the flow. This intuitive argument—for which hyperbolicity is a crucial ingredient—is the reason why a formal expansion works for exactly the hyperbolic trajectory and for no other solution.

The goal now is to determine the control velocity c, which in terms of its Taylor expansion in ε reads

$$c_i(x,t,\varepsilon)=\varepsilon c_i^{(1)}(x,t)+\varepsilon^2 c_i^{(2)}(x,t)+\cdots+\varepsilon^r c_i^{(r)}(x,t)+\frac{\varepsilon^{r+1}}{(r+1)!}\frac{\partial^{r+1}c_i(x,t,\tilde{\varepsilon})}{\partial\varepsilon^{r+1}} \tag{5.13}$$

within $\mathscr{V}$, in component form. There is no ε^0 term since $c(x,t,0)$ is assumed to be zero; "turning on" the parameter ε turns on the control. The question now is to determine the coefficients of (5.13), which shall be done in a step-by-step fashion in terms of the known quantities, namely, $u(x,t)$ and the coefficients $a_i^{(0)}(t)$, $a_i^{(1)}(t)$, $a_i^{(2)}(t)$, etc.

Using the component notation proves convenient in the expressions to be derived. Substituting the Taylor expansions (5.12) and (5.13) into the ith component of (5.9) gives

$$\begin{aligned}
&\dot a_i^{(0)}+\varepsilon\dot a_i^{(1)}+\varepsilon^2\dot a_i^{(2)}+\cdots+\varepsilon^r\dot a_i^{(r)}+\mathcal{O}(\varepsilon^{r+1})=u_i\\
&+\left(\varepsilon a_j^{(1)}\frac{\partial u_i}{\partial x_j}+\varepsilon^2 a_j^{(2)}\frac{\partial u_i}{\partial x_j}+\frac{\varepsilon^2}{2}a_j^{(1)}a_k^{(1)}\frac{\partial^2 u_i}{\partial x_j\partial x_k}\right)\\
&+\left(\varepsilon c_i^{(1)}+\varepsilon^2 a_j^{(1)}\frac{\partial c_i^{(1)}}{\partial x_j}\right)+\varepsilon^2 c_i^{(2)}+\mathcal{O}(\varepsilon^3)
\end{aligned}\tag{5.14}$$

for each component i. Here, each of the quantities on the right is evaluated on the hyperbolic trajectory $\left(a^{(0)}(t),t\right)$ of (5.9) when $\varepsilon=0$, and the Einstein summation convention has been used: when a sub/superscript is repeated, it is assumed that these are summed over for each term. Since the terms become increasingly complicated, the right-hand side has been written out accurately to $\mathcal{O}\left(\varepsilon^2\right)$ only. Setting the order ε^0 term to zero gives the fact that $x\doteq a_i^{(0)}(t)=u_i\left(x^{(0)}(t),t\right)$, which is Hypothesis 5.5(c). The $\mathcal{O}(\varepsilon)$ term of (5.14) gives

$$c_i^{(1)}\left(a^{(0)}(t),t\right)=\dot a_i^{(1)}(t)-a_j^{(1)}(t)\frac{\partial u_i}{\partial x_j}\left(a^{(0)}(t),t\right),\tag{5.15}$$

which specifies the first-order term of the control velocity (5.13) on the unperturbed hyperbolic trajectory $\left(a^{(0)}(t),t\right)$. Taking the $\mathcal{O}\left(\varepsilon^2\right)$ term of (5.14), one gets

$$\begin{aligned}
c_i^{(2)}\left(a^{(0)}(t),t\right)&=\dot a_i^{(2)}(t)-a_j^{(2)}(t)\frac{\partial u_i}{\partial x_j}\left(a^{(0)}(t),t\right)-\frac{1}{2}a_j^{(1)}(t)a_k^{(1)}(t)\frac{\partial^2 u_i}{\partial x_j\partial x_k}\left(a^{(0)}(t),t\right)\\
&-a_j^{(1)}(t)\frac{\partial c_i^{(1)}}{\partial x_j}\left(a^{(0)}(t),t\right).
\end{aligned}\tag{5.16}$$

A difficulty has emerged: (5.16) does not express $c_i^{(2)}$ in terms of known quantities because of the spatial derivative of the control velocity appearing on the right. This can be resolved by using (5.15). Think of setting

$$c_i^{(1)}(x,t)=\dot a_i^{(1)}(t)-a_j^{(1)}(t)\frac{\partial u_i}{\partial x_j}(x,t)\tag{5.17}$$

in a neighborhood around $\left(a^{(0)}(t),t\right)$, that is, in the conceptual tubular neighborhood $\mathcal{V}$. This is consistent with the first-order condition (5.15) and retains the option of extending c beyond $\mathcal{V}$ in any fashion. Differentiating (5.17) with respect to a component x_k gives

$$\frac{\partial c_i^{(1)}}{\partial x_k}(x,t)=-a_j^{(1)}\frac{\partial^2 u_i}{\partial x_j\partial x_k}(x,t).$$

This can be evaluated at $\left(a^{(0)}(t),t\right)$ and substituted into the troublesome term of (5.16). The last two terms in (5.16) are then seen to be of the same type and can be combined. Now, this results in an equation only on the unperturbed hyperbolic trajectory as before, and this can be once again extended to $\mathcal{V}$ by the simple strategem of replacing $a^{(0)}(t)$ with x in the argument. This leads to the expression

$$c_i^{(2)}(x,t)=\dot a_i^{(2)}(t)-a_j^{(2)}(t)\frac{\partial u_i}{\partial x_j}(x,t)+\frac{1}{2}a_j^{(1)}(t)a_k^{(1)}(t)\frac{\partial^2 u_i}{\partial x_j\partial x_k}(x,t)\tag{5.18}$$

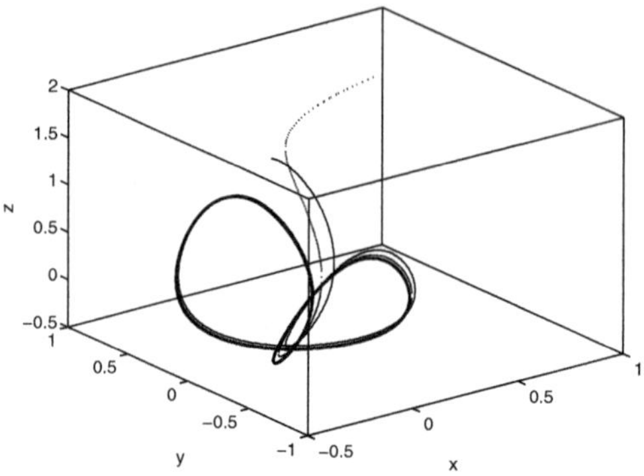

Figure 5.2. *Two initial conditions of the Lorenz system (5.20) subject to the control velocity $c_{(p)}$, showing the approach with high accuracy to the specified periodic attracting trajectory $a_{(p)}$ as given in (5.21). Here, $\varepsilon = 0.6$. [53] Reproduced with permission from The American Physical Society.*

in $\mathcal{V}$, for the second-order term of the control velocity. It is clear that the errors resulting from this procedure would be $\mathcal{O}\left(\varepsilon^3\right)$, with coefficient bounded because of the conditions in Hypotheses 5.5 and 5.7. That is, one can take $r = 2$ here, with the bounds on the third derivative furnishing the requirement for the error to be $\mathcal{O}\left(\varepsilon^3\right)$. If the procedure is continued for one more order, the condition for the third-order term in the control velocity turns out to be

$$c_i^{(3)}(x,t) = \dot{a}_i^{(3)}(t) - a_j^{(3)}(t)\frac{\partial u_i}{\partial x_j}(x,t) + a_j^{(1)}(t)a_k^{(2)}(t)\frac{\partial^2 u_i}{\partial x_j \partial x_k}(x,t) \\ -\frac{1}{6}a_j^{(1)}(t)a_k^{(1)}(t)a_l^{(1)}(t)\frac{\partial^3 u_i}{\partial x_j \partial x_k \partial x_l}(x,t) \tag{5.19}$$

in $\mathcal{V}$. (The calculations for this are not shown since they are straightforward though tedious.) So if the control velocity (5.13) obeys (5.17), (5.18), and (5.19), there will be a hyperbolic trajectory of (5.9) which is $\mathcal{O}\left(\varepsilon^3\right)$-close to $a(t,\varepsilon)$. Clearly, this process can be continued to as high an order (i.e., rth order) as desired, and in general the error for the hyperbolic trajectory will be $\mathcal{O}\left(\varepsilon^{r+1}\right)$.

Remark 5.8 (incompressibility is preserved). If (5.9) is associated with a fluid flow in which the fluid is incompressible, one would like to choose the control velocity in such a way that the velocity field in (5.9) is divergence-free. Fortunately, if $\nabla \cdot u = 0$, then the expressions derived for c in (5.17), (5.18), and (5.19) will also be divergence-free, thereby ensuring consistency.

Example 5.9 (Lorenz system). Here, some numerical validation results first presented in [53] for the celebrated Lorenz system [250] are discussed. The Lorenz system is given by

$$\dot{x} = \sigma(y - x), \quad \dot{y} = x(\rho - z) - y, \quad \dot{z} = xy - \beta z, \tag{5.20}$$

in Cartesian coordinates $\underline{r} = (x,y,z)^{\mathsf{T}}$. When $\rho = 0.5$, $\sigma = 10$, and $\beta = 8/3$, (5.20) is nonchaotic and has a *globally* attracting fixed point at the origin. A control will be applied to ensure that the long-term behavior of (5.20) follows a prescribed nonautonomous trajectory. A simple example would be to try to make (5.20) settle in the long-term to the periodic trajectory

$$a_{(p)}(t)=\begin{pmatrix}0\\0\\0\end{pmatrix}+\begin{pmatrix}\sinh[\varepsilon\sin t\cos t]\\ \sin[\varepsilon\cos t]\\ e^{\varepsilon\cos 2t}-1\end{pmatrix}. \tag{5.21}$$

Following the procedure outlined here (see [53] for details), the control velocity expansion terms (5.17), (5.18), and (5.19) yield

$$c^{(1)}_{(p)}(\underline{r},t)=\begin{pmatrix}\cos 2t+\sigma\cos t(\sin t-1)\\ -(1+(\rho-z)\cos t)\sin t+\cos t+x\cos 2t\\ -2\sin 2t-y\sin t\cos t-x\cos t+\beta\cos 2t\end{pmatrix},$$

$$c^{(2)}_{(p)}(\underline{r},t)=\begin{pmatrix}0\\ \frac{x}{2}\cos^2 2t-\sin t\cos t\cos 2t\\ -2\sin 2t\cos 2t+\frac{\beta}{2}\cos^2 2t+\sin t\cos^2 t\end{pmatrix},\text{ and}$$

$$c^{(3)}_{(p)}(\underline{r},t)=\frac{1}{2}\begin{pmatrix}\sin^2 t\cos^2 t\cos 2t\\ \sin t\cos t(\cos t-\cos^2 2t)\\ -2\cos^2 2t\sin 2t\end{pmatrix}+\frac{1}{6}\begin{pmatrix}\sigma\cos^3 t(\sin^3 t+1)\\ -\left[(\rho-z)\sin^3 t+1\right]\cos^3 t+x\cos^2 2t\\ \cos^3 t\left(x-y\sin^3 t\right)+\beta\cos^3 2t\end{pmatrix}$$

in a neighborhood around the origin. To verify the accuracy of these expressions, two arbitrary initial conditions (dotted and solid curves) were evolved with the controlled system (with $\varepsilon=0.6$) as shown in Figure 5.2; both approached the periodic trajectory (5.21) with an accuracy of $\mathcal{O}\left(\varepsilon^4\right)$ as verified in [53]. In this case, *all* initial conditions are eventually attracted to this trajectory, and global flow control has been achieved. This can be done for any specification of $a(t,\varepsilon)$, and not just the one presented here, for the parameter ε "suitably" small. Exactly how small ε needs to be is unclear from this process, but the fact that it can be extended to any order of ε indicates that large values are probably admissible.

At the classical parameter values $\sigma=10$, $\beta=8/3$, and $\rho=28$, the system (5.20) possesses a chaotic attractor [250, 402]. The origin is now a saddle point and possesses a two-dimensional stable manifold and a one-dimensional unstable manifold. Unlike in classical chaos control methods which attempt to stabilize the saddle point by pushing trajectories onto its stable manifold [301, 213, 435, e.g.], the approach in [53] is to move the hyperbolic fixed point to a quasi-periodic hyperbolic trajectory

$$a_{(q)}(t)=\begin{pmatrix}0\\0\\0\end{pmatrix}+\varepsilon\begin{pmatrix}\sin t\\ \cos\omega t+\cos t\\ \cos 2t\end{pmatrix}, \tag{5.22}$$

which too is unstable. The required control velocity can be determined by using (5.17), (5.18), and (5.19) to be

$$c^{(1)}_{(q)}(\underline{r},t)=\begin{pmatrix}(1-\sigma)\cos t+\sigma(\sin t-\cos\omega t)\\ -\omega\sin\omega t-(\rho-z+1)\sin t+x\cos 2t+\cos\omega t+\cos t\\ -2\sin 2t-y\sin t-x(\cos\omega t+\cos t)+\beta\cos 2t\end{pmatrix},$$

$$c^{(2)}_{(q)}(\underline{r},t)=\begin{pmatrix}0\\ -\sin t\cos 2t\\ \sin t(\cos\omega t+\cos t)\end{pmatrix}\quad\text{and}\quad c^{(3)}_{(q)}(\underline{r},t)=\begin{pmatrix}0\\0\\0\end{pmatrix}.$$

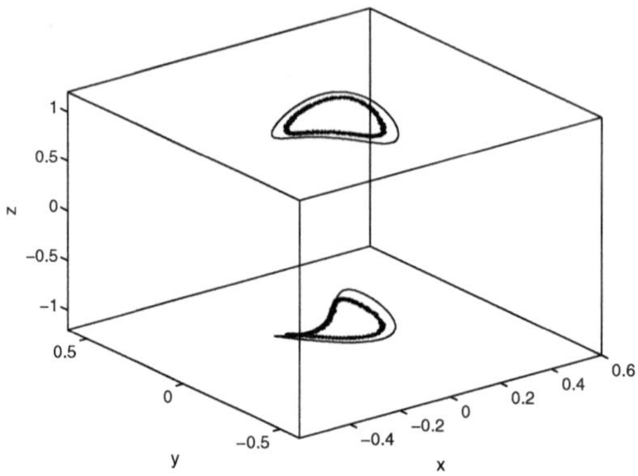

Figure 5.3. *The desired (thick) and numerically verified (thin) hyperbolic trajectories for the controlled Hadamard-Rybczynski droplet as described in Example 5.10 with $\omega_z = 1$ and $\varepsilon = 0.2$. [53] Reproduced with permission from The American Physical Society.*

It was shown in [53] that using this control velocity resulted in the hyperbolic trajectory (5.22) to an accuracy of better than 10^{-4}, even when setting $\varepsilon = 1$. Thus, the technique works even when hyperbolic trajectory is not attracting (and is indeed difficult to numerical determine since it is within a chaotic region [53]). ■

It is sometimes the case that the control of not just one, but several, hyperbolic trajectories simultaneously is of interest. This can be achieved easily with this method by defining the control velocity in each neighborhood $\mathcal{V}_i$ around the hyperbolic trajectories and then patching these together to form a global control velocity in some reasonable way. The simplest approach might be to set the velocity to zero outside each $\mathcal{V}_i$, possibly using some C^∞-bump function to smoothen the transition to zero. This process is illustrated in the next example.

Example 5.10 (Hadamard-Rybczynski droplet). A classical solution to the Stokes equations (for highly viscous flow) in three dimensions is the Hadamard-Rybczynski solution [381, 229, 380, 98, 97, 51] in which a spherical fluid droplet travels at constant speed within another fluid with which it does not mix. This droplet solution been analyzed by many authors in the quest for improving mixing within microdroplets [38, 381, 229, 380, 98, 97]. The kinematic structure of the droplet is similar to the classical Hill spherical vortex [234, 195, 233, 370, 20, 356, 249, 11, 87, 94, 50] for Euler flows as pictured in Figure 1.1(e). Here, a Hadamard-Rybczynski solution with an additive solid rotation around the axis of symmetry is considered. In Cartesian coordinates $\underline{r} = (x, y, z)^\top$, particle trajectories within the sphere obey

$$\dot{x} = zx - \omega_z y, \quad \dot{y} = zy + \omega_z x, \quad \dot{z} = 1 - 2r^2 + z^2, \tag{5.23}$$

where $r = |\underline{r}|$ and ω_z is a constant representing the solid rotation [229]. The z-axis for $1 < z < -1$ is the one-dimensional stable manifold of the north-pole saddle fixed point $a_{(n)} = (0, 0, 1)^\top$ and simultaneously the one-dimensional unstable manifold of the south-pole saddle fixed point $a_{(s)} = (0, 0, -1)^\top$. Now, if mixing is to be achieved within the droplet, the symmetric and autonomous structure within it must be broken. This is to be achieved by moving the north-pole saddle point around as a hyperbolic trajectory, and at the same time moving the south-pole around as a different hyperbolic trajectory. Each will retain its stable and unstable manifold, but these no longer coincide, and the motions of each of these one-dimensional manifolds in relation to one another will engender transport within the droplet. The first step in achieving this is to move the north and south hyperbolic trajectories around, and here it is demonstrated how each can be moved according to

$$a_{(n)}(t) = \begin{pmatrix} 0 \\ 0 \\ 1 \end{pmatrix} + \varepsilon \begin{pmatrix} \sin t \\ \cos t \\ \sin t \cos t \end{pmatrix} \quad \text{and} \quad a_{(s)}(t) = \begin{pmatrix} 0 \\ 0 \\ -1 \end{pmatrix} + \varepsilon \begin{pmatrix} \sin t \\ \sin t \cos t \\ \cos t \end{pmatrix},$$

respectively, as investigated in [53]. The corresponding first-order control velocities are, from (5.17),

$$c_{(n)}^{(1)}(\underline{r}, t) = \begin{pmatrix} 2\cos t - z\sin t - x\cos t\sin t \\ -2\sin t - z\cos t - y\cos t\sin t \\ \cos 2t + 4x\sin t + 4y\cos t + 2z\cos t\sin t \end{pmatrix}$$

and

$$c_{(s)}^{(1)}(\underline{r}, t) = \begin{pmatrix} \cos t - z\sin t + \cos t\sin t - x\cos t \\ \cos 2t - \sin t - z\cos t\sin t - y\cos t \\ -\sin t + 4x\sin t + 4y\cos t\sin t + 2z\cos t \end{pmatrix}.$$

Both these can be achieved in a localized fashion by tempering each with a Gaussian function centered at the relevant pole, as described in [53]. The result of implementing this control strategy is shown in Figure 5.3. The accuracy is not as great as in the previous example since only a first-order control was used. However, this process controls the motion of the "foot" of the one-dimensional stable and unstable manifolds emanating into the droplet. As numerically verified in [53], the nonautonomous motion of these manifolds in relation to each other is complicated and results in complicated mixing within the droplet. To improve mixing, one can experiment with different ways of moving the "foot" of each manifold, for example, by having them spinning in opposite directions. ∎

In this standard perturbation process, c will only acquire conditions on this hyperbolic trajectory, and the extension of c to Ω in a reasonable fashion remains open. For example, one might want to extend c in a way which minimizes energy at each instance in time (such as in the approaches of [246, 42]), in which case the control velocities obtained from this procedure would be specified as constraints in the problem of minimizing $\int_\Omega c(x,t,\varepsilon)^2\,\mathrm{d}x$, and a numerical method would need to be used to determine c globally on Ω. For example, in the above example there would be two such constraints. It is not clear whether this would be an easy problem to solve, and possibly a more reasonable approach would be to minimize $\int_{-T}^{T}\int_\Omega c(x,t,\varepsilon)^2\,\mathrm{d}x\mathrm{d}t$ over a given time-interval $[-T, T]$. Issues such as these in relation to the *global* extension of c are the subject of future investigation.

This section has outlined how it is possible to control hyperbolic trajectories, be they attracting, repelling, or saddle-like, in any dimension, to any order of accuracy. In order to achieve this, full information on the uncontrolled system is needed. This may be impractical in some applications, but it is possible that by, say, using particle image velocimetry (PIV) measurements in an experiment or using numerical data from a computational simulation, the quantity $u(x,t)$ could be approximated in some way. This information could then be fed into the control velocity equations to find the necessary velocity to achieve specified variation in $a(t,\varepsilon)$.

The control methodology is not only useful in fluid mechanical systems in which the phase space is the same as physical space. As illustrated through Example 5.9, there is a strong connection between this method and the controlling of chaotic systems in which the phase space may be something different. Unlike in the standard chaos control approaches, the idea here is not to push trajectories towards an unstable entity (such as a fixed point or periodic orbit) but rather to establish some specified *nonautonomous* behavior (which need not be periodic) on key entities in the phase space. If these are attractors, then the procedure will ensure that the eventual behavior of the system follows this specified nonautonomous variation. If these are not attractors (for example, the origin of the Lorenz system in the classical chaotic parameter regime), one might still be able to achieve this long-term behavior by combining the method described here with a form of the OGY method [301] or other methods [428, 393, e.g.] to repeatedly push trajectories onto the stable manifold of the hyperbolic trajectory. This will result in the system eventually following the (unstable) hyperbolic trajectory, whose time-variation can be specified according to our wishes.

Hyperbolic trajectories which are not attracting or repelling (that is, they are saddle-like in the sense that $0 < m < n$) are at the "foot" of stable and unstable manifolds at any instance in time. Given the importance of stable and unstable manifolds in governing phase space transport (i.e., these are nonautonomous flow barriers), controlling the location of hyperbolic trajectories is a first step in controlling phase space transport. The next step would be to be able to control not just the "foot" of these flow barriers but also their directions of emanation from the hyperbolic point. This can loosely be thought of as "local" control of flow barriers and is the subject of Section 5.3. Pushing this further, one would like to control not just the direction of emanation of these flow barriers but also their global extent as well; this is addressed in Section 5.4.

5.3 ▪ Control of local stable/unstable manifolds

Section 5.1 showed how the time-varying *locations* of hyperbolic trajectories could be controlled to first-order in two-dimensional nearly autononous flows, using a rigorous approach. In contrast, Section 5.2 was able to relax all of these conditions, while also achieving nonautonomous hyperbolic trajectory control, but did so by following a *formal* process. This section attempts to take the next step in flow barrier control by not just controlling the location of the "foot" of the flow barrier (i.e., the hyperbolic trajectory) as done in Section 5.1 but also controlling the direction of emanation of the flow barrier. Thus, this setting is once again that of Section 5.1 (two-dimensional, nearly autonomous).

The flow, then, satisfies (5.1) and Hypothesis 5.1. When the control velocity $c(x,t)$ is set equal to zero, the uncontrolled system $\dot{x} = u(x)$ possesses a saddle fixed point a, with eigenvalues $\lambda_{s,u}$ satisfying $\lambda_s < 0 < \lambda_u$, and with corresponding normalized eigenvectors $v_{s,u}$. Thus, before the control is imposed, the directions of emanation of the stable and unstable manifolds from the hyperbolic trajectory location a are given, respectively, by v_s and v_u. Note that there is a $\pm$ ambiguity in the definition of $v_{s,u}$, which can be tied to the fact that there are *two* branches of each of the one-dimensional stable and unstable manifolds emanating from a, and these occur in opposite directions. Imagine that one branch each has been chosen, and the directions $v_{s,u}$ represent the local directions of the corresponding manifold. These direction are *local* directions of the emanation of the manifolds in that the manifolds may be curved. Alternative characterizations are as the local tangent directions to the manifolds at a, the basis for the Oseledets space associated with the trajectory (a,t), or in terms of the range associated with the projection operator of exponential dichotomies; see Section 2.5 for more details. For a picture of the geometry described so far for the uncontrolled system, the reader is referred to Figure 2.3.

Once an additional control velocity is added to the velocity field, the hyperbolic trajectory itself becomes nonautonomous and will be at a location $a(t)$ rather than at a. The local tangent vectors to the stable and unstable manifolds in a time-slice will also change from the constant directions v_s and v_u to the time-varying directions $v_s(t)$ and $v_u(t)$, as shown in red in Figure 2.12(a). The question that is to be addressed here is: for *specified* $v_{s,u}(t)$, is it possible to find the required control velocity $c(x,t)$?

Remark 5.11. One can first control the location of the hyperbolic trajectory $\tilde{a}(t)$ by using Theorem 5.2. If, for example, the hyperbolic trajectory needs to be kept at a for all time, one can then use the condition $c(a,t)=0$ for all t. *Having decided on this*, the next question is the determination of additional conditions on c to make sure that the required time-varying tangents $v_{s,u}(t)$ to the local stable/unstable manifolds to $\tilde{a}(t)$ are achieved. These directions can be thought of at counterclockwise time-dependent rotations by $\tilde{\theta}_{s,u}(t)$ of $v_{s,u}$, as shown in Figure 2.12(a).

Theorem 5.12 (controlling local manifold directions [39]). *Consider the flow (5.1), subject additionally to Hypothesis 5.1. Let the required counterclockwise rotation angles of $v_{s,u}$ be given by $\tilde{\theta}_{s,u}(t)$ for each time $t\in\mathbb{R}$, defined such that*

$$\left|\tilde{\theta}_{s,u}(t)\right|+\left|\dot{\tilde{\theta}}_{s,u}(t)\right|<\varepsilon \quad \textit{for all } t\in\mathbb{R}. \tag{5.24}$$

Suppose c is chosen subject to the existence of $\tilde{C}$ such that

$$|c(x,t)|+\|Dc(x,t)\|+\left|\frac{\partial c}{\partial t}(x,t)\right|\leq\tilde{C}\varepsilon \quad \textit{for all } (x,t)\in\Omega\times\mathbb{R}. \tag{5.25}$$

If $c(x,t)$ satisfies the velocity shear conditions

$$v_s\cdot\nabla\left[c(x,t)\cdot v_s^{\perp}\right]\Big|_{x=a}=\dot{\tilde{\theta}}_s(t)-(\lambda_u-\lambda_s)\tilde{\theta}_s(t) \tag{5.26}$$

and

$$v_u\cdot\nabla\left[c(x,t)\cdot v_u^{\perp}\right]\Big|_{x=a}=\dot{\tilde{\theta}}_u(t)+(\lambda_u-\lambda_s)\tilde{\theta}_u(t), \tag{5.27}$$

then if the actual rotational angles of the stable and unstable manifolds at the hyperbolic trajectory location $a_\varepsilon(t)$ in a general time-slice t of (5.1) are $\theta_{s,u}(t)$, then there exists a constant K such that

$$\left|\theta_{s,u}(t)-\tilde{\theta}_{s,u}(t)\right|\leq\varepsilon^2K.$$

Proof. Given (5.25), c is $\mathcal{O}(\varepsilon)$, and thus let $c(x,t)=\varepsilon g(x,t)$, where

$$|g(x,t)|+\|Dg(x,t)\|+\left|\frac{\partial g(x,t)}{\partial t}\right|\leq C \quad \text{for all}(x,t)\in\Omega\times\mathbb{R}.$$

This is stronger than the condition under which Theorem 2.27—which quantifies how a perturbation affects the rotation of the local tangent vectors to the stable/unstable manifolds—is valid. Thus, for any g satisfying the above, the *actual* rotations $\theta_{s,u}(t)$

are given by (2.79) and (2.78). The first of these equations, rewritten for convenience, is

$$\begin{aligned}\theta_s(t) &= -\varepsilon\, e^{(\lambda_u-\lambda_s)t}\int_t^{\infty} e^{(\lambda_s-\lambda_u)\tau} v_s \cdot \nabla\left[g \cdot v_s^{\perp}\right](a,\tau)\,\mathrm{d}\tau + \varepsilon^2 K_s(t)\\ &= -e^{(\lambda_u-\lambda_s)t}\int_t^{\infty} e^{(\lambda_s-\lambda_u)\tau} v_s \cdot \nabla\left[c(x,\tau)\cdot v_s^{\perp}\right]\Big|_{x=a}\mathrm{d}\tau + \varepsilon^2 K_s(t),\end{aligned}$$

where $|K_s(t)|$ is bounded for $t \in [T_s, \infty)$. Now, this means that

$$e^{(\lambda_s-\lambda_u)t}\theta_s(t) = \int_{\infty}^{t} e^{(\lambda_s-\lambda_u)\tau} v_s \cdot \nabla\left[c(x,\tau)\cdot v_s^{\perp}\right]\Big|_{x=a}\mathrm{d}\tau + \varepsilon^2 K_s(t) e^{(\lambda_s-\lambda_u)t}.$$

Differentiating with respect to t, and bearing in mind that condition (5.25) ensures that the higher-order term in ε remains so after differentiation, leads to

$$e^{(\lambda_s-\lambda_u)t}\left[\dot{\theta}_s(t) - (\lambda_u - \lambda_s)\theta_s(t)\right] = e^{(\lambda_s-\lambda_u)t} v_s \cdot \nabla\left[c(x,t)\cdot v_s^{\perp}\right]\Big|_{x=a} + \varepsilon^2 K_s(t) e^{(\lambda_s-\lambda_u)t},$$

and thus

$$v_s \cdot \nabla\left[c(x,t)\cdot v_s^{\perp}\right]\Big|_{x=a} = \dot{\theta}_s(t) - (\lambda_u - \lambda_s)\theta_s(t) + \varepsilon^2 K_s(t). \tag{5.28}$$

Now, (5.28) looks like (5.26) except for two things: the higher-order term and the fact that this has θ_s rather than $\tilde{\theta}_s$. The point is that $\tilde{\theta}_s$ is the quantity that is *prescribed*, and θ_s is the *actual* rotation that results from using a control velocity satisfying (5.26). It is required to show that the actual rotation is within $\mathcal{O}\left(\varepsilon^2\right)$ of the desired rotation. Defining $\eta_s(t) := \theta_s(t) - \tilde{\theta}_s(t)$ and subtracting (5.28) from (5.26) yields

$$\dot{\eta}_s(t) - (\lambda_u - \lambda_s)\eta_s(t) = -\varepsilon^2 K_s(t).$$

Now, multiplying the above by $e^{(\lambda_s-\lambda_u)t}$, bearing in mind that $\lambda_s - \lambda_u < 0$, and integrating from a general value of t to ∞ gives

$$-e^{(\lambda_s-\lambda_u)t}\eta_s(t) = -\varepsilon^2\int_t^{\infty} K_s(t) e^{(\lambda_s-\lambda_u)\tau}\mathrm{d}\tau,$$

and therefore there exists a K such that

$$e^{(\lambda_s-\lambda_u)t}\,|\eta_s(t)| \le K\varepsilon^2\int_t^{\infty} e^{(\lambda_s-\lambda_u)\tau}\mathrm{d}\tau = K\varepsilon^2\frac{e^{(\lambda_s-\lambda_u)t}}{\lambda_u - \lambda_s}.$$

Canceling out the exponential factor shows that $|\eta_s(t)| = \mathcal{O}\left(\varepsilon^2\right)$, as required, for the rotation of the stable manifold. The proof for the unstable manifold is similar, and will be skipped. □

Theorem 5.12 shows that the time-variation of the local tangent vectors to the stable/unstable manifolds can be controlled via imposing a prescribed velocity shear as given in (5.26) and (5.27). A validation of this process is given in the following example.

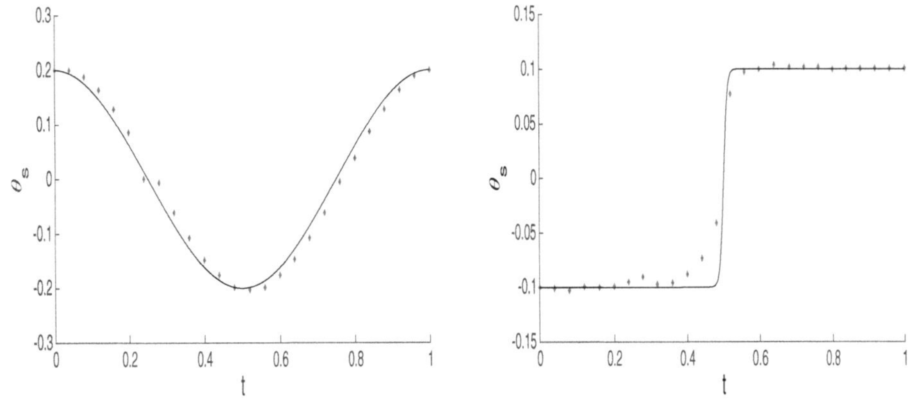

(a) Periodic (b) Abruptly switching

Figure 5.4. *Numerical results [red diamonds] reported in [39] for the implementation of the control (5.30) in (5.29) for two different protocols for* $\tilde{\theta}_s(t)$ *[black curves]. Reproduced with permission from Springer Science+Business Media.*

Example 5.13 (Taylor-Green flow cont.). The Taylor-Green flow described in Examples 3.41 and 4.9 is now re-examined, but with a slightly different parametrization which is compatible with the double-gyre. The controlled Taylor-Green system shall be

$$\left.\begin{aligned} \dot{x}_1 &= -\pi A \sin(\pi x_1)\cos(\pi x_2) + c_1(x_1,x_2,t) \\ \dot{x}_2 &= \pi A \cos(\pi x_1)\sin(\pi x_2) + c_2(x_1,x_2,t) \end{aligned}\right\}, \tag{5.29}$$

where A is a parameter. With the control switched off, this system has a stable manifold coming vertically downwards to the saddle fixed point $(1,0)$, which is simultaneously a branch of the unstable manifold emanating downwards from the point $(1,1)$. This formulation of the Taylor-Green flow is identical to the double-gyre (2.64) with the nonautonomous parameter set equal to zero, and thus the uncontrolled geometry associated with (5.29) is shown in Figure 2.9. (It is also equivalent to the system (4.16) in Example 4.9 with the choice of parameters $V = -\pi A$, $L = 1$.) The red line is the heteroclinic manifold of interest in this example as well. The relevant eigenvalues are $\pm\pi^2 A$ at both saddle points $(1,0)$ and $(1,1)$. It should be borne in mind that when a time-dependent control is included as in (5.29), these saddle points becomes hyperbolic trajectories which, while possessing a time-variation, remain $\mathcal{O}(\varepsilon)$-close to the original fixed points. Rather than focusing on one of these saddle points and attempting to rotate its stable and unstable manifolds, the focus shall be on the endpoints of this red line: the local stable manifold of $(1,0)$ and the local unstable manifold of $(1,1)$. The reason is that by controlling each of these, the transport across the previously impermeable red line of Figure 2.9 is affected. Thus, the local control of these manifolds impacts the transfer of fluid between the left and right gyres in Figure 2.9. Suppose, then, that the local directions of emanation of these stable and unstable manifolds are prescribed as $\tilde{\theta}_s(t)$ and $\tilde{\theta}_u(t)$, each of which are bounded by ε as given in Theorem 5.12. This leads to the conditions

$$\frac{\partial c_1}{\partial x_2}(1,0,t) = -\left[\dot{\tilde{\theta}}_s(t) - 2\pi^2 A \tilde{\theta}_s(t)\right] \text{ and } \frac{\partial c_1}{\partial x_2}(1,1,t) = -\left[\dot{\tilde{\theta}}_u(t) + 2\pi^2 A \tilde{\theta}_u(t)\right]$$

on the control velocity. Since c_2 does not contribute to leading-order, it shall be simply set to zero. The conditions above can be implemented by setting

$$\begin{aligned} c_1(x_1,x_2,t) = &-\left[\dot{\tilde{\theta}}_s(t) - 2\pi^2 A\tilde{\theta}_s(t)\right] x_2 I_{(1,0)}(x_1,x_2) \\ &-\left[\dot{\tilde{\theta}}_u(t) + 2\pi^2 A\tilde{\theta}_u(t)\right](x_2-1) I_{(1,1)}(x_1,x_2), \end{aligned} \tag{5.30}$$

where the "patch" function

$$I_{(\tilde{x}_1,\tilde{x}_2)}(x_1,x_2) = \frac{1}{2}\left[\tanh\frac{\sqrt{(x_1-\tilde{x}_1)^2+(x_2-\tilde{x}_2)^2}+\varepsilon}{\varepsilon^2} - \tanh\frac{\sqrt{(x_1-\tilde{x}_1)^2+(x_2-\tilde{x}_2)^2}-\varepsilon}{\varepsilon^2}\right]$$

is a smoothened version of the indicator function of the ball of radius ε centered at $(\tilde{x}_1,\tilde{x}_2)$. A test of how well this method works is presented in [39]. For specified choices of $\tilde{\theta}_{s,u}(t)$, the system (5.29) with the control (5.30) is numerically run. Backward- and forward-time FTLEs are then numerically calculated; as explained in Section 1.4, the ridges of these scalar fields form proxies for the unstable and stable manifolds, respectively. In this case, by focusing near $(1,0)$ and $(1,1)$, it is possible to identify these ridges with no ambiguity and moreover to determine their tangent line slopes. This procedure can be implemented in each time-slice and the results compared to the desired tangent vector rotations as specified by $\tilde{\theta}_{s,u}(t)$. The results of two different numerical experiments are shown in Figure 5.4, with the choice $A=1$ and the smallness parameter $\varepsilon = 0.2$, which is not too small. For a periodic specification, as shown by the black curve in (a), the FTLE-based numerical validation (red dots) demonstrates excellent agreement with the desired time-variation of the local tangent vector to the stable manifolds. For an abruptly changing specification as shown in Figure 5.4(b), the results are still very good, in spite of the fact that an abruptly changing $\tilde{\theta}_s(t)$ does not satisfy the hypotheses of Theorem 5.12. Indeed, such an abrupt change can only result if there are impulsive perturbations in the velocity, and in this case the stable manifold would need to be a stable pseudomanifold for which the theory of Section 2.7 would be required. There is some uncertainty in the numerical calculations near the abrupt change (at $t = 0.5$), which results also from the fact that the FTLE ridges are more difficult to identify when they are moving rapidly. The reader is referred to [39] for more detail on the results shown in Figure 5.4, including an error analysis. In the main, however, the strategy of Theorem 5.12 is highly successful in controlling the local tangent vector directions of the manifolds, as desired. ■

The results shown above are very promising, demonstrating the potential for utilizing these methods for controlling *transport* through this process. It is the intermingling of the stable and unstable manifolds which causes transport between the gyres in Figure 2.9, and by "aiming" the stable and unstable manifolds emanating from the points $(1,0)$ and $(1,1)$ in various directions in a time-varying way, it will be possible to influence their intersection pattern and hence transport. For example, suppose at some instance in time the local manifolds of the controlled Taylor-Green flow are enforced to have the configuration shown in Figure 5.5(a). (The arrows shown on the local manifolds are to help with identifying which is the stable and which is the unstable manifold; the actual velocity directions will be a combination of flow along the relevant manifold in the direction shown *and* the rotation of the manifold.) This scenario corresponds to $\theta_s(t) > 0$ and $\theta_u(t) < 0$ at this instance in time, as is clear from

(a) Pointing into different gyres

(b) Pointing into same gyre

Figure 5.5. *Controlling local directions of Taylor-Green manifolds to influence transport: possible pictures in a time-slice t.*

considering the rotations necessary from the unperturbed manifolds which lie along $x_1 = 1$. At this t, there is greater potential for the *primary* segments of the manifolds (i.e., segments close to the line $\{(1, x_2) : 0 \leq x_2 \leq 1\}$, along which both manifolds lie in the absence of a control) to be in the left gyre. Primary intersections between them are therefore more likely to be in this gyre. Transport between the gyres will occur in the sense of Sections 3.2 and 3.3. Suppose a gate $\mathcal{G}$ is drawn at the midpoint of the unperturbed flow barrier along the line $x_2 = 0.5$ as shown by the dotted line. The gate $\hat{\mathcal{G}}$ *through which flux occurs* would be the segment of this line lying between the stable and the unstable manifold (see Section 3.3), which in this scenario is *likely* to lie entirely within the left gyre. By thinking of extending the local manifold segments of Figure 5.5(a) further, $\hat{\mathcal{G}}$ is likely to be small, with a relatively small transport occurring. Contrast this with the situation shown in Figure 5.5(b), in which the manifolds emanate into *different gyres*, with this picture corresponding to a situation in which $\theta_s(t) < 0$ and $\theta_u(t) < 0$. The gate $\hat{\mathcal{G}}$ across which flux occurs will generically be much larger, since the manifolds are likely to be on different sides of the nominal flow separator lying along $x_1 = 1$. In the pictured situation there is likely to be a relatively large flux from the right to the left gyre at this instance in time, as can be imagined by extending the manifolds until they intersect $\mathcal{G}$. One might indeed consider two time-periodic specifications which would lead to the pictures in Figure 5.5 at time 0: (a) $\tilde{\theta}_s(t) = \varepsilon \cos \omega t$ and $\tilde{\theta}_u(t) = -\varepsilon \cos \omega t$, and (b) $\tilde{\theta}_s(t) = \tilde{\theta}_u(t) = -\varepsilon \cos \omega t$. Since the control velocity resulting from these is time-periodic, the standard Poincaré map methods are applicable, and generically a heteroclinic tangle in which the global manifolds intersect at infinitely many locations (i.e., a displaced version of Figure 2.10) will occur. The situation (b) will in general give a much larger time-periodic flux amplitude because of the reasons outlined. The point is that by *specifying* $\theta_{s,u}$ in certain ways and determining the relevant control velocity, transport can be influenced strongly.

The above argument was based on thinking of simply extending the local manifolds in Figure 5.5 in a "reasonable" way. In reality, however, the global extent of these manifolds will be influenced by global velocity conditions and *cannot* be predetermined by local conditions near the heteroclinic trajectories. This is the subject of the next section, in which methods for controlling the *global* manifolds (subject to some restrictions) will be pursued. If purely *localized* controls are used, then once the local manifolds emanate into regions in which there is no additional control velocity, their global extent will be governed by the *unperturbed* velocity. If this was the strategy used in Figure 5.5(a), then, for example, the trajectories that will "feed into" the local stable manifold would be those coming from the outer closed curves in the left gyre in Figure 2.9. Each point on the local manifold Γ^s pictured in Figure 5.5(a) would in backwards time go towards a different closed loop, and the union of these entities, when thought of in the augmented phase space, will give the nonautonomous stable manifold. While difficult to draw, these intuitions can help understand how the global manifolds will look if the control velocities are localized near $(1,0)$ and $(1,1)$. Thus, using some careful intuition, the local control methodology can be utilized for more global transport control.

5.4 ▪ Controlling global stable/unstable manifolds

The previous section outlined how it was possible to control the time-varying directions of emanation of the stable and unstable manifolds from a hyperbolic trajectory. In each time-slice t, this controls the tangent vector to the manifolds drawn at the hyperbolic trajectory, in other words, the *local* stable/unstable manifold. This section describes a method for controlling the *global* extent of these manifolds (suitably restricted), following the development in [54]. This is the *only* currently existing theory for controlling stable and unstable manifolds nonautonomously [54], and its results are modified slightly in this presentation. Since the theory is quite new and highly restrictive (two-dimensional, nearly autonomous, primary segments of manifolds only), there is considerable scope for the improvement of the results. Even within these restrictions, the current method of defining the control is somewhat difficult to implement, and thus more effective characterizations of this method would be desirable.

The basic setup is exactly as in Sections 5.1 and 5.3, with mild variations in the smoothness requirements. However, to ensure that this section remains self-contained, the assumptions shall be stated explicitly. The system to be controlled takes the form

$$\dot{x} = f(x) \tag{5.31}$$

for $x \in \Omega$, a two-dimensional open connected set. (The notation f, as opposed to u as used in previous sections of this chapter, shall be used here in order to make connections to Chapter 2.)

Hypothesis 5.14 (uncontrolled system). *The quantities associated with (5.31) satisfy the following:*

(a) $f \in C^2(\Omega)$ *and is such that there exists a constant* C_f *satisfying*

$$\|f\|_{2,\infty} \leq C_f\,; \tag{5.32}$$

that is, all spatial derivatives up to second-order are uniformly bounded.

(b) *The system (5.31) possesses a saddle fixed point* a, *that is,* $f(a) = 0$, *and* $Df(a)$ *possesses real eigenvalues* λ_s *and* λ_u *such that* $\lambda_s < 0 < \lambda_u$.

The focus shall be on one branch each of the stable and unstable manifolds of a (whose existence is guaranteed by Hypothesis 5.14(b)), parametrized in the form

$$\Gamma^s := \{\bar{x}^s(p) : p \in [S, \infty)\} \quad \text{and} \quad \Gamma^u := \{\bar{x}^u(p) : p \in (-\infty, U]\},$$

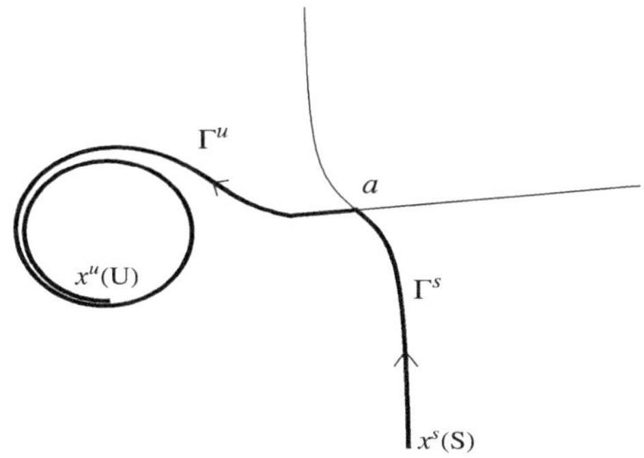

Figure 5.6. *Clipped branches of the stable Γ^s and unstable Γ^u manifolds (thick curves) of a in the uncontrolled system (5.31).*

in which $\bar{x}^{s,u}(p)$ are solutions to (5.31) satisfying $\bar{x}^s(p) \to a$ as $p \to \infty$ and $\bar{x}^u(p) \to a$ as $p \to -\infty$. The large quantities $-S$ and U will "clip" these manifold branches at some point. Thus, while each of these one-dimensional manifold branches may exhibit various types of behavior at its "other end" (connecting to another—or the same—fixed point to form a heteroclinic, spiraling towards a limit cycle or fixed point, escaping to infinity, etc.), this clipping process ensures that each of $\Gamma^{s,u}$ is well-defined as a parametric curve. A generic picture of these entities (thick curves) is presented in Figure 5.6.

Since after a control is introduced the system becomes nonautonomous, it pays to understand these entitites in the augmented phase space $\Omega \times \mathbb{R}$ (i.e., when thinking of writing (5.31) in the form $\dot{x} = f(x)$, $\dot{t} = 1$). The clipped manifolds $\Gamma^{s,u}$ are *two*-dimensional in this phase space. It proves advantageous to represent these parametrically in the form

$$\Gamma^s := \{(\bar{x}^s(t - T_s + p), t) \ : \ (p,t) \in [S, \infty) \times [T_s, \infty)\}, \tag{5.33}$$

$$\Gamma^u := \{(\bar{x}^u(t - T_u + p), t) \ : \ (p,t) \in (-\infty, U] \times (-\infty, T_u]\}, \tag{5.34}$$

in which the notation $\Gamma^{s,u}$ is retained with an abuse of notation. The parametrization in (5.34) is chosen such that t is the time-slice, and p identifies a specific trajectory, as illustrated in Figure 5.7. To explain the form of these expressions, consider the clipped stable manifold Γ^s. The temporal clipping is necessitated by the spatial clipping: any point on the clipped curve Γ^s in Figure 5.6 will in backwards time eventually reach $\bar{x}^s(S)$, after which there is no information on how the manifold behaves. The time at which this occurs would be the lowest time-value for which the corresponding trajectory is "admissible," and it is exactly this which is captured in (5.34) by $t \geq T_s$. To further understand this parametrization in the augmented phase space, the reader is referred to Figure 5.7. The fullest extent of the clipped stable manifold, ranging from $x^s(S)$ to a, is retained at the least admissible time value, T_s. For each point $\bar{x}^s(p)$ chosen in the time-slice $t = T_s$, $(\bar{x}^s(t - T_s + p), t)$ represents the corresponding forward trajectory on Γ^s as it evolves with time t; four such trajectories are pictured in Figure 5.7. The outermost one has $p = P$, which when inserted into the parametric representation of (5.34) yields the value $x_s(S)$ at $t = T_s$. The hyperbolic trajectory, indicated by the dashed line, corresponds to $p = \infty$. The two-dimensional surface in Figure 5.7 bounded by these two curves is the clipped stable manifold, represented

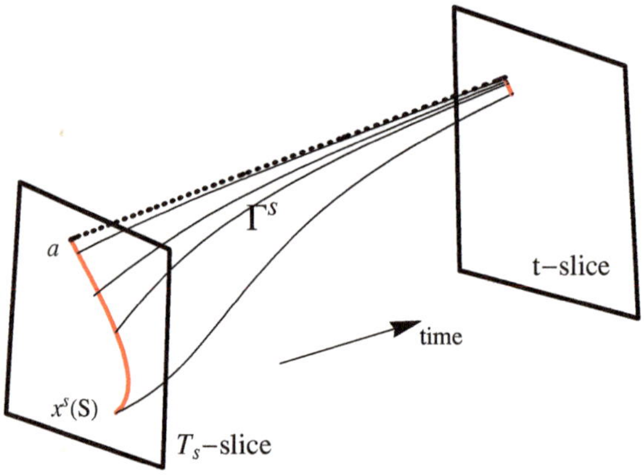

Figure 5.7. *The uncontrolled clipped stable manifold Γ^s as given in (5.34), shown with several trajectories with different p-values (thin curves). The hyperbolic trajectory (a,t) is shown by the dashed line, and the clipped manifolds visible in the time-slices T_s and t are the red curves.*

parametrically by (5.34). In the time-slice $t = T_s$, the full clipped stable manifold of Figure 5.6 is present, shown as the thick red curve going from $\bar{x}^s(S)$ to a. As time evolves, the clipped stable manifold gets smaller in length. The definition for Γ^u in (5.34) is analogous, with T_u now being a time-value below which the unstable manifold is defined.

The goal now is to move the uncontrolled stable manifold to a *prescribed* nonautonomous stable manifold $\tilde{\Gamma}^s$ (the unstable manifold will be dealt with thereafter). To specify this carefully, $\tilde{\Gamma}^s$ will be considered in parametric form

$$\tilde{\Gamma}^s := \{(x^s(p,t),t) \, : \, (p,t) \in [S,\infty) \times [T_s,\infty)\}\,, \tag{5.35}$$

where x^s is assumed *given*. As in (5.34), in the parametric representation $x^s(p,t)$, the quantity t shall be time and p shall be a parameter identifying *which* trajectory on the perturbed manifold is being considered. So $\tilde{\Gamma}^s$ is itself a clipped stable manifold. Thus, $x_\varepsilon(p,t)$ for each fixed p is to be thought of as a solution to (5.41). For this to occur, and to be able to think of $\tilde{\Gamma}^s$ as being a nonautonomous perturbation of the clipped stable manifold Γ^s, x^s cannot be specified completely arbitrarily but must satisfy the following hypothesis.

Hypothesis 5.15 (stable manifold requirements). *The quantity x^s which parametrizes $\tilde{\Gamma}^s$ satisfies the following:*

(a) *[Closeness] There exists a constant ε such that for all $(p,t) \in [S,\infty) \times [T_s,\infty)$,*

$$|x^s(p,t) - \bar{x}^s(t - T_s + p)| + \left| \frac{\partial}{\partial p}(x^s(p,t) - \bar{x}^s(t - T_s + p)) \right| \leq \varepsilon\,. \tag{5.36}$$

(b) *[Smoothness] There exists a constant $K_s > 0$ such that for all $(p,t) \in [S,\infty) \times [T_s,\infty)$,*

$$|x^s(p,t)| + \left| \frac{\partial}{\partial t} x^s(p,t) \right| + \left| \frac{\partial}{\partial p} x^s(p,t) \right| < K_s\,. \tag{5.37}$$

(c) *[Limit] For each* $t \geq T_s$, $\lim_{p\to\infty} x^s(p,t)$ *is well defined.*

(d) *[Mappability] For each* $t \geq T_s$, *there exist intervals* $[S_1(t), S_2(t)) \subseteq [S, \infty)$ *and* $[\tilde{S}_1(t), \tilde{S}_2(t)) \subseteq [S, \infty)$, *and a scalar function* $r^s(\cdot, t)$ *defined on* $[S_1(t), S_2(t))$ *which satisfies*

$$x^s(\tilde{p}, t) = \bar{x}^s(t - T_s + p) + r^s(p, t) \frac{f^{\perp}(\bar{x}^s(t - T_s + p))}{|f(\bar{x}^s(t - T_s + p))|}, \tag{5.38}$$

such that the mapping $p \to \tilde{p}$ *from* $[S_1(t), S_2(t))$ *to* $[\tilde{S}_1(t), \tilde{S}_2(t))$ *defined through (5.38) is a diffeomorphism.*

(e) *[Initial parametrization] For all* $p \in [S_1(T_s), S_2(T_s))$,

$$[f(\bar{x}^s(p))]^{\top} [x^s(p, T_s) - \bar{x}^s(p)] = 0. \tag{5.39}$$

It will help to expand on some of these hypotheses. Hypothesis 5.15(a) gives an estimate of the closeness between $x^s(p,t)$ and $\bar{x}^s(t - T_s + p)$, i.e., between $\tilde{\Gamma}^s$ and Γ^s. These are specifically within ε of one another. The errors to be determined will be stated in terms of this parameter ε, which is derived from the closeness of the prescribed $\tilde{\Gamma}^s$ to the uncontrolled Γ^s. The reason for this parametrization of $\bar{x}^s$ is to ensure that in the beginning time-slice T_s (i.e., the minimum time value of the clipped manifold that is permissible), $x^s(p, T_s)$ is associated with $\bar{x}^s(p)$. Hypothesis 5.15(b) is a straightforward smoothness assumption on $\tilde{\Gamma}^s$. The limit $p \to \infty$ in condition (c) means going along the stable manifold curve in that time slice to the perturbed hyperbolic trajectory and is therefore a statement to ensure that this hyperbolic trajectory (of which $\tilde{\Gamma}^s$ is the stable manifold) is well-defined. In each time-slice t, the ability of "mapping" $\bar{x}^s(t - T_s - p)$ to $x^s(\tilde{p}, t)$ by proceeding in the normal direction at $\bar{x}^s(t - T_s - p)$ by a signed distance r^s is the requirement given in (d), for which an illustration appears in Figure 5.8 [left]. This prevents, for example, $\tilde{\Gamma}^s$ having self-intersections or twists which would give problems in being able to associate each uncontrolled trajectory $\bar{x}^s(t - T_s + p)$ unambiguously with a controlled trajectory $x^s(p, t)$. It may be that such issues occur when considering $p \in [S, \infty)$, but the requirement is that there is a suitable restriction $p \in [S_1(t), S_2(t))$ for which this mapping works, and it will be *only for this restriction* that the control strategy will be legitimate. For the situation pictured in the left subfigure of Figure 5.8, the uncontrolled and controlled manifolds are defined in terms of the parameters $p \in [S, \infty)$ and $\tilde{p} \in [S, \infty)$, but in this situation they overlap only for $p \in [S_1(t), S_2(t))$ with $S_1(t) > S$ and $S_2(t) = \infty$. Define

$$\Xi_s := \{(p, t) \ : \ t \geq T_s \text{ and } S_1(t) \leq p < S_2(t)\}, \tag{5.40}$$

in which it shall be assumed that the largest possible interval $[S_1(t), S_2(t))$ has been chosen for each t in order to fulfill the mappability condition of Hypothesis 5.15. Next, condition (e) is merely a choice of parametrization p on the controlled nonautonomous manifold. The idea is that the p-parametrization of $\tilde{\Gamma}^s$ should be chosen such that it is identical to the point on the uncontrolled manifold Γ^s to which it is mapped by the mapping of condition (d), in the "initial" time-slice T_s. Thus, the p-value of $x^s(p, T_s)$ should be the same as the p-value in $\bar{x}^s(p)$, as shown in the right subfigure of Figure 5.8. It should be emphasized that as trajectories evolve to a general

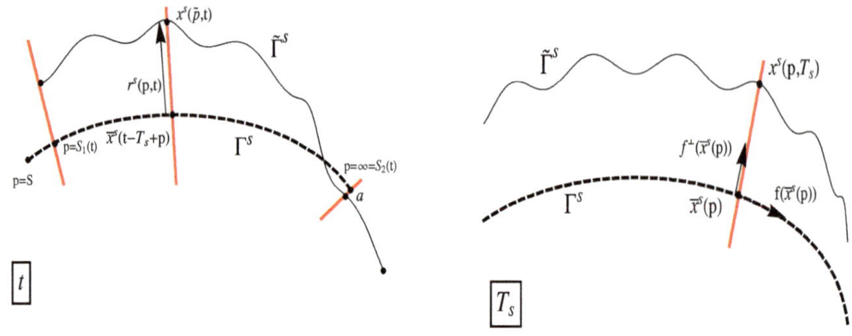

(a) Condition (d) (b) Condition (e)

Figure 5.8. *Illustrations of the conditions of Hypothesis 5.15 for the controlled stable manifold* $\tilde{\Gamma}^s$.

time t, there is no necessity for a particular $x^s(p,t)$ to lie on the normal line drawn to Γ^s at the point $\bar{x}^s(t-T_s+p)$. Hypothesis 5.15(e) makes no such restrictions on the evolution of trajectories but is simply a statement that the curve $\tilde{\Gamma}^s \cap \{t=T_s\}$ is best parametrized consistently with the parametrization on $\Gamma^s \cap \{t=T_s\}$.

The goal is to now determine the control velocity $c(x,t)$ which achieves the desired stable manifold $\tilde{\Gamma}^s$. Thus, the controlled system is

$$\dot{x} = f(x) + c(x,t). \tag{5.41}$$

Clearly, control shall be possible only for the *primary* manifold, that is, the manifold segment which remains close to the unperturbed one. Once the manifold segment has progressed "beyond" the endpoint $\bar{x}^s(S)$, it will no longer be possible to control it. It will be convenient to consider (5.41) in its augmented form

$$\left.\begin{aligned} \dot{x} &= f(x) + c(x,t) \\ \dot{t} &= 1 \end{aligned}\right\} \tag{5.42}$$

with phase space now being $\Omega \times \mathbb{R}$. It is for (5.42) with $c \equiv 0$ that the geometry of Figure 5.7 applies, with Γ^s to be moved to a prescribed nearby entity $\tilde{\Gamma}^s$ by turning on the control velocity c.

For $(p,t) \in \Xi_s$, define the quantities

$$M^s(p,t) := \left[f^\perp(\bar{x}^s(t-T_s+p))\right]^\top \left[x^s(p,t) - \bar{x}^s(t-T_s+p)\right] \tag{5.43}$$

and

$$B^s(p,t) := \left[f(\bar{x}^s(t-T_s+p))\right]^\top \left[x^s(p,t) - \bar{x}^s(t-T_s+p)\right]. \tag{5.44}$$

These definitions are inspired by the normal and tangential Melnikov functions (see Theorem 2.23), but there is a difference here: both M^s and B^s are now $\mathcal{O}(\varepsilon)$ entities since $|x^s(p,t) - \bar{x}^s(t-T_s+p)|$ is uniformly bounded by ε. In other words, the ε-perturbation framework of Theorem 2.23 has been rescaled, absorbing the small parameter into the definitions of the Melnikov functions. Based on the above definitions, the control velocity is now specified.

Hypothesis 5.16 (control velocity conditions for stable manifold). *The control velocity c in (5.41) obeys the following conditions:*

(a) $c(\cdot,t) \in C^2(\Omega)$ *for each* $t \in \mathbb{R}$.

(b) $c(x,\cdot) \in C^1(\mathbb{R})$ *for each* $x \in \Omega$.

(c) *There exists a constant* C_c *satisfying*

$$\|c(\cdot,t)\|_{1,\infty} + \left\| \frac{\partial c}{\partial t}(\cdot,t) \right\|_{0,\infty} \leq \varepsilon C_c \text{ for all } t \in \mathbb{R}. \tag{5.45}$$

(d) *The normal component of the control velocity,* $c^n := c \cdot \hat{f}^\perp$, *is chosen to satisfy*

$$c^n(\bar{x}^s(t-T_s+p),t) = \frac{\frac{\partial M^s}{\partial t}(p,t) - \mathrm{Tr}\left[Df(\bar{x}^s(t-T_s+p))\right] M^s(p,t)}{|f(\bar{x}^s(t-T_s+p))|}. \tag{5.46}$$

(e) *The tangential component of the control velocity,* $c^t := c \cdot \hat{f}$, *is chosen according to*

$$c^t(\bar{x}^s(t-T_s+p),t) = \frac{|f|^2 \frac{\partial B^s(p,t)}{\partial t} - f^\top \left[(Df)+(Df)^\top\right]\left[f^\perp M^s(p,t) + f B^s(p,t)\right]}{|f|^3}, \tag{5.47}$$

in which the dependence of f and Df on the argument $\bar{x}^s(t-T_s+p)$ has been suppressed for brevity.

While the first three conditions on the control velocity are straightforward smoothness ones, conditions (d) and (e) are the crucial conditions which will ensure that the prescribed stable manifold is achieved to leading order. Together, they specify what the control velocity needs to be on the clipped uncontrolled stable manifold. Quantifying the effectiveness of using this control velocity requires a characterization of the *actual* stable manifold, $\widehat{\Gamma}^s$, resulting from the inclusion of a control velocity as specified in Hypothesis 5.16 in (5.41). This will have the form

$$\widehat{\Gamma}_s := \{(\widehat{x}^s(p,t),t) \; : \; (p,t) \in \Xi_s\}, \tag{5.48}$$

in which the $\widehat{x}^s(p,t)$ is an exact trajectory of (5.41) for fixed p, and the parametrization shall be be chosen to be consistent with that of the desired stable manifold $\tilde{\Gamma}^s$; i.e., Hypothesis 5.15(d) and (e) are met for $\widehat{x}^s$ in just the same way that they were satisfied for x^s. Now, since c is $\mathcal{O}(\varepsilon)$, Theorem 2.23 ensures that the actual stable manifold deforms no more than $\mathcal{O}(\varepsilon)$ from the *unperturbed* one, and thus there exists a constant C_s such that

$$|\widehat{x}^s(p,t) - \bar{x}^s(t-T_s+p)| + \left| \frac{\partial}{\partial p}(\widehat{x}^s(p,t) - \bar{x}^s(t-T_s+p)) \right| \leq \varepsilon C_s \tag{5.49}$$

for $(p,t) \in \Xi_s$. The goal, however, is to show that the choice specified in Hypothesis 5.16 ensures that the *desired* stable manifold $\tilde{\Gamma}^s$ is close to the *actual* stable manifold $\widehat{\Gamma}^s$. This can be characterized through the error

$$e_s(p,t) := \widehat{x}^s(p,t) - x^s(p,t), \tag{5.50}$$

with normal and tangential components

$$e_s^n(p,t) := e_s(p,t)\cdot \hat{f}^{\perp}(\bar{x}^s(t-T_s+p)) \quad \text{and} \quad e_s^t(p,t) := e_s(p,t)\cdot \hat{f}(\bar{x}^s(t-T_s+p)).$$

Theorem 5.17 (control of stable manifold [54]). *Assume the control velocity c satisfies Hypothesis 5.16. Then, the error $e_s(p,t) = \mathcal{O}(\varepsilon^2)$. More specifically,*

$$|e_s^n(p,t)| \leq \varepsilon^2 \left(C_s C_c + \frac{C_s^2 C_f}{2} \right) \frac{\int_t^\infty |f(\bar{x}^s(\tau - T_s + p))| \exp\left[\int_\tau^t \mathrm{Tr}[Df(\bar{x}^s(\xi - T_s + p))]\mathrm{d}\xi\right] \mathrm{d}\tau}{|f(\bar{x}^s(t - T_s + p))|} \tag{5.51}$$

and

$$|e_s^t(p,t)| \leq \varepsilon^2 \left(C_s C_c + \frac{C_s^2 C_f}{2} \right) |f(\bar{x}^s(t - T_s + p))|$$
$$\times \int_{T_s}^t \frac{|f(\bar{x}^s(\tau - T_s + p))| + 2C_f \int_\tau^\infty |f(\bar{x}^s(\zeta - T_s + p))| \exp\left[\int_\zeta^\tau \mathrm{Tr}[Df(\bar{x}^s(\xi - T_s + p))]\mathrm{d}\xi\right] \mathrm{d}\zeta}{|f(\bar{x}^s(\tau - T_s + p))|^2} \mathrm{d}\tau \tag{5.52}$$

for $(p,t) \in \Xi_s$.

Proof. See Section 5.4.1. □

Remark 5.18. It can be shown that the improper integrals in (5.51) and (5.52) are convergent by using essentially the argument presented in (2.43) and its adjacent discussion, justifying the claim that $e_s = \mathcal{O}(\varepsilon^2)$.

Remark 5.19. Subject to remaining in Ξ_s, the normal and tangential bounds for the error $e_s(p,t)$ as given in Theorem 5.17 can be shown to obey the limiting conditions [54]

$$\lim_{t \to \infty} |e_s^n(p,t)| \leq \frac{\varepsilon^2 \left[C_s C_c + \frac{C_s^2 C_f}{2} \right]}{-\lambda_s}, \quad \lim_{p \to \infty} |e_s^n(p,t)| \leq \varepsilon^2 \frac{\left[C_s C_c + \frac{C_s^2 C_f}{2} \right]}{\lambda_u}, \tag{5.53}$$

and

$$\lim_{t \to \infty} |e_s^t(p,t)| = 0, \ \lim_{p \to \infty} |e_s^t(p,t)| \leq \varepsilon^2 \frac{\left(2C_c C_s + C_s^2 C_f\right)\left(\lambda_u + 2C_f\right)}{-2\lambda_s \lambda_u} \left[1 - e^{\lambda_s (t - T_s)}\right], \tag{5.54}$$

as is detailed in Section 5.4.1.

Next, results for the unstable manifold are presented. Suppose the desired unstable manifold was specified by

$$\tilde{\Gamma}^u := \{(x^u(p,t), t) \ : \ (p,t) \in (-\infty, U] \times (-\infty, T_u]\} \tag{5.55}$$

and is subject to the following conditions.

Hypothesis 5.20 (unstable manifold requirements). *The quantity x^u which parametrizes $\tilde{\Gamma}^u$ satisfies the following:*

(a) *[Closeness] There exists a constant* ε *such that for all* $(p,t) \in (-\infty, U] \times (-\infty, T_u]$,

$$|x^u(p,t) - \bar{x}^u(t - T_u + p)| + \left| \frac{\partial}{\partial p}(x^u(p,t) - \bar{x}^u(t - T_u + p)) \right| \leq \varepsilon. \quad (5.56)$$

(b) *[Smoothness] There exists a constant* $K_u > 0$ *such that for all* $(p,t) \in (-\infty, U] \times (-\infty, T_u]$,

$$|x^u(p,t)| + \left| \frac{\partial}{\partial t} x^u(p,t) \right| + \left| \frac{\partial}{\partial p} x^u(p,t) \right| < K_u. \quad (5.57)$$

(c) *[Limit] For each* $t \leq T_u$, $\lim_{p \to -\infty} x^u(p,t)$ *is well defined.*

(d) *[Mappability] For each* $t \leq T_u$, *there exist intervals* $(U_1(t), U_2(t)] \subseteq (-\infty, U]$ *and* $(\tilde{U}_1(t), \tilde{U}_2(t)] \subseteq (-\infty, U]$, *and a scalar function* $r^u(\cdot, t)$ *defined on* $(U_1(t), U_2(t)]$ *which satisfies*

$$x^u(\tilde{p}, t) = \bar{x}^u(t - T_u + p) + r^u(p,t) \frac{f^\perp(\bar{x}^u(t - T_u + p))}{|f(\bar{x}^u(t - T_u + p))|}, \quad (5.58)$$

such that the mapping $p \to \tilde{p}$ *from* $(U_1(t), U_2(t)]$ *to* $(\tilde{U}_1(t), \tilde{U}_2(t)]$ *defined through (5.58) is a diffeomorphism.*

(e) *[Initial parametrization] For all* $p \in (U_1(T_u), U_2(T_u)]$,

$$[f(\bar{x}^u(p))]^\top [x^u(p, T_u) - \bar{x}^u(p)] = 0. \quad (5.59)$$

Next, define

$$\Xi_u := \{(p,t) \; : \; t \leq T_u \text{ and } U_1(t) < p \leq U_2(t)\}, \quad (5.60)$$

based on the largest interval from Hypothesis 5.20(f), and for $(p,t) \in \Xi_u$, define the quantities

$$M^u(p,t) := \left[f^\perp(\bar{x}^u(t - T_u + p)) \right]^\top [x^u(p,t) - \bar{x}^u(t - T_u + p)] \quad (5.61)$$

and

$$B^u(p,t) := [f(\bar{x}^u(t - T_u + p))]^\top [x^u(p,t) - \bar{x}^u(t - T_u + p)]. \quad (5.62)$$

Assume that the control velocity is chosen according to the smoothness conditions of Hypothesis 5.16(a–c), but with (d) and (e) replaced by the following:

(d)′ *The normal component of the control velocity,* $c^n := c \cdot \hat{f}^\perp$ *is chosen to satisfy*

$$c^n(\bar{x}^u(t - T_u + p), t) = \frac{\frac{\partial M^u}{\partial t}(p,t) - \mathrm{Tr}\,[Df(\bar{x}^u(t - T_u + p))] M^u(p,t)}{|f(\bar{x}^u(t - T_u + p))|}. \quad (5.63)$$

(e)′ *The tangential component of the control velocity, $c^t := c \cdot \hat{f}$, is chosen according to*

$$c^t\left(\bar{x}^u(t-T_u+p),t\right) = \frac{|f|^2 \frac{\partial B^u(p,t)}{\partial t} - f^\top\left[(Df)+(Df)^\top\right]\left[f^\perp M^u(p,t)+f B^u(p,t)\right]}{|f|^3}, \quad (5.64)$$

in which the dependence of f and Df on the argument $\bar{x}^u(t-T_u+p)$ has been suppressed for brevity.

Suppose the true unstable manifold under such a control velocity c is given by

$$\widehat{\Gamma}_u := \left\{(\widehat{x}^u(p,t),t) \,:\, (p,t)\in \Xi_u\right\}, \quad (5.65)$$

in which $\widehat{x}^u(p,t)$ is an exact trajectory for each fixed p, and for which there exists C_u such that

$$|\widehat{x}^u(p,t)-\bar{x}^u(t-T_u+p)| + \left|\frac{\partial}{\partial p}\left(\widehat{x}^u(p,t)-\bar{x}^u(t-T_u+p)\right)\right| \leq \varepsilon C_u \quad (5.66)$$

for $(p,t)\in\Xi_u$. The relevant error to be characterized is

$$e_u(p,t) := \widehat{x}^u(p,t) - x^u(p,t), \quad (5.67)$$

with normal and tangential components

$$e_u^n(p,t) := e_u(p,t)\cdot \hat{f}^\perp(\bar{x}^u(t-T_u+p)) \quad \text{and} \quad e_u^t(p,t) := e_u(p,t)\cdot \hat{f}(\bar{x}^u(t-T_u+p)).$$

Theorem 5.21 (control of unstable manifold [54]). *Assume the control velocity c satisfies Hypothesis 5.16, with the modifed conditions (d)′ and (e)′ above. Then, the error $e_u(p,t)=\mathcal{O}(\varepsilon^2)$. More specifically,*

$$|e_u^n(p,t)| \leq \varepsilon^2\left(C_u C_c + \frac{C_u^2 C_f}{2}\right) \frac{\int_{-\infty}^t |f(\bar{x}^u(\tau-T_u+p))| \exp\left[\int_\tau^t \mathrm{Tr}[Df(\bar{x}^u(\xi-T_u+p))]\mathrm{d}\xi\right]\mathrm{d}\tau}{|f(\bar{x}^u(t-T_u+p))|} \quad (5.68)$$

and

$$|e_u^t(p,t)| \leq \varepsilon^2\left(C_u C_c + \frac{C_u^2 C_f}{2}\right)|f(\bar{x}^u(t-T_u+p))| \times \int_t^{T_u} \frac{|f(\bar{x}^u(\tau-T_u+p))| + 2C_f\int_{-\infty}^{\tau} |f(\bar{x}^u(\zeta-T_u+p))| \exp\left[\int_\zeta^\tau \mathrm{Tr}[Df(\bar{x}^u(\xi-T_u+p))]\mathrm{d}\xi\right]\mathrm{d}\zeta}{|f(\bar{x}^u(\tau-T_u+p))|^2}\mathrm{d}\tau \quad (5.69)$$

for $(p,t)\in\Xi_u$.

Proof. This is fairly similar to the proof of Theorem 5.17 and will be omitted. □

Example 5.22 (Taylor-Green flow cont.). The uncontrolled velocity field shall be taken to be of Taylor-Green form, previously addressed in Examples 3.41 and 5.13, and which is a special case of the double-gyre. Here, though, a slightly different parametrization which highlights the dimensional parameters will be chosen. Consider the version given by

$$\left.\begin{aligned} \dot{x}_1 &= -\pi U \sin\left(\tfrac{\pi x_1}{L}\right)\cos\left(\tfrac{\pi x_2}{L}\right) \\ \dot{x}_2 &= \pi U \cos\left(\tfrac{\pi x_1}{L}\right)\sin\left(\tfrac{\pi x_2}{L}\right) \end{aligned}\right\}, \quad (5.70)$$

in which U and L are dimensional positive parameters representing velocity and length scales. The branch of the stable manifold of the saddle point $(L,0)$ (which emanates from (L,L) as its unstable manifold) shall be the focus of the control; this is illustrated in Figure 2.9 for the choice $L=1$. This stable manifold can be represented by $\bar{x}_1^s(t)=L$ and

$$\bar{x}_2^s(t)=\frac{2L}{\pi}\tan^{-1}e^{-\pi^2Ut/L}$$

as a solution trajectory of (5.70). In generalizing the particular example examined in [54], suppose this stable manifold is to instead behave according to

$$x^s(p,t)=\begin{pmatrix} L \\ \frac{2L}{\pi}\tan^{-1}e^{-\pi^2U(t-T_s+p)/L} \end{pmatrix}+\varepsilon\begin{pmatrix} h(p,t) \\ 0 \end{pmatrix}, \tag{5.71}$$

for some smooth function $h(p,t)$. The requirement of being within ε of the stable manifold is encapsulated by the condition $|h(p,t)|<1$. The above is written to be consistent with (5.36) with the first term being $\bar{x}^s(t-T_s+p)$. Only the x_1-part of the manifold is "moved" with the x_2 value being kept as it is, and thus $\hat{f}^{\perp}$ is purely in the x_1-direction at all points and all instances in time. Consequently, it is possible to take the quantities $S_1(t)=S$ and $S_2(t)=\infty$ (that is, for any given segment of the uncontrolled manifold, if the controlled manifold obeys (5.71), then that entire segment can be controlled). Since this parametric representation is not transparent, the primary stable manifold can also be written in a more standard form: x_2 represented as a function of x_1, with t appearing as a parameter. To do this, note that $x_1=L+\varepsilon h(p,t)$ and

$$x_2=\frac{2L}{\pi}\tan^{-1}e^{-\pi^2U(t-T_s+p)/L}$$

from (5.71), enabling the parameter p to be represented by

$$p(x_2,t)=\frac{L}{\pi^2U}\ln\left[\cot\frac{\pi x_2}{2L}\right]+T_s-t\,. \tag{5.72}$$

Substituting for p in the x_1 expression gives the stable manifold representation

$$x_1=L+\varepsilon h\left(p(x_2,t),t\right) \tag{5.73}$$

in terms of the definition (5.72). By specifying h, various forms of moving the stable manifold off the line $x_1=L$ can be achieved, with some prescribed time variation. This can be done only in a "clipped" fashion in the domain $t\in[T_s,\infty)$ and $p\in[S,\infty)$ (for any chosen T_s and S). The latter condition translates to

$$0<x_2\leq x_2^{\max}(t):=\frac{2L}{\pi}\cot^{-1}\left[\exp\frac{\pi^2U(t-T_s+S)}{L}\right] \tag{5.74}$$

by using (5.72), which demonstrates that segments of the manifold that can be controlled shorten as t increases, as indicated in Figure 5.7, as a consequence of the point $\bar{x}^s(S)$ approaching the saddle point when advected by the flow from the "initial" time-slice $t=T_s$.

For the choice of controlled stable manifold (5.71), it is clear that

$$x^s(p,t)-\bar{x}^s(t-T_s+p)=\begin{pmatrix} \varepsilon h(p,t) \\ 0 \end{pmatrix}.$$

Moreover, the uncontrolled velocity field, given by (5.70), on the stable manifold is

$$f\left(\bar{x}^s(p)\right)=\begin{pmatrix} 0 \\ -\pi U \sin\left[2\tan^{-1} e^{-\pi^2 U p/L}\right] \end{pmatrix}=\begin{pmatrix} 0 \\ -\pi U \operatorname{sech} \frac{\pi^2 U p}{L} \end{pmatrix}.$$

Consequently, from (5.44) $B^s(p,t)=0$, and the [more difficult] tangential control condition can be discarded. Thus, $c^t=0$, and so it is not necessary to impose any control velocity in the x_2-direction. Since $f^\perp(\bar{x}^s(p))=\left(\pi U \operatorname{sech}\frac{\pi^2 U p}{L},0\right)^\top$, (5.43) gives the Melnikov-like function as

$$M^s(p,t)=\varepsilon\pi U \operatorname{sech}\frac{\pi^2 U(t-T_s+p)}{L}\,h(p,t),$$

and thus

$$\frac{\partial M^s(p,t)}{\partial t}$$
$$=\varepsilon\pi U \operatorname{sech}\frac{\pi^2 U(t-T_s+p)}{L}\left[\frac{\partial h(p,t)}{\partial t}-\tanh\frac{\pi^2 U(t-T_s+p)}{L}\frac{\pi^2 U}{L}h(p,t)\right].$$

In determining c^n, the fact that $\operatorname{Tr} Df=0$ for (5.70) helps to simplify (5.46), and so using the above expressions,

$$c^n\left(\bar{x}^s(t-T_s+p),t\right)=\varepsilon\left[\frac{\partial h(p,t)}{\partial t}-\frac{\pi^2 U}{L}\tanh\frac{\pi^2 U(t-T_s+p)}{L}h(p,t)\right].$$

While this is the required control, it may be more convenient to express it in terms of the (x_1,x_2) coordinates and its unit vectors rather than in terms of the parameter p. This can be done using (5.72), and thus the control velocity in order to achieve the stable manifold (5.71) is

$$c(L,x_2,t)=\begin{pmatrix} \varepsilon\frac{\partial h(p(x_2,t),t)}{\partial t}-\frac{\varepsilon\pi^2 U}{L}\cos\frac{\pi x_2}{L}h\left(p(x_2,t),t\right) \\ 0 \end{pmatrix}. \tag{5.75}$$

The efficacy of this control strategy was investigated in [54] for a particular choice of $h(p,t)$, and excellent results were obtained. The error characterizations of Theorem 5.17 were also assessed. Rather than presenting similar results here, a different issue will be tackled. In the results in [54], a difficulty was that everything needed to be defined in terms of (p,t) rather than the more physical (x_2,t); this is reflected in the continuing presence of p in the formulae (5.73) and (5.75). Thus, define

$$\tilde{h}(x_2,t):=h\left(p(x_2,t),t\right)$$

and note from (5.73) that the expectation then is for the clipped controlled manifold to be specified in the more standard form

$$x_1=L+\varepsilon\tilde{h}\left(x_2,t\right) \tag{5.76}$$

of a graph at each t. The question now is to rewrite (5.75) in this more natural coordinate system. Note that

$$\frac{\partial\tilde{h}(x_2,t)}{\partial t}=\frac{d}{dt}\left[h\left(p(x_2,t),t\right)\right]=\frac{\partial h}{\partial p}\frac{\partial p}{\partial t}+\frac{\partial h}{\partial t}=\frac{\partial h}{\partial t}-\frac{\partial h}{\partial p} \quad \text{and}$$

$$\frac{\partial\tilde{h}(x_2,t)}{\partial x_2}=\frac{\partial h}{\partial p}\frac{\partial p}{\partial x_2}=-\frac{1}{\pi U}\csc\frac{\pi x_2}{L}\frac{\partial h}{\partial p},$$

and hence

$$\frac{\partial h}{\partial t} = \frac{\partial \tilde{h}}{\partial t} + \frac{\partial h}{\partial p} = \frac{\partial \tilde{h}}{\partial t} - \pi U \sin \frac{\pi x_2}{L} \frac{\partial \tilde{h}}{\partial x_2}.$$

Substituting the above into (5.75) yields

$$c(L, x_2, t) = \begin{pmatrix} \varepsilon \frac{\partial \tilde{h}(x_2,t)}{\partial t} - \varepsilon \pi U \sin \frac{\pi x_2}{L} \frac{\partial \tilde{h}(x_2,t)}{\partial x_2} - \frac{\varepsilon \pi^2 U}{L} \cos \frac{\pi x_2}{L} \tilde{h}(x_2, t) \\ 0 \end{pmatrix}, \tag{5.77}$$

which expresses the control velocity required to achieve (5.76) to leading-order, in the physical (x_1, x_2, t) coordinates.

As a special case, suppose the stable manifold is to be *pinned*, so that it is not varying in time. Now, given a choice of T_s and S, there is an initial primary manifold segment that is being considered, and as shown in Figure 5.7, this inevitably shortens as t progresses. Thus, this pinning is only possible subject to the fact that the manifold shortens, as further specified in (5.74). That is, if considering a range of time $t \in [T_s, t^{\max}]$ and a fixed S, pinning can be achieved only for $x_2 \in \left[0, x_2^{\max}(t^{\max})\right]$. For such an x_2, the x_1 value on the stable manifold must not change with t, which implies that $\tilde{h}(x_2, t)$ must be independent of t. The corresponding x_1-component of the control velocity must therefore be

$$c^n(L, x_2, t) = -\varepsilon \pi U \sin \frac{\pi x_2}{L} \tilde{h}'(x_2) - \frac{\varepsilon \pi^2 U}{L} \cos \frac{\pi x_2}{L} \tilde{h}(x_2). \tag{5.78}$$

The forms for the control velocity arising from this analysis do not seem to possess any intuitively obvious characteristics, and hence the full analysis would seem to be necessary to determine them. ∎

The example above shows how more implementable control conditions than in the original article [54] are possible to obtain through a process of transforming the trajectory-identifying parameter p to a physical length parameter. This process should be adaptable to other situations relatively easily and is the subject of future investigation.

The methods of this section allows the control of stable or unstable manifolds according to user specifications. In flow barrier situations, it is the *relative* positioning of a stable and unstable manifold that has an impact on transport, as has been addressed in detail in Chapter 3. When a heteroclinic separator breaks apart into a stable and an unstable manifold, an interesting question is whether one can independently control each of these manifolds. This is also under investigation.

5.4.1 ▪ Proof of Theorem 5.17 (control of stable manifold)

The notation

$$y(t) := \bar{x}^s(t - T_s + p) \tag{5.79}$$

will prove useful in keeping this proof brief. Now, $\widehat{x}^s(p, t)$ lies on the true perturbed manifold $\widehat{\Gamma}_s$ for each p, with the t parametrizing the time evolution. In keeping with Hypothesis 5.15, the parametrization shall be chosen such that

$$[f(\bar{x}^s(p))]^\top [\widehat{x}^s(p, T_s) - \bar{x}^s(p)] = 0. \tag{5.80}$$

Let $a(t)$ be the actual hyperbolic trajectory to which a modifies under the inclusion of the control velocity. Thus, as $t \to \infty$, $|\widehat{x}^s(p,t) - a(t)| \to 0$ for any p. The goal is to show that $e_s(p,t) = \mathcal{O}(\varepsilon^2)$. Now for $(p,t) \in \Xi_s$,

$$\begin{aligned}|e_s(p,t)| &= |\widehat{x}^s(p,t) - x^s(p,t)| \\ &= |\widehat{x}^s(p,t) - a(t) + a(t) - a + a - y(t) + y(t) - x^s(p,t)| \\ &\leq |\widehat{x}^s(p,t) - a(t)| + |a(t) - a| + |a - y(t)| + |y(t) - x^s(p,t)| \,.\end{aligned}$$

As $t \to \infty$, the first and third terms above go to zero because each represents the difference between a trajectory on a stable manifold and its corresponding hyperbolic trajectory. The second and fourth terms are each $\mathcal{O}(\varepsilon)$ by virtue of Theorem 2.7 and Hypothesis 5.15(a), respectively. Thus, $e_s(p,t) = \mathcal{O}(\varepsilon)$ and is bounded on Ξ_s. Extending this to $\mathcal{O}(\varepsilon^2)$ now requires quite a bit of work. Define on Ξ_s

$$\begin{aligned}\widehat{M}^s(p,t) &:= \left[f^\perp(\bar{x}^s(t - T_s + p))\right]^\top \left[\widehat{x}^s(p,t) - \bar{x}^s(t - T_s + p)\right] \\ &= \left[f^\perp(y(t))\right]^\top \left[x^s(p,t) + e_s(p,t) - y(t)\right],\end{aligned} \tag{5.81}$$

which is the normal displacement associated with the *true* stable manifold as opposed to the desired one (compare with (5.43)). Differentiating with respect to time t yields

$$\begin{aligned}\frac{\partial \widehat{M}^s}{\partial t}(p,t) &= \left[f^\perp(y(t))\right]^\top \left[f(x^s(p,t) + e_s(p,t)) + c(x^s(p,t) + e_s(p,t),t) - f(y(t))\right] \\ &\quad + \left[Df^\perp(y(t))\frac{\partial y(t)}{\partial t}\right]^\top \left[x^s(p,t) - y(t) + e_s(p,t)\right],\end{aligned} \tag{5.82}$$

where the fact that $x^s(p,t) + e_s(p,t)$ is an exact solution to (5.41) is used in writing its derivative in terms of $f + c$, and similarly since $y(t)$ obeys (5.31), its derivative is $f(y(t))$. Now, Taylor's theorem guarantees the presence of points $\xi_{1,2} \in \Omega$ such that

$$\begin{aligned}f(x^s(p,t) + e_s(p,t)) &= f(y(t)) + Df(y(t))(x^s(p,t) + e_s(p,t) - y(t)) \\ &\quad + \frac{1}{2}(x^s(p,t) + e_s(p,t) - y(t))^\top D^2 f(\xi_1)(x^s(p,t) + e_s(p,t) - y(t))\end{aligned} \tag{5.83}$$

and

$$c(x^s(p,t) + e_s(p,t),t) = c(y(t),t) + Dc(\xi_2,t)(x^s(p,t) + e_s(p,t) - y(t)). \tag{5.84}$$

Substituting the above into (5.82) and rearranging,

$$\begin{aligned}&\frac{\partial \widehat{M}^s}{\partial t}(p,t) \\ &= \left[f^\perp(y(t))\right]^\top c(y(t),t) + \left[f^\perp(y(t))\right]^\top Df(y(t))\left[x^s(p,t) + e_s(p,t) - y(t)\right] \\ &\quad + \left[Df^\perp(y(t))f(y(t))\right]^\top \left[x^s(p,t) - y(t)\right] + \left[Df^\perp(y(t))f(y(t))\right]^\top e_s(p,t) \\ &\quad + \left[f^\perp(y(t))\right]^\top E_s(p,t),\end{aligned} \tag{5.85}$$

where $E_s(p,t)$ is a higher-order term satisfying

$$|E_s(p,t)| \leq \varepsilon^2 \left[C_s C_c + \frac{C_s^2 C_f}{2}\right], \tag{5.86}$$

by virtue of the bounds in (5.32), (5.36), (5.45), and (5.49), valid for $(p,t)\in\Xi_s$. Now, using (2.39) with the choice $A=Df(y(t))$, $b=f(y(t))$, and $c=x^s(p,t)-y(t)$ gives the identity

$$[Dff]^{\perp}\cdot[x^s(p,t)-y(t)]+f^{\perp}\cdot(Df[x^s(p,t)-y(t)])=(\mathrm{Tr}Df)f^{\perp}\cdot[x^s(p,t)-y(t)],$$

but it is easy to verify that $[Dff]^{\perp}=Df^{\perp}f$, and hence the above combination appears on the right-hand side of (5.85). Replacing with the trace form from above (but retaining the matrix multiplication notation as opposed to the dot product) gives

$$\begin{aligned}\frac{\partial\widehat{M}^s}{\partial t}(p,t)=&\left[f^{\perp}(y(t))\right]^{\mathsf{T}}c(y(t),t)+\mathrm{Tr}[Df(y(t))]\left[f^{\perp}(y(t))\right]^{\mathsf{T}}[x^s(p,t)-y(t)]\\&+\mathrm{Tr}[Df(y(t))]\left[f^{\perp}(y(t))\right]^{\mathsf{T}}e_s(p,t)+\left[f^{\perp}(y(t))\right]^{\mathsf{T}}E_s(p,t).\end{aligned}$$

Substituting into this the condition

$$\widehat{M}^s(p,t)=M^s(p,t)+\left[f^{\perp}(y(t))\right]^{\mathsf{T}}e_s(p,t),$$

which arises from (5.81) and (5.43), yields

$$\begin{aligned}&\frac{\partial M^s}{\partial t}(p,t)+\frac{\partial}{\partial t}\left\{\left[f^{\perp}(y(t))\right]^{\mathsf{T}}e_s(p,t)\right\}=\left[f^{\perp}(y(t))\right]^{\mathsf{T}}E_s(p,t)+\left[f^{\perp}(y(t))\right]^{\mathsf{T}}c(y(t),t)\\&+\mathrm{Tr}[Df(y(t))]M^s(p,t)+\mathrm{Tr}[Df(y(t))]\left[f^{\perp}(y(t))\right]^{\mathsf{T}}e_s(p,t).\end{aligned}\tag{5.87}$$

Now, under condition (5.46) of Hypothesis 5.16(d), which is a specification for the normal component of the control velocity, it is clear that

$$\frac{\partial M^s}{\partial t}(p,t)=\left[f^{\perp}(y(t))\right]^{\mathsf{T}}c(y(t),t)+\mathrm{Tr}[Df(y(t))]M^s(p,t).$$

After deleting this combination of terms in (5.87), the remainder is

$$\frac{\partial}{\partial t}\left\{\left[f^{\perp}(y(t))\right]^{\mathsf{T}}e_s(p,t)\right\}-\mathrm{Tr}[Df(y(t))]\left[f^{\perp}(y(t))\right]^{\mathsf{T}}e_s(p,t)=\left[f^{\perp}(y(t))\right]^{\mathsf{T}}E_s(p,t).$$

By multiplying through by the integrating factor

$$\mu(p,t):=\exp\left[\int_t^{T_s}\mathrm{Tr}[Df(y(\xi))]\mathrm{d}\xi\right],$$

the expression

$$\frac{\partial}{\partial t}\left\{\mu(p,t)\left[f^{\perp}(y(t))\right]^{\mathsf{T}}e_s(p,t)\right\}=\mu(p,t)\left[f^{\perp}(y(t))\right]^{\mathsf{T}}E_s(p,t)$$

results. In integrating this from a general t-value to ∞, it is first required to check that the limit $\lim_{L\to\infty}|\mu(p,L)f^{\perp}(y(L))e_s(p,L)|$ arising from the boundary term is zero. Coupled with the already established fact that e_s is bounded, this is analogous to the calculation that has been done in relation to (2.43) and therefore shall be skipped. Integration yields

$$-\mu(p,t)\left[f^{\perp}(y(t))\right]^{\mathsf{T}}e_s(p,t)=\int_t^{\infty}\mu(p,\tau)\left[f^{\perp}(y(\tau))\right]^{\mathsf{T}}E_s(p,\tau)\mathrm{d}\tau.$$

Dividing by $\mu(p,t)|f(y(\tau))|$ and utilizing the bound on E_s furnished by (5.86) leads to the bound for the normal component of the error (5.51), as required.

Before proceeding in analyzing the tangential component, the limits for the normal bound given in Remark 5.19 will be tackled. By applying L'Hôpital's rule to (5.51),

$$\lim_{t\to\infty} |e_s^n(p,t)| \le \left[C_s C_c + \frac{C_s^2 C_f}{2} \right] \varepsilon^2 \lim_{t\to\infty} \frac{-|f(y(t))|}{\frac{\partial}{\partial t}|f(y(t))|},$$

but since $|f(y(t))| \sim e^{\lambda_s(t-T_s+p)}$, the limit above is $1/\lambda_s$, which gives (5.53). The limit $p\to\infty$ at each fixed t is easiest computed with the formal replacements $|f(y(t))| \sim e^{\lambda_s(t-T_s+p)}$ and $\operatorname{Tr} Df(y(\xi)) \sim \lambda_u + \lambda_s$. Thus,

$$\lim_{p\to\infty} |e_s^n(p,t)| \le \left[C_s C_c + \frac{C_s^2 C_f}{2} \right] \varepsilon^2 \int_t^\infty \frac{e^{\lambda_s(\tau-T_s+p)}}{e^{\lambda_s(t-T_s+p)}} \exp\left[\int_\tau^{\mathsf{T}} (\lambda_s+\lambda_u)\, d\xi \right] d\tau,$$

which leads to the second limit in (5.53).

The next order of business is to bound the tangential component of the error. This is done by defining

$$\begin{aligned}\widehat{B}^s(p,t) &:= [f(\bar{x}^s(t-T_s+p))]^\top [\widehat{x}^s(p,t) - \bar{x}^s(t-T_s+p)] \\ &= [f(y(t))]^\top [x^s(p,t)+e_s(p,t)-y(t)]\end{aligned} \tag{5.88}$$

and then taking the t-derivative, leading to

$$\begin{aligned}\frac{\partial \widehat{B}^s}{\partial t}(p,t) = {} & [f(y(t))]^\top c(x^s(p,t)+e_s(p,t),t) \\ & + [f(y(t))]^\top [f(x^s(p,t)+e_s(p,t)) - f(y(t))] \\ & + [Df(y(t))f(y(t))]^\top [x^s(p,t)+e_s(p,t)-y(t)].\end{aligned}$$

Applying the expansions (5.83) and (5.84) and then using the relationship between $\widehat{B}^s$ and B^s (as done for M^s previously in this proof) eventually leads to

$$\begin{aligned}\frac{\partial B^s}{\partial t}(p,t) + \frac{\partial}{\partial t}\left\{ [f(y(t))]^\top e_s(p,t) \right\} = [f(y(t))]^\top E_s(p,t) + [f(y(t))]^\top c(y(t),t) \\ + f^\top \left[Df + (Df)^\top \right]\Big|_{y(t)} [x^s(p,t)-y(t)] + f^\top \left[Df + (Df)^\top \right]\Big|_{y(t)} e_s(p,t),\end{aligned} \tag{5.89}$$

where $E_s(p,t)$ once again satisfies (5.86). It is possible to write

$$x^s(p,t) - y(t) = \frac{f^\perp(y(t))}{|f(y(t))|^2} M^s(p,t) + \frac{f(y(t))}{|f(y(t))|^2} B^s(p,t)$$

in terms of the orthogonal coordinate system with unit vectors $(\hat{f}^\perp, \hat{f})$ by using (2.62) and (2.63). Substituting this into (5.89) yields

$$\begin{aligned}\frac{\partial B^s}{\partial t}(p,t) + \frac{\partial}{\partial t}\left\{ [f(y(t))]^\top e_s(p,t) \right\} = [f(y(t))]^\top E_s(p,t) + [f(y(t))]^\top c(y(t),t) \\ + \left. \frac{f^\top \left[Df + (Df)^\top \right] f^\perp}{|f|^2} \right|_{y(t)} M^s(p,t) + \left. \frac{f^\top \left[Df + (Df)^\top \right] f}{|f|^2} \right|_{y(t)} B^s(p,t) \\ + f^\top \left[Df + (Df)^\top \right]\Big|_{y(t)} e_s(p,t).\end{aligned} \tag{5.90}$$

By applying the imposed tangential component of the control velocity (5.47) (i.e., Hypothesis 5.16(e)), the condition

$$[f(y(t))]^\top c(y(t),t) = \frac{\partial B^s}{\partial t}(p,t) - \left.\frac{f^\top\left[Df + (Df)^\top\right]f^\perp}{|f|^2}\right|_{y(t)} M^s(p,t)$$
$$- \left.\frac{f^\top\left[Df + (Df)^\top\right]f}{|f|^2}\right|_{y(t)} B^s(p,t)$$

arises, and hence (5.90) reduces to

$$\frac{\partial}{\partial t}\left\{[f(y(t))]^\top e_s(p,t)\right\}$$
$$= [f(y(t))]^\top E_s(p,t) + f^\top\left[Df + (Df)^\top\right]\Big|_{y(t)} e_s(p,t). \tag{5.91}$$

Writing $e_s(p,t)$ in terms of the $(\hat{f}^\perp, \hat{f})$ coordinate system and rearranging gives

$$\frac{\partial}{\partial t}\left[f^\top e_s\right] - \frac{f^\top\left[Df + (Df)^\top\right]f}{|f|^2}\left[f^\top e_s\right] = f^\top E_s + \frac{f^\top\left[Df + (Df)^\top\right]f^\perp}{|f|}\frac{(f^\perp)^\top}{|f|}e_s,$$

where e_s's and E_s's (p,t)-dependence and f's $y(t)$-dependence have been suppressed for brevity. The left-hand side above can be simplified by using the ideas of (2.53) and its subsequent discussion to enable the representation

$$\frac{\partial}{\partial t}\left[f^\top e_s\right] - \frac{\partial}{\partial t}\left[\ln|f|^2\right]\left[f^\top e_s\right] = f^\top E_s + \frac{f^\top\left[Df + (Df)^\top\right]f^\perp}{|f|}e_s^n,$$

which when multiplied through by the integrating factor $|f(y(t))|^{-2}$ and integrated from T_s to a general t-value yields

$$\frac{f^\top(y(t))}{|f(y(t))|^2}e_s(p,t) = \int_{T_s}^t \frac{f^\top E_s(p,\tau) + \left.\frac{f^\top\left[Df+(Df)^\top\right]f^\perp}{|f|}\right|_{y(\tau)} e_s^n}{|f(y(\tau))|^2}\,d\tau.$$

The boundary term at T_s has been eliminated because of the choice of parameters as expressed in (5.39) and (5.80). Therefore,

$$e_s^t(p,t) = \frac{f^\top(y(t))e_s(p,t)}{|f(y(t))|} = |f(y(t))|\int_{T_s}^t \frac{f^\top E_s(p,\tau) + \left.\frac{f^\top\left[Df+(Df)^\top\right]f^\perp}{|f|}\right|_{y(\tau)} e_s^n}{|f(y(\tau))|^2}\,d\tau. \tag{5.92}$$

To bound e_s^t, the integrand of (5.92) is first bounded using (5.86):

$$\left|\frac{f^\top E_s(p,\tau) + \frac{f^\top\left[Df+(Df)^\top\right]f^\perp}{|f|}e_s^n}{|f|^2}\right| \le \frac{\left|f^\top\right|}{|f|^2}\left|\left|E_s\right| + \left|Df + (Df)^\top\right|\left|e_s^n\right|\right|$$
$$\le |f|^{-1}\left|\varepsilon^2\left(C_s C_c + \frac{C_s^2 C_f}{2}\right) + 2C_f\left|e_s^n\right|\right|.$$

Now, inserting the already established bound (5.51) for e_s^n into the above and then into the integrand in (5.92) gives the required result (5.52).

The limiting behavior of (5.52), as given in Remark 5.19, is next established. Consider the integrand of (5.52) as a quotient; its denominator goes to zero as $\tau \to \infty$ since $|f(y(\tau))| \to |f(a)| = 0$, while its numerator (which shall be called $H(p,\tau)$) remains bounded. Thus, by applying L'Hôpital's rule,

$$\lim_{t\to\infty} \frac{\int_{T_s}^{t} \frac{H(p,\tau)}{|f(y(\tau))|} \mathrm{d}\tau}{|f(y(t))|^{-1}} = \lim_{t\to\infty} \frac{\frac{H(p,t)}{|f(y(t))|}}{-|f(y(t))|^{-2}} = -\lim_{t\to\infty} H(p,t)|f(y(t))| = 0,$$

and hence $e_s^t(p,t) \to 0$ as $t \to \infty$. The $p \to \infty$ limit is established as before by replacing each term with its limiting behavior:

$$\begin{aligned}
&\lim_{p\to\infty} |e_s^t(p,t)| \\
&\leq \varepsilon^2 \left(C_c C_s + \frac{C_s^2 C_f}{2} \right) e^{\lambda_s(t-T_s+p)} \int_{T_s}^{t} \frac{e^{\lambda_s(\tau-T_s+p)} + 2C_f \int_\tau^\infty e^{\lambda_s(\zeta-T_s+p)} e^{(\lambda_s+\lambda_u)(\tau-\zeta)} \mathrm{d}\zeta}{e^{2\lambda_s(\tau-T_s+p)}} \mathrm{d}\tau \\
&= \varepsilon^2 \left(C_c C_s + \frac{C_s^2 C_f}{2} \right) e^{\lambda_s(t-T_s+p)} \int_{T_s}^{t} e^{-\lambda_s(\tau-T_s+p)} \left[1 + 2C_f e^{\lambda_u \tau} \int_\tau^\infty e^{-\lambda_u \zeta} \mathrm{d}\zeta \right] \mathrm{d}\tau \\
&= \varepsilon^2 \left(C_c C_s + \frac{C_s^2 C_f}{2} \right) \left(1 + \frac{2C_f}{\lambda_u} \right) e^{\lambda_s t} \frac{e^{-\lambda_s T_s} - e^{-\lambda_s t}}{\lambda_s},
\end{aligned}$$

which simplifies to (5.54) as required.

Bibliography

[1] S.S. ABDULLAEV, *Structure of motion near saddle points and chaotic transport in Hamiltonian systems*, Phys. Rev. E, 62 (2000), pp. 3508–3528. (Cited on pp. 4, 140, 184)

[2] R. ABERNATHY, J. MARSHALL, M. MAZLOFF, AND E. SHUCKBURGH, *Enhancement of mesoscale eddy stirring at steering levels in the Southern Ocean*, J. Phys. Oceanogr., 40 (2010), pp. 170–184. (Cited on p. 12)

[3] D. ACHESON, *Elementary Fluid Dynamics*, Oxford University Press, Oxford, UK, 1990. (Cited on pp. 5, 14, 151, 152, 160, 161, 164)

[4] A. ADJARI, *Pumping liquids using asymmetric electrode arrays*, Phys. Rev. E, 61 (2000), pp. R45–R48. (Cited on pp. 14, 187)

[5] T. AHN AND S. KIM, *Separatrix map analysis of chaotic transport in planar periodic vortical flows*, Phys. Rev. E, 49 (1994), pp. 2900–2912. (Cited on pp. 4, 140, 184)

[6] A. ALAM, A. AFZAL, AND K.-Y. KIM, *Mixing performance of a planar micromixer with circular obstructions in a curved microchannel*, Chem. Eng. Res. Design, 92 (2014), pp. 423–434. (Cited on pp. 5, 13, 14, 15, 28)

[7] B. ALDRIDGE, G. HALLER, P. SORGER, AND D. LAUFFENBURGER, *Direct Lyapunov exponent analysis enables parametric study of transient signaling governing cell behaviour*, IEE Proc.-Sys. Biol., 153 (2006), pp. 425–432. (Cited on p. 198)

[8] J. ALLEN, R. SAMELSON, AND P. NEWBERGER, *Chaos in a model of forced quasi-geostrophic flow over topography—an application of Melnikov method*, J. Fluid Mech., 226 (1991), pp. 511–547. (Cited on p. 9)

[9] K.T. ALLIGOOD, T.D. SAUER, AND J.A. YORKE, *Chaos: An Introduction to Dynamical Systems*, Springer-Verlag, New York, 1996. (Cited on pp. 13, 28, 35, 36, 103)

[10] M.R. ALLSHOUSE AND J.-L. THIFFEAULT, *Detecting coherent structures using braids*, Phys. D, 241 (2012), pp. 95–105. (Cited on p. 24)

[11] J.R. ANGILELLA AND J.P. BRANCHER, *Note on chaotic advection in an oscillating drop*, Phys. Fluids, 15 (2003), pp. 261–264. (Cited on pp. 5, 14, 210)

[12] M. ANSARI, K.-Y. KIM, K. ANWAR, AND S. KIM, *A novel passive micromixer based on unbalanced splits and collisions of fluid streams*, J. Micromech. Microeng., 20 (2010), p. 055007. (Cited on pp. 5, 13, 14, 15, 28)

[13] F. ARAI, A. ICHIKAWA, T. FUKUDA, K. HORIO, AND K. ITOIGAWA, *Stagnation point control by pressure balancing in microchannel for high speed and high purity separation of microobject*, in Proceedings of the 2001 IEEE/RSJ International Conference on Intelligent Robots and Systems, 2001, pp. 1343–1348. Maui, HI, Oct 29–Nov 3, 2001. (Cited on p. 198)

[14] H. AREF, *Stirring by chaotic advection*, J. Fluid Mech., 143 (1984), pp. 1–21. (Cited on pp. 8, 13)

[15] H. AREF ET AL., *Frontiers of chaotic advection*, submitted (2015), arXiv:1403.2953. (Cited on p. 13)

[16] V.I. ARNOLD, *Sur la topologie des écoulements stationnaires des fluides parfaits*, C. R. Acad. Sci. Paris, 261 (1965), pp. 17–20. (Cited on p. 13)

[17] D.K. ARROWSMITH AND C.M. PLACE, *An Introduction to Dynamical Systems*, Cambridge University Press, Cambridge, UK, 1990. (Cited on pp. 13, 36, 38, 39, 40, 49, 50, 61, 116, 117, 129, 130, 131)

[18] D. AUERBACH, C. GREBOGI, E. OTT, AND J.A. YORKE, *Controlling chaos in high dimensional systems*, Phys. Rev. Lett., 69 (1992), pp. 3479–3482. (Cited on p. 204)

[19] D. BAINOV AND P. SIMEONOV, *Systems with Impulsive Effects: Stability Theory and Applications*, Ellis Horwood, Chichester, UK 1989. (Cited on pp. 75, 81, 143)

[20] R. BALASUBRAMANIAM AND R.S. SUBRAMANIAN, *The migration of a drop in a uniform temperature gradient at large Marangoni numbers*, Phys. Fluids, 12 (2000), pp. 733–743. (Cited on pp. 5, 14, 210)

[21] S. BALASURIYA, *Meridional and zonal wavenumber dependence in tracer flux in Rossby waves*, Fluids, 1 (2016), 30. (Cited on pp. 136, 137)

[22] ——, *Vanishing viscosity in the barotropic β-plane*, J. Math. Anal. Appl., 214 (1997), pp. 128–150. (Cited on pp. 134, 159, 161, 162, 163, 166, 167)

[23] ——, *Gradient evolution in potential vorticity flows*, Nonlin. Proc. Geophys., 8 (2001), pp. 253–263. (Cited on pp. 10, 12, 13, 198)

[24] ——, *An analytical study of general hyper-diffusivity and barotropic eddies*, Geophys. Astrophys. Fluid Dyn., 98 (2004), pp. 39–62. (Cited on pp. 4, 10, 12, 159, 161, 162)

[25] ——, *Approach for maximizing chaotic mixing in microfluidic devices*, Phys. Fluids, 17 (2005), 118103. (Cited on pp. 5, 13, 15, 62, 183, 184, 187, 198)

[26] ——, *Direct chaotic flux quantification in perturbed planar flows: General time-periodicity*, SIAM J. Appl. Dyn. Sys., 4 (2005), pp. 282–311. (Cited on pp. 4, 62, 129, 130, 133, 137, 138, 139, 140, 142, 175, 184)

[27] ——, *Optimal perturbation for enhanced chaotic transport*, Phys. D, 202 (2005), pp. 155–176. (Cited on pp. 4, 15, 133, 140, 175, 183)

[28] ——, *Cross-separatrix flux in time-aperiodic and time-impulsive flows*, Nonlinearity, 19 (2006), pp. 282–311. (Cited on pp. 71, 78, 81, 113, 121, 127, 129, 139, 147, 149)

[29] ——, *Separatrix splitting through high-frequency non-smooth perturbations*, Dyn. Continuous Discrete Imulsive Sys. Ser. A, 14 (2007), pp. 333–343. (Cited on p. 139)

[30] ——, *Invasions with density-dependent ecological parameters*, J. Theor. Biol., 266 (2010), pp. 657–666. (Cited on p. 54)

[31] ——, *Optimal frequency for microfluidic mixing across a fluid interface*, Phys. Rev. Lett., 105 (2010), p. 064501. (Cited on pp. ix, 15, 123, 155, 178, 180, 181, 193, 198)

[32] ——, *A tangential displacement theory for locating perturbed saddles and their manifolds*, SIAM J. Appl. Dyn. Sys., 10 (2011), pp. 1100–1126. (Cited on pp. 43, 48, 49, 50, 53, 54, 56, 60, 116)

[33] ——, *Explicit invariant manifolds and specialised trajectories in a class of unsteady flows*, Phys. Fluids, 24 (2012), p. 127101. (Cited on pp. 8, 26, 31, 33)

[34] S. BALASURIYA, *Flow barriers in realistic flows, and their relationship to invariant manifolds*, in Workshop on Mixing, Transport and Coherent Structures, Oberwolfach Reports 11, European Mathematical Society, Zürich, Switzerland, 2014, pp. 270–272. (Cited on p. 24)

[35] S. BALASURIYA, *Nonautonomous flows as open dynamical sytems: characterising escape rates and time-varying boundaries*, in Ergodic Theory, Open Dynamics and Structures, W. Bahsoun, C. Bose, and G. Froyland, eds., Springer Proceedings in Mathematics and Statistics, Springer, New York, 2014, ch. 1, pp. 1–30. (Cited on pp. 118, 129, 130, 175)

[36] ——, *Dynamical systems techniques for enhancing microfluidic mixing*, J. Micromech. Microeng., 25 (2015), p. 094005. (Cited on pp. 15, 28, 178, 180, 181, 187, 189)

[37] ——, *Impulsive perturbations to differential equations: stable/unstable pseudo-manifolds, heteroclinic connections, and flux* (2015), submitted. (Cited on pp. 75, 80, 82, 85, 143, 144, 146, 149)

[38] ——, *Quantifying transport within a two-cell microdroplet induced by circular and sharp channel bends*, Phys. Fluids, 27 (2015), p. 052005. (Cited on pp. 5, 13, 14, 15, 210)

[39] S. BALASURIYA, *Local stable and unstable manifolds and their control in nonautonomous finite-time flows*, J. Nonlin. Sci., 26 (2016), pp. 895–927. (Cited on pp. viii, x, 16, 67, 68, 94, 96, 213, 215, 216)

[40] S. BALASURIYA, *Transport between two fluids across their mutual flow interface: the streakline approach* (2016), submitted. (Cited on pp. 150, 152, 154, 188)

[41] S. BALASURIYA AND B.J. BINDER, *Nonautonomous analysis of steady Korteweg-de Vries waves under nonlocalised forcing*, Phys. D, 285 (2014), pp. 28–41. (Cited on p. 53)

[42] S. BALASURIYA AND M.D. FINN, *Energy constrained transport maximization across a fluid interface*, Phys. Rev. Lett., 108 (2012), p. 244503. (Cited on pp. 5, 13, 15, 62, 108, 181, 183, 184, 187, 198, 211)

[43] S. BALASURIYA AND G. FROYLAND, *Minimum advective+diffusive flux curves and surfaces in unsteady flows* (2016), in preparation. (Cited on pp. 24, 112, 124)

[44] S. BALASURIYA, G. FROYLAND, AND N. SANTITISSADEEKORN, *Absolute flux optimising curves of flows on a surface*, J. Math. Anal. Appl., 409 (2014), pp. 119–139. (Cited on pp. 24, 112, 124, 198)

[45] S. BALASURIYA AND G.A. GOTTWALD, *Wavespeed in reaction-diffusion systems, with applications to chemotaxis and population pressure*, J. Math. Biol., 61 (2010), pp. 377–399. (Cited on p. 54)

[46] S. BALASURIYA, G.A. GOTTWALD, J. HORNIBROOK, AND S. LAFORTUNE, *High Lewis number combustion wavefronts: A perturbative Melnikov analysis*, SIAM J. Appl. Math., 67 (2007), pp. 464–486. (Cited on p. 54)

[47] S. BALASURIYA AND C.K.R.T. JONES, *Diffusive draining and growth of eddies*, Nonlin. Proc. Geophys., 8 (2001), pp. 241–251. (Cited on pp. 10, 12, 27, 161, 168)

[48] S. BALASURIYA, C.K.R.T. JONES, AND B. SANDSTEDE, *Viscous perturbations of vorticity-conserving flows and separatrix splitting*, Nonlinearity, 11 (1998), pp. 47–77. (Cited on pp. 4, 5, 10, 12, 27, 100, 117, 161, 163, 165, 168, 171)

[49] S. BALASURIYA, R. KALAMPATTEL, AND N. OUELLETTE, *Hyperbolic neighborhoods as organizers of finite-time exponential stretching*, J. Fluid Mech., accepted (2016). (Cited on pp. 20, 21, 24)

[50] S. BALASURIYA, I. MEZIĆ, AND C.K.R.T. JONES, *Weak finite-time Melnikov theory and 3D viscous perturbations of Euler flows*, Phys. D, 176 (2003), pp. 82–106. (Cited on pp. 5, 6, 13, 14, 39, 117, 164, 210)

[51] S. BALASURIYA AND N.-T. NGUYEN, *Mixing within floating liquid marbles due to unsteady translation*, in preparation. (Cited on pp. 5, 13, 14, 15, 210)

[52] S. BALASURIYA AND K. PADBERG-GEHLE, *Controlling the unsteady analogue of saddle stagnation points*, SIAM J. Appl. Math., 73 (2013), pp. 1038–1057. (Cited on pp. 44, 198, 201, 202, 203)

[53] ——, *Accurate control of hyperbolic trajectories in any dimension*, Phys. Rev. E, 90 (2014), 032903. (Cited on pp. x, 44, 204, 208, 209, 210, 211)

[54] ——, *Nonautonomous control of stable and unstable manifolds in two-dimensional flows*, Phys. D, 276 (2014), pp. 48–60. (Cited on pp. 4, 10, 62, 140, 218, 224, 226, 227, 228, 229)

[55] S. BALASURIYA AND V.A. VOLPERT, *Wavespeed analysis: Approximating Arrhenius kinetics with step-function kinetics*, Combust. Theor. Model., 12 (2008), pp. 643–670. (Cited on p. 54)

[56] E. BARNES AND D. HARTMANN, *Rossby wave scales, propagation, and the variability of eddy-driven jets*, J. Atmos. Sci., 68 (2011), pp. 2893–2908. (Cited on p. 134)

[57] L. BARREIRA, M. FAN, C. VALLS, AND J. ZHANG, *Invariant manifolds for impulsive equations and nonuniform polynomial dichotomies*, J. Stat. Phys., 141 (2010), pp. 179–200. (Cited on pp. 75, 81, 143)

[58] L. BARREIRA AND C. VALLS, *Stability of Nonautonomous Differential Equations*, Springer, New York, 2008. (Cited on pp. 32, 75, 81, 143)

[59] F. BATTELLI AND M. FECKAN, *Homoclinic trajectories in discontinuous systems*, J. Dyn. Diff. Equat., 20 (2008), pp. 337–376. (Cited on pp. 39, 76, 144)

[60] ——, *Melnikov theory for nonlinear implicit ODEs*, J. Differential Equations, 256 (2014), pp. 1157–1190. (Cited on p. 39)

[61] F. BATTELLI AND C. LAZZARI, *Exponential dichotomies, heteroclinic orbits, and Melnikov functions*, J. Differential Equations, 86 (1990), pp. 342–366. (Cited on pp. 30, 39, 117, 162, 163)

[62] M.Z. BAZANT AND T.M. SQUIRES, *Induced-charge electrokinetic phenomena: Theory and microfluidic applications*, Phys. Rev. Lett., 92 (2004), 066101. (Cited on pp. 14, 187)

[63] D. BEIGIE, A. LEONARD, AND S. WIGGINS, *Invariant manifold templates for chaotic advection*, Chaos Solitons Fractals, 4 (1994), pp. 749–868. (Cited on p. 108)

[64] R. BELLMAN AND K. COOKE, *Differential-Difference Equations*, Academic Press, New York, 1963. (Cited on pp. 89, 90)

[65] B.J. BENTLEY AND L.G. LEAL, *A computer-controlled four-roll mill for investigations of particle and drop dynamics in two-dimensional shear flows*, J. Fluid Mech., 167 (1986), pp. 219–240. (Cited on pp. 198, 200)

[66] A. BERGER, T.S. DOAN, AND S. SIEGMUND, *Nonautonomous finite-time dynamics*, Discrete Cont. Dyn. Sys. B, 9 (2008), pp. 483–492. (Cited on pp. 17, 32, 99, 100)

[67] P BERLOFF, J. MCWILLIAMS, AND A. BRACCO, *Material transport in oceanic gyres. Part I: Phenomenology*, J. Phys. Oceanogr., 32 (2002), pp. 764–796. (Cited on p. 10)

[68] F. BERON-VERA, M. OLASCOAGA, G. HALLER, J. TRINANES, AND M. ISKANDARANI, *Drifter motion in the Gulf of Mexico constrained by altimetric Lagrangian coherent structures*, Geophys. Res. Lett., 40 (2013), pp. 6171–6175. (Cited on pp. 11, 24)

[69] A. BERTOZZI, *Heteroclinic orbits and chaotic dynamics in planar fluid flows*, SIAM J. Math. Anal., 19 (1988), pp. 1271–1294. (Cited on pp. 4, 140, 168, 169, 177)

[70] J.H. BETTENCOURT, C. LÓPEZ, AND E. HERNÁNDEZ-GARCÍA, *Oceanic three-dimensional Lagrangian coherent structures: A study of a mesoscale eddy in the Benguela upwelling region*, Ocean Modelling, 51 (2012), pp. 73–83. (Cited on p. 19)

[71] B. BINDER, M. BLYTH, AND S. BALASURIYA, *Steady free-surface flow over spatially periodic topography*, J. Fluid Mech., 781 (2015), p. R3. (Cited on p. 53)

[72] D. BLAZEVSKI AND R. DE LA LLAVE, *Time-dependent scattering theory for ODEs and applications to reaction dynamics*, J. Phys. A: Math. Theor., 44 (2011), 195101. (Cited on p. 94)

[73] D. BLAZEVSKI AND J. FRANKLIN, *Using scaterring theory to compute invariant manifolds and numerical results for the laser-driven Hénon-Heiles system*, Chaos, 22 (2012), 043138. (Cited on p. 94)

[74] D. BLAZEVSKI AND G. HALLER, *Hyperbolic and elliptic transport barriers in three-dimensional unsteady flows*, Phys. D, 273–274 (2014), pp. 46–64. (Cited on pp. 23, 24)

[75] S. BOCCALETTI, C. GREBOGI, Y.C. LAI, H. MANCINI, AND D. MAZA, *The control of chaos: Theory and applications*, Phys. Reports, 329 (2000), pp. 103–197. (Cited on p. 204)

[76] E. BOLAND, E. SHUCKBURGH, P. HAYNES, J. LEDWELL, M. MESSIAS, AND A. WATSON, *Estimating a submesoscale diffusivity using a roughness measure applied to a tracer release experiment in the southern ocean*, J. Phys. Oceanogr., 45 (2015), pp. 1610–1631. (Cited on p. 12)

[77] E. BOLLT, *Combinatorial control of global dynamics in a chaotic differential equation*, Internat. J. Bifurc. Control, 11 (2001), pp. 2145–2162. (Cited on p. 197)

[78] J.B. BOREYKO, P. MRUETUSATORN, S.T. RETTERER, AND C.P. COLLIER, *Aqueous two-phase microdroplets with reversible phase transitions*, Lab Chip, 13 (2013), pp. 1295–1301. (Cited on pp. 5, 14, 15)

[79] F. BOTTAUSCI, C. CARDONNE, C. MEINHART, AND I. MEZIĆ, *An ultrashort mixing length micromixer: The shear superposition micromixer*, Lab Chip, 7 (2007), pp. 396–398. (Cited on pp. x, 5, 13, 14, 155, 173, 177, 179, 186, 187, 196)

[80] F. BOTTAUSCI, I. MEZIĆ, C. MEINHART, AND C. CARDONNE, *Mixing in the shear superposition micromixer: Three-dimensional analysis*, Proc. Trans. R. Soc. Lond. A, 362 (2004), pp. 1001–1018. (Cited on pp. 155, 179, 186, 187)

[81] A. BOWER, *A simple kinematic model for mixing fluid parcels across a meandering jet*, J. Phys. Oceanogr., 21 (1991), pp. 173–180. (Cited on pp. 9, 10)

[82] A. BOWER AND M. LOZIER, *A closer look at particle exchange in the Gulf Stream*, J. Phys. Oceanogr., 24 (1994), pp. 1399–1418. (Cited on pp. 9, 11, 161)

[83] A. BOWER AND T. ROSSBY, *Evidence of cross-frontal exchange processes in the Gulf Stream based on isopycnal RAFOS float data*, J. Phys. Oceanogr., 19 (1989), pp. 1177–1190. (Cited on pp. 9, 11)

[84] K. BOWMAN, *Rossby wave phase speeds and mixing barriers in the stratosphere. Part I: Observations*, J. Atmos. Sci., 53 (1996), pp. 905–916. (Cited on p. 9)

[85] A. BOZORGMAGHAM, S. ROSS, AND D. SCHMALE, *Real-time prediction of atmospheric Lagrangian coherent structures based on forecast data: An application and error analysis*, Phys. D, 258 (2013), pp. 47–60. (Cited on p. 24)

[86] M. BRANICKI, A.M. MANCHO, AND S. WIGGINS, *A Lagrangian description of transport associated with front-eddy interaction: Application to data from the North-Western Mediterranean Sea*, Phys. D, 240 (2011), pp. 282–304. (Cited on p. 198)

[87] M. BRANICKI AND S. WIGGINS, *An adaptive method for computing invariant manifolds in non-autonomous, three-dimensional dynamical systems*, Phys. D, 238 (2009), pp. 1625–1657. (Cited on pp. 5, 14, 20, 210)

[88] M. BRANICKI AND S. WIGGINS, *Finite-time Lagrangian transport analysis: Stable and unstable manifolds of hyperbolic trajectories and finite-time Lyapunov exponents*, Nonlin. Proc. Geophys., 17 (2010), pp. 1–36. (Cited on p. 20)

[89] A. BRESSAN, *Impulsive control of Lagrangian systems and locomotion in fluids*, Discrete Cont. Dyn. Sys., 20 (2008), pp. 1–35. (Cited on p. 75)

[90] M.R. BRINGER, C.J. GERDTS, H. SONG, J.D. TICE, AND R.F. ISMAGILOV, *Microfluidic systems for chemical kinetics that rely on chaotic mixing in droplets*, Phil. Trans. R. Soc. Lond. A, 362 (2004), pp. 1087–1104. (Cited on pp. 5, 14, 15, 27)

[91] M. BROWN AND R. SAMELSON, *Particle conservation in vorticity-conserving, 2-dimensional flows*, Phys. Fluids, 6 (1994), pp. 2875–2876. (Cited on pp. 10, 13)

[92] S.L. BRUNTON AND C.W. ROWLEY, *Fast computation of finite-time Lyapunov exponent fields for unsteady flows*, Chaos, 20 (2010), 017503. (Cited on p. 20)

[93] M. BUDISIĆ, R. MOHR, AND I. MEZIĆ, *Applied Koopmanism*, Chaos, 22 (2012), 047510. (Cited on p. 23)

[94] M. BUDIŠIĆ AND I. MEZIĆ, *Geometry of ergodic quotient reveals coherent structures in flows*, Phys. D, 241 (2012), pp. 1255–1269. (Cited on pp. 5, 14, 23, 24, 210)

[95] A. CALAMAI AND M. FRANCA, *Mel'nikov methods and homoclinic orbits in discontinuous systems*, J. Dyn. Diff. Equat., 25 (2013), pp. 733–764. (Cited on pp. 39, 76, 144)

[96] R. CHABREYRIE, C. CHANDRE, AND N. AUBRY, *Complete chaotic mixing in an electro-osmotic flow by destabilization of key periodic pathlines*, Phys. Fluids, 23 (2011), 072002. (Cited on p. 20)

[97] R. CHABREYRIE, D. VAINCHTEIN, C. CHANDRE, P. SINGH, AND N. AUBRY, *Tailored mixing inside a translating droplet*, Phys. Rev. E, 77 (2008), 036314. (Cited on pp. 5, 13, 14, 15, 210)

[98] ——, *Using resonances to control chaotic mixing within a translating and rotating droplet*, Commun. Nonlinear Sci. Numer. Simulat., 15 (2010), pp. 2124–2132. (Cited on p. 210)

[99] R. CHAĆON, *Melnikov method approach to control homoclinic/heteroclinic chaos by weak harmonic excitations*, Phil. Trans. R. Soc. A, 364 (2006), pp. 2335–2351. (Cited on p. 38)

[100] S. CHANDRASESEKHAR, *Hydrodynamics and Hydrodynamic Stability*, Dover, New York, 1961. (Cited on pp. 62, 184)

[101] S.R. CHANNON AND J.L. LEBOWITZ, *Numerical experiments in stochasticity and homoclinic oscillations*, Annals New York Acad. Sci., 357 (1980), pp. 108–118. (Cited on p. 104)

[102] Z. CHE, N.-T. NGUYEN, AND T.N. WONG, *Analysis of chaotic mixing in plugs moving in meandering microchannels*, Phys. Rev. E, 84 (2011), 066309. (Cited on pp. 5, 13, 14, 15, 28)

[103] Z. CHE, T.N. WONG, AND N.-T. NGUYEN, *An analytical model for a liquid plug moving in curved microchannels*, Intern. J. Heat Mass Transfer, 53 (2010), pp. 1977–1985. (Cited on pp. 13, 15)

[104] A. CHIAN, E. REMPEL, G. AULANIER, B. SCHMEISTER, S. SHADDEN, B. WELSCH, AND A. YEATES, *Detection of coherent structures in turbulent photospheric flows*, Astrophys. J., 786 (2014), p. 51. (Cited on p. 20)

[105] L. CHIERCHIA AND G. GALLAVOTTI, *Drift and diffusion in phase space*, Ann. Inst. Henri Poincaré Physique Theorique, 60 (1994), pp. 1–144. (Cited on p. 138)

[106] S-N. CHOW, J. K. HALE, AND J. MALLET-PARET, *An example of bifurcation to homoclinic orbits*, J. Differential Equations, 37 (1980), pp. 351–373. (Cited on pp. 39, 117, 162, 163)

[107] C.-K. CHUNG, C. LAI, T. SHIH, E. CHANG, AND S. CHEN, *Simulation and fabrication of capillary-driven meander micromixer for short-distance mixing*, Micro. Nano. Lett., 8 (2013), pp. 567–570. (Cited on pp. 14, 15, 27, 28)

[108] T. CLOPEAU, A. MIKELIC, AND R. ROBERT, *On the vanishing viscosity limit for the 2D incompressible Navier-Stokes equations with the friction type boundary conditions*, Nonlinearity, 11 (1998), pp. 1625–1636. (Cited on pp. 159, 162)

[109] K. CONLISK AND G.M. O'CONNOR, *Analysis of passive microfluidic mixers incorporating 2D and 3D baffle geometries fabricated using an excimer laser*, Microfluid. Nanofluid., 12 (2012), pp. 941–951. (Cited on pp. 27, 28)

[110] P. CONSTANTIN, *On the Euler equations of incompressible fluids*, Bull. Amer. Math. Soc., 44 (2007), pp. 603–621. (Cited on pp. 159, 162)

[111] P. CONSTANTIN, I. KUCAVICA, AND V. VICOL, *On the inviscid limit of the Navier-Stokes equations*, J. Differential Equations, 143 (2015), pp. 3075–3090. (Cited on pp. 159, 162)

[112] P. CONSTANTIN AND J. WU, *Inviscid limit for vortex patches*, Nonlinearity, 8 (1995), pp. 735–742. (Cited on pp. 159, 162)

[113] C. CONTI, D. ROSSINELLI, AND P. KOUMOUTSAKOS, *GPU and APU computations of finite time Lyapunov exponent fields*, J. Comp. Phys., 231 (2012), pp. 2229–2244. (Cited on p. 20)

[114] C. COOKE AND J. KROLL, *The existence of periodic solutions to certain impulsive differential equations*, Computers Math. Appl., 44 (2002), pp. 667–676. (Cited on p. 75)

[115] W. A. COPPEL, *Dichotomies in Stability Theory*, Lecture Notes in Mathematics 629, Springer-Verlag, Berlin, 1978. (Cited on pp. 30, 99)

[116] M.L. CORDERO, H.O. ROLFSNES, D.R. BURNHAM, P.A. CAMPBELL, D. MCGLOIN, AND C.N. BAROUD, *Mixing via thermocapillary generation of flow patterns inside a microfluidic drop*, New J. Phys., 11 (2009), 075033. (Cited on pp. 5, 14, 15)

[117] L. CORTELEZZI, A. ADROVER, AND M. GIONA, *Feasibility, efficiency and transportability of short-horizon optimal mixing protocols*, J. Fluid Mech., 597 (2008), pp. 199–231. (Cited on pp. 28, 181, 198)

[118] C. CORTEZ-QUIROZ, A. AZARBADEGAN, M. ZANGENEH, AND A. GOTO, *Analysis and multi-criteria design optimization of geometric characteristics of grooved micromixer*, Chem. Eng. J., 160 (2010), pp. 852–864. (Cited on pp. 5, 13, 15, 28)

[119] C. COTTÉ, F. D'OVIDIO, A. CHAIGNEUA, M. LÉVY, I. TAUPIER-LETAGE, B. MATE, AND C. GUINET, *Scale-dependent interactions of Mediterranean whales with marine dynamics*, Limnol. Oceanography, 106 (2011), pp. 219–232. (Cited on p. 10)

[120] B. CUSHMAN-ROISIN, *Trajectories in Gulf-stream meanders*, J. Geophys. Res. Oceans, 98 (1993), pp. 2543–2554. (Cited on p. 9)

[121] P. DANCKWERTS, *The definition and measurement of some characteristics of mixtures*, Appl. Sci. Res. A, A3 (1952), pp. 279–296. (Cited on p. 28)

[122] R. DAVIS, *Lagrangian ocean studies*, Annu. Rev. Fluid Mech., 23 (1991), pp. 43–64. (Cited on p. 9)

[123] J. DE JONG, R.G.H. LAMMERTINK, AND M. WESSLING, *Membranes and microfluidics: A review*, Lab Chip, 6 (2006), pp. 1125–1139. (Cited on pp. 14, 186, 187, 196)

[124] S. DE-MONTE, F. D'OVIDIO, C. COTTÉ, M. LÉVY, M. LE CORRE, AND F. WEMERSKIRSH, *Frigatebird behaviour at the ocean-atmosphere interface: Integrating animal behaviour with multisatellite data*, J. R. Soc. Interface, 9 (2012), pp. 3351–3358. (Cited on p. 10)

[125] D. DEL CASTILLO NEGRETE AND P.J. MORRISON, *Chaotic transport by Rossby waves in shear flow*, Phys. Fluids A, 5 (1993), pp. 948–965. (Cited on pp. 4, 5, 9, 10, 12, 100, 135, 161, 168, 198)

[126] A. DELSHAMS AND T. SEARA, *Splitting of separatrices in Hamiltonian systems with one and a half degrees of freedom*, Math. Phys. Electron. J., 3 (1997), p. 4. (Cited on p. 138)

[127] T.S. DOAN, D. KARRASCH, T.Y. NGUYEN, AND S. SIEGMUND, *A unified approach to finite-time hyperbolicity which extends finite-time Lyapunov exponents*, J. Differential Equations, 252 (2012), pp. 5535–5554. (Cited on pp. 17, 32, 93, 99, 100)

[128] H. DOGAN, S. NAS, AND M. MURADOGLU, *Mixing of miscible liquids in gas-segmented serpentine channels*, Int. J. Multiphase Flow, 35 (2009), pp. 1149–1158. (Cited on pp. 15, 28)

[129] F. D'OVIDIO, A. DELLA PENNA, T. TRULL, F. NENCIOLI, M.-I. PUJOL, M.-H. RIO, Y.-H. PARK, C. COTTÉ, M. ZHOU, AND S. BLAIN, *The biogeochemical structuring role of horizontal stirring: Lagrangian perspectives on iron delivery downstream of the Kerguelen plateau*, Biogeosciences, 12 (2015), pp. 5567–5581. (Cited on pp. 11, 12)

[130] F. D'OVIDIO, J. ISERN, C. LÓPEZ, C. HERNÁNDEZ-GARCÍA, AND E. GARCÍA-LADONA, *Comparison between Eulerian diagnostics and finite-size Lyapunov exponents computed from the altimetry in the Algerian basin*, Deep-Sea Res., 56 (2009), pp. 15–31. (Cited on pp. 19, 20)

[131] F. D'OVIDIO, S. DE MONTE, S. ALVAIN, Y. DANDONNEAU, AND M. LÉVY, *Fluid dynamical niches of phytoplankton types*, Proc. Natl. Acad. Sci. USA, 107 (2010), pp. 18366–18370. (Cited on pp. 10, 20)

[132] F. D'OVIDIO, S. DE MONTE, A. PENNA, C. COTTÉ, AND C. GUINET, *Ecological implications of eddy rtentionin the open ocean: A Lagrangian approach*, J. Phys. A: Math. Theor., 46 (2013), p. 254023. (Cited on p. 20)

[133] Z. DU AND W. ZHANG, *Melnikov method for homoclinic bifurcations in nonlinear impact oscillators*, Computers Math. Appl., 50 (2005), pp. 445–458. (Cited on pp. 39, 76, 143, 144)

[134] J. DUAN AND S. WIGGINS, *Fluid exchange across a meandering jet with quasiperiodic variability*, J. Phys. Oceanogr., 26 (1996), pp. 1176–1188. (Cited on p. 10)

[135] L.H. DUC AND S. SIEGMUND, *Hyperbolicity and invariant manifolds for planar nonautonomous systems on finite time intervals*, Int. J. Bifurcation Chaos, 18 (2008), pp. 641–674. (Cited on pp. 17, 99, 100)

[136] S. DUTKIEWICZ, A. GRIFFA, AND D. OLSON, *Particle diffusion in a meandering jet*, J. Geophys. Res. Oceans, 98 (1993), pp. 16487–16500. (Cited on p. 9)

[137] S. DUTKIEWICZ AND N. PALDOR, *On the mixing enhancement in a meandering jet due to interaction with an eddy*, J. Phys. Oceanogr., 24 (1994), pp. 2418–2423. (Cited on p. 9)

[138] D. EBIN AND J. MARSDEN, *Groups of diffeomorphisms and the motion of an incompressible fluid*, Ann. Math., 92 (1970), pp. 102–163. (Cited on pp. 159, 162)

[139] K. EL RIFAI, G. HALLER, AND A. BAJAJ, *Global dynamics of an autoparametric spring-masspendulum system*, Nonlinear Dyn., 49 (2007), pp. 105–116. (Cited on p. 198)

[140] M. FARAZMAND AND G. HALLER, *Computing Lagrangian coherent structures from variational LCS theory*, Chaos, 22 (2012), 013128. (Cited on pp. 23, 24)

[141] N. FENICHEL, *Persistence and smoothness of invariant manifolds for flows*, Indiana Univ. Math. J., 21 (1971/1972), pp. 193–226. (Cited on p. 42)

[142] J.L. FENNER AND S. SIEGMUND, *An (h,k) dichotomy spectrum for systems of linear differential equations with impulse effect*, Dyn. Cont. Discr. Impul. Sys. A, 11 (2004), pp. 1–17. (Cited on p. 75)

[143] H. FIGUEROA AND D. OLSON, *Eddy resolution versus eddy diffusion in a double-gyre GCM. 1. The Lagrangian and Eulerian description*, J. Phys. Oceanogr., 24 (1994), pp. 371–386. (Cited on p. 9)

[144] M.D. FINN, S.M. COX, AND H.M. BYRNE, *Mixing measures for a two-dimensional chaotic stokes flow*, J. Eng. Math., 48 (2004), pp. 129–155. (Cited on p. 28)

[145] D.P.G. FOURES, C.P. CAULFIELD, AND P.J. SCHMID, *Optimal mixing in two-dimensional plane Poiseuille flow at finite Peclet number*, J. Fluid Mech., 748 (2014), pp. 241–277. (Cited on p. 198)

[146] A. FRADKOV, R. EVANS, AND B. ANDRIEVSKY, *Control of chaos: Methods and applications in mechanics*, Phil. Trans. R. Soc. A, 364 (2006), pp. 2279–2307. (Cited on p. 197)

[147] A. L. FRADKOV AND R. J. EVANS, *Control of chaos: Methods and applications in engineering*, Ann. Rev. Control, 29 (2005), pp. 33–56. (Cited on p. 197)

[148] E. FRANCO, D.N. PEKEREK, J. PENG, AND J.O. DABIRI, *Geometry of unsteady fluid transport during fluid-structure interactions*, J. Fluid Mech., 589 (2007), pp. 125–145. (Cited on p. 108)

[149] L. FRENZ, A. EL HARRACK, M. PAULY, S. BÉGIN-COLIN, D. GRIFFITHS, AND J.-C. BARET, *Droplet-based microreactors for the synthesis of magnetic iron oxide nanoparticles*, Angew. Chem. Int. Ed., 47 (2008), pp. 6817–6820. (Cited on pp. 5, 14, 15)

[150] G. FROYLAND, *An analytic framework for identifying finite-time coherent sets in time-dependent dynamical systems*, Phys. D, 250 (2013), pp. 1–19. (Cited on pp. 16, 23, 61, 67, 93, 126)

[151] G. FROYLAND, C. GONZÁLEZ-TOKMAN, AND T.M. WATSON, *Optimal mixing enhancement by local perturbation*, SIAM Review, 58 (2016), pp. 494–513. (Cited on p. 198)

[152] G. FROYLAND, C. HORENKAMP, V. ROSSI, N. SANTITISSADEEKORN, AND A. SEN GUPTA, *Three-dimensional characterization and tracking of an Agulhas Ring*, Ocean Model., 52–53 (2012), pp. 69–75. (Cited on p. 12)

[153] G. FROYLAND, S. LLOYD, AND A. QUAS, *Coherent structures and isolated spectrum for Perron–Frobenius cocycles*, Ergod. Th. Dynam. Sys., 30 (2010), pp. 729–756. (Cited on pp. 16, 67)

[154] G. FROYLAND AND K. PADBERG, *Almost invariant sets and invariant manifolds: Connecting probablistic and geometric descriptions of coherent structures in flows*, Phys. D, 238 (2009), pp. 1507–1523. (Cited on pp. 23, 93)

[155] G. FROYLAND AND K. PADBERG-GEHLE, *Finite-time entropy: A probabilistic approach for measuring nonlinear stretching*, Phys. D, 241 (2012), pp. 1612–1628. (Cited on p. 24)

[156] G. FROYLAND, K. PADBERG, M. ENGLAND, AND A.-M. TREGUIER, *Detection of coherent structures via transfer operators*, Phys. Rev. Lett., 98 (2007), 224503. (Cited on pp. 61, 126)

[157] G. FROYLAND AND K. PADBERG-GEHLE, *Almost-invariant and finite-time coherent sets: Directionality, duration, and diffusion*, in Ergodic Theory, Open Dynamics, and Coherent Structures, W. Bahsoun, C. Bose, and G. Froyland, eds., Springer, New York, 2014, pp. 171–216. (Cited on p. 23)

[158] G. FROYLAND AND K. PADBERG-GEHLE, *A rough-and-ready cluster-based approach for extracting finite-time coherent sets from sparse and incomplete trajectory data*, Chaos, 25 (2015), 087406. (Cited on p. 24)

[159] G. FROYLAND, N. SANTITISSADEEKORN, AND A. MONAHAN, *Transport in time-dependent dynamical systems: Finite-time coherent sets*, Chaos, 20 (2010), 043116. (Cited on pp. 23, 93)

[160] G. FROYLAND, R. STUART, AND E. VAN SEBILLE, *How well-connected is the global ocean?*, Chaos, 24 (2014), 033126. (Cited on p. 12)

[161] L.-M. FU, R.-J. YANG, C.-H. LIN, AND Y.-S. CHIEN, *A novel microfluidic mixer utilizing electrokinetic driving forces under low switching frequency*, Electrophoresis, 5 (2005), pp. 1814–1824. (Cited on pp. 15, 173, 177)

[162] D. GALLOWAY AND U. FRISCH, *Dynamo action in a family of flows with chaotic streamlines*, Geophys. Astrophys. Fluid Dyn., 36 (1986), pp. 53–83. (Cited on pp. 6, 8)

[163] P. GARG, J. RICARDO, AND S. PUSHPAVANAM, *Chaotic mixing in a planar, curved channel using periodic slip*, Phys. Fluids, 27 (2015), 032004. (Cited on p. 28)

[164] A. GARGETT, *Vertical eddy diffusivity in the ocean interior*, J. Marine Res., 42 (1984), pp. 359–393. (Cited on p. 12)

[165] F. GAROFALO AND M. GIONA, *On the modelling of time-interleaved sequential lamination micromixers*, Microfluid. Nanofluid., 12 (2012), pp. 175–187. (Cited on pp. 14, 27)

[166] P.H. GASKELL, M.D. SAVAGE, AND H.M. THOMPSON, *Stagnation-saddle points and flow patterns in Stokes flow between contra-rotating cylinder*, J. Fluid Mech., 370 (1998), pp. 221–247. (Cited on p. 198)

[167] L. GAULTIER, B. DJATH, J. VERRON, J.M. BRANKART, P. BRASSEUR, AND A. MELET, *Inversion of submesoscale patterns from a high-resolution Solomon Sea model: Feasibility assessment*, J. Geophys. Res. Oceans, 119 (2014), pp. 4520–4541. (Cited on p. 20)

[168] L. GAULTIER, J. VERRON, J.-M. BRANKART, O. TITAUD, AND P. BRASSEUR, *On the inversion of submesoscale tracer fields to estimate the surface ocean circulation*, J. Marine Sys., 126 (2013), pp. 33–42. (Cited on p. 20)

[169] P. GIESL AND M. RASMUSSEN, *Areas of attraction for nonautonomous differential equations on finite time intervals*, J. Math. Anal. Appl., 390 (2012), pp. 27–46. (Cited on p. 32)

[170] F. GINELLI, H. CHATÉ, R. LIVI, AND A. POLITI, *Covariant Lyapunov vectors*, J. Phys. A: Math. Theor., 46 (2013), 254005. (Cited on pp. 16, 67)

[171] M. GIONA, P. ANDERSON, AND F. GAROFALO, *Short-time behavior of advecting-diffusing scalar fields in Stokes flows*, Phys. Rev. E, 87 (2013), 063011. (Cited on pp. 13, 14, 27)

[172] I. GLASGOW AND N. AUBRY, *Enhancement of microfluidic mixing using time pulsing*, Lab Chip, 3 (2003), pp. 114–120. (Cited on pp. 123, 173, 177)

[173] H. GOLOVKIN, *Vanishing viscosity in Cauchy's problem for hydrodynamics equations*, Mem. Steklov. Inst., 92 (1966), pp. 33–53. (Cited on pp. 159, 162)

[174] O. GORODETSKYI, M. GIONA, AND P. ANDERSON, *Spectral analysis of mixing in chaotic flows via the mapping matrix formalism: Inclusion of molecular diffusion and quantitative eigenvalue estimate in the purely convective limit*, Phys. Fluids, 24 (2012), 073603. (Cited on pp. 14, 27)

[175] E. GRENIER, C. JONES, F. ROUSSET, AND B. SANDSTEDE, *Viscous perturbations of marginally stable Euler flow and finite-time Melnikov theory*, Nonlinearity, 18 (2005), pp. 465–483. (Cited on pp. 5, 159, 162)

[176] R.O. GRIGORIEV, *Chaotic mixing in thermocapillary-driven microdroplets*, Phys. Fluids, 17 (2005), 033601. (Cited on pp. 5, 14, 15)

[177] R.O. GRIGORIEV, M.F. SCHATZ, AND V. SHARMA, *Chaotic mixing in microdroplets*, Lab Chip, 6 (2006), pp. 1369–1372. (Cited on pp. 5, 13, 14, 15, 27)

[178] R. GRIMSHAW AND X. TIAN, *Periodic and chaotic behaviour in a reduction of the perturbed Korteweg-de Vries equation*, Proc. R. Soc. Lond. A, 445 (1994), pp. 1–21. (Cited on pp. 38, 131)

[179] J. GRIZZLE, G. ABBA, AND F. PLESTAN, *Asymptotically stable walking for biped robots: Analysis via systems with impulse effects*, IEEE Trans. Auto. Contol, 46 (2001), pp. 51–64. (Cited on p. 75)

[180] J. GRUENDLER, *The existence of homoclinic orbits and the method of Melnikov for systems in R^n*, SIAM J. Math. Anal., 16 (1985), pp. 907–931. (Cited on p. 39)

[181] ——, *Homoclinic solutions and chaos in ordinary differential equations with singular perturbations*, Trans. Amer. Math. Soc., 350 (1998), pp. 3797–3814. (Cited on p. 39)

[182] O. GUBANOV AND L. CORTELEZZI, *On the cost efficiency of mixing optimization*, J. Fluid Mech., 692 (2012), pp. 112–136. (Cited on p. 198)

[183] J. GUCKENHEIMER AND P. HOLMES, *Nonlinear Oscillations, Dynamical Systems and Bifurcations of Vector Fields*, Springer, New York, 1983. (Cited on pp. 13, 30, 35, 36, 38, 39, 40, 49, 50, 54, 61, 67, 99, 103, 113, 116, 117, 119, 129, 130, 131, 133, 198, 202)

[184] J. HADAMARD, *Mouvement permanent lent d'une sphère liquide et visqueuse dans un liquide visqueux*, C.R. Acad. Sci. Paris, 152 (1911), pp. 1735–1738. (Cited on pp. 5, 14)

[185] A. HADJIGHASEM, D. KARRASCH, H. TERAMOTO, AND G. HALLER, *Spectral clustering approach to Lagrangian vortex detection*, Phys. Rev. E, 93 (2016), 063107. (Cited on p. 24)

[186] J. HALE, *Integral manifolds of perturbed differential systems*, Ann. Math., 73 (1961), pp. 496–531. (Cited on p. 41)

[187] G. HALLER, *Lagrangian coherent structures from approximate velocity data*, Phys. Fluids, 14 (2002), pp. 1851–1861. (Cited on p. 20)

[188] ——, *A variational theory for Lagrangian Coherent Structures*, Phys. D, 240 (2011), pp. 574–598. (Cited on pp. 20, 21, 24)

[189] G. HALLER, *Lagrangian coherent structures*, Annu. Rev. Fluid Mech., 47 (2015), pp. 137–162. (Cited on pp. 8, 11, 20, 21, 23, 24, 25)

[190] G. HALLER AND F. BERON-VERA, *Geodesic theory for transport barriers in two-dimensional flows*, Phys. D, 241 (2012), pp. 1680–1702. (Cited on p. 24)

[191] G. HALLER, A. HADJIGHASEM, M. FARAZMAND, AND F. HUHN, *Defining coherent vortices objectively from the vorticity*, J. Fluid Mech., 795 (2016), pp. 136–173. (Cited on p. 24)

[192] G. HALLER AND A.C. POJE, *Finite time transport in aperiodic flows*, Phys. D, 119 (1998), pp. 352–380. (Cited on pp. 110, 112)

[193] G. HALLER AND G.-C. YUAN, *Lagrangian coherent structures and mixing in two-dimensional turbulence*, Phys. D, 147 (2000), pp. 352–370. (Cited on pp. 4, 8, 10, 24)

[194] X. HAN AND J. LU, *Impulsive control induced effects on dynamics of single and coupled ODE systems*, Nonlin. Dyn., 59 (2010), pp. 101–111. (Cited on pp. 75, 81, 143)

[195] J.F. HARPER AND D.W. MOORE, *The motion of a spherical liquid drop at high Reynolds number*, J. Fluid Mech., 32 (1968), pp. 367–391. (Cited on pp. 5, 14, 210)

[196] Y. HATTORI AND K. HIJIYA, *Short-wavelength stability analysis of Hill's vortex with/without swirl*, Phys. Fluids, 22 (2010), 074104. (Cited on pp. 5, 14)

[197] P. HAYNES, D. POET, AND E. SHUCKBURGH, *Transport and mixing in kinematic and dynamically consistent flows*, J. Atmos. Sci., 64 (2007), pp. 3640–3651. (Cited on p. 10)

[198] C. HERNANDEZ, Y. BERNARD, AND A. RAZEK, *A global assessment of piezoelectric actuated micro-pumps*, Eur. Phys. J. Appl. Phys., 51 (2010), 20101. (Cited on pp. 14, 186, 187, 196)

[199] M.W. HIRSCH, C.C. PUGH, AND M. SHUB, *Invariant manifolds*, Bull. Amer. Math. Soc., 76 (1970), pp. 1015–1019. (Cited on p. 42)

[200] P.J. HOLMES, *Averaging and chaotic motions in forced oscillations*, SIAM J. Appl. Math., 38 (1980), pp. 65–80. (Cited on pp. 38, 54, 56, 116)

[201] P. HOLMES, *Some remarks on chaotic particle paths in time-periodic, three-dimensional swirling flows*, Contemp. Math., 28 (1984), pp. 393–404. (Cited on pp. 5, 14)

[202] P. HOLMES AND J. MARSDEN, *Melnikov's method and Arnold diffusion for perturbations of integrable Hamiltonian systems*, J. Math. Phys., 23 (1982), pp. 669–675. (Cited on pp. 38, 39)

[203] P.J. HOLMES AND D. WHITLEY, *On the attracting set for Duffing's equation*, Phys. D, 7 (1983), pp. 111–123. (Cited on pp. 8, 45)

[204] A. HOLTSLAG AND B. BOVILLE, *Local versus nonlocal boundary-layer diffusion in a global climate model*, J. Climate, 6 (1993), pp. 1825–1842. (Cited on p. 12)

[205] M. HOLZER AND E. SIGGIA, *Turbulent mixing of a passive scalar*, Phys. Fluids, 6 (1994), pp. 1820–1837. (Cited on p. 12)

[206] J. HRDLICKA, P. CERVENKA, M. PRIBYL, AND D. SNITA, *Zig-zag arrangement of four electrodes for ac electro-osmotic micropumps*, Phys. Rev. E, 84 (2011), 016307. (Cited on pp. 14, 187)

[207] S. HU, D.E. LACROIX, AND A.R. MCFARLAND, *Stagnation point displacement: Control of losses on a conically shaped aerosol distributor*, Aerosol Sci. Tech., 43 (2009), pp. 311–321. (Cited on p. 198)

[208] Y.T. HU, D.J. PINE, AND L.G. LEAL, *Drop deformation, breakup, and coalescence with compatibilizer*, Phys. Fluids, 12 (2000), pp. 484–489. (Cited on p. 200)

[209] A. HÜBLER AND E. LÜSCHER, *Resonant stimulation and control of conlinear oscillators*, Naturwissenschaften, 76 (1989), pp. 67–69. (Cited on p. 204)

[210] S.D. HUDSON, F.R. PHELAN, M.D. HANDLER, J.T. CABRAL, K.B. MIGLER, AND E.J. AMIS, *Microfluidic analog of the four-roll mill*, Appl. Phys. Lett., 85 (2004), pp. 335–337. (Cited on p. 200)

[211] F. HUHN, A. VON KAMEKE, V. PÉREZ-MUNUZURI, M. OLASCOAGA, AND F. BERON-VERA, *The impact of advective transport by the South Indian Ocean Countercurrent on the Madagascar plankton bloom*, Geophys. Res. Lett., 39 (2012), L06602. (Cited on pp. 10, 11)

[212] K. IDE, D. SMALL, AND S. WIGGINS, *Distinguished hyperbolic trajectories in time dependent fluid flows: Analytical and computational approach for velocity fields defined as data sets*, Nonlin. Proc. Geophys., 9 (2002), pp. 237–263. (Cited on pp. 10, 32)

[213] E.A. JACKSON, *Controls of dynamic flows with attractors*, Phys. Rev. A, 44 (1991), pp. 4839–4853. (Cited on pp. 204, 209)

[214] G. JIANG AND Q. LU, *Impulsive state feedback control of a predator-prey model*, J. Comp. Appl. Math., 200 (2007), pp. 193–207. (Cited on p. 75)

[215] J.A.J. JIMENEZ-MADRID AND A.M. MANCHO, *Distinguished trajectories in time dependent vector fields*, Chaos, 19 (2009), 013111. (Cited on p. 45)

[216] B. JOSEPH AND B. LEGRAS, *Relation between kinematic boundaries, stirring, and barriers for the antarctic polar vortex*, J. Atmos. Sci., 59 (2002), pp. 1198–1212. (Cited on pp. 4, 10, 19)

[217] E. KAI, V. ROSSI, J. SUDRE, H. WEIMERSKIRCH, C. LOPEZ, E. HÉRNANDEZ-GARCIA, F. MARSAC, AND V. GARCON, *Top marine predators track Lagrangian coherent structures*, Proc. Natl. Acad. Sci. USA, 106 (2009), pp. 8245–8250. (Cited on p. 10)

[218] I. KAMENKOVICH, I. RYPINA, AND P. BERLOFF, *Properties and origins of the anisotropic eddy-induced transport in the North Atlantic*, J. Phys. Oceanogr., 45 (2015), pp. 778–791. (Cited on p. 12)

[219] D. KARRASCH, *Linearization of hyperbolic finite-time processes*, J. Differential Equations, 254 (2013), pp. 256–282. (Cited on pp. 17, 32, 93, 99, 100)

[220] D. KARRASCH AND G. HALLER, *Do finite-size Lyapunov exponents detect coherent structures?*, Chaos, 23 (2013), 043126. (Cited on pp. 19, 20)

[221] T. KATO AND G. PONCE, *Well-posedness of the Euler and Navier-Stokes equations in the Lebesgue spaces $L_s^p(R^2)$*, Rev. Mat. Iberoamericana, 2 (1986), pp. 73–88. (Cited on pp. 159, 162)

[222] A. KATZ AND E. DOWELL, *From single-well chaos to cross-well chaos: A detailed explanation in terms of manifold intersections*, Intern. J. Bifurc. Chaos, 4 (1994), pp. 933–941. (Cited on pp. 198, 202)

[223] D. KELLEY, M. ALLSHOUSE, AND N. OUELLETTE, *Lagrangian coherent structures separate dynamically distinct regions in fluid flow*, Phys. Rev. E, 88 (2013), 013017. (Cited on p. 20)

[224] P. KLOEDEN AND M. RASMUSSEN, *Nonautonomous Dynamical Systems*, American Mathematical Society, Providence, RI, 2011. (Cited on p. 206)

[225] E. KNOBLOCH AND J. WEISS, *Chaotic advection by modulated traveling waves*, Phys. Rev. A, 36 (1987), pp. 1522–1524. (Cited on pp. 5, 9, 12, 38, 39, 50, 100, 134, 135, 168)

[226] T.-Y. KOH AND B. LEGRAS, *Hyperbolic lines and the stratospheric polar vortex*, Chaos, 12 (2002), pp. 382–394. (Cited on pp. 4, 10, 12)

[227] K. KOSHEL AND S. PRANTS, *Chaotic advection in the ocean*, Physics-Uspekhi, 49 (2006), pp. 1151–1178. (Cited on p. 9)

[228] G. KOVACIC, *Lobe area via an action formalism in a class of Hamiltonian systems*, Phys. D, 51 (1991), pp. 226–233. (Cited on pp. 127, 128)

[229] D. KROUJILINE AND H.A. STONE, *Chaotic streamlines in steady bounded three-dimensional Stokes flows*, Phys. D, 130 (1999), pp. 105–132. (Cited on pp. 5, 13, 14, 15, 210)

[230] P. KUKUCKA, *Mel'nikov method for discontinuous planar systems*, Nonlin. Anal., 66 (2007), pp. 2698–2719. (Cited on pp. 39, 76, 144)

[231] M. KUNZE AND T. KUPPER, *Non-smooth dynamical systems: An overview*, in Ergodic Theory, Analysis, and Efficient Simulation of Dynamical Systems, B. Fiedler, ed., Springer-Verlag, New York, 2001, pp. 431–452. (Cited on p. 76)

[232] O.A. LADYZHENSKAYA, *The Mathematical Theory of Viscous Incompressible Flow*, Gordon & Breach, New York, 1969. (Cited on pp. 159, 162)

[233] T.S. LAKER AND S.M. GHIAASIAAN, *Monte-Carlo simulation of aerosol transport in rising spherical bubbles with internal circulation*, J. Aerosol Sci., 35 (2004), pp. 473–488. (Cited on pp. 5, 14, 210)

[234] H. LAMB, *Hydrodynamics*, Cambridge University Press, Cambridge, UK, 1879. (Cited on pp. 5, 14, 210)

[235] T.N. LE, Y.K. SUH, AND S. KANG, *A numerical study on the flow and mixing in a microchannel using magnetic particles*, J. Mech, Sci. Tech., 24 (2010), pp. 441–450. (Cited on pp. 173, 177)

[236] D. LEE AND Y. CHEN, *Mixing in tangentially crossing microchannels*, AIChE J., 57 (2011), pp. 571–580. (Cited on pp. 5, 13, 28)

[237] J.S. LEE, R. DYLLA-SPEARS, N.P TECLEMARIAM, AND S.J. MULLER, *Microfluidic four-roll mill for all flow types*, Appl. Phys. Lett., 90 (2007), 074103. (Cited on pp. 5, 13, 14, 15, 20, 27, 28, 155, 179, 180, 186, 187, 189, 196, 200)

[238] S. LEIBOVICH, *The structure of vortex breakdown*, Annu. Rev. Fluid Mech., 10 (1978), pp. 221–246. (Cited on pp. 5, 14)

[239] S. LENCI AND G. REGA, *Optimal numerical control of single-well to cross-well chaos transition in mechanical systems*, Chaos, Solitons and Fractals, 15 (2003), pp. 173–186. (Cited on pp. 198, 202)

[240] C. LIANG, G. LIAO, AND W. SUN, *A note on approximating properties of the Oseledets splitting*, Proc. Amer. Math. Soc., 142 (2014), pp. 3825–3838. (Cited on pp. 16, 20, 67)

[241] A. LIFSCHITZ, *Instabilities of ideal fluids and related topics*, Z. Angew. Math. Mech., 75 (1995), pp. 411–422. (Cited on pp. 5, 14)

[242] C.Y. LIM, Y.C. LAM, AND C. YANG, *Mixing enhancement in microfluidic channel with a constriction under periodic electro-osmotic flow*, Biomicrofluidics, 4 (2010), 014101. (Cited on pp. 14, 15, 123, 173, 177, 187)

[243] R. LIMA AND M. PETTINI, *Parametric resonant control of chaos*, Int. J. Bifurc. Chaos, 8 (1998), pp. 1675–1684. (Cited on pp. 197, 198, 202)

[244] C.-H. LIN, L.-M. FU, AND Y.-S. CHIEN, *Microfluidic T-form mixer utilizing switching electroosmotic flow*, Anal. Chem., 76 (2004), pp. 5265–5272. (Cited on pp. 15, 123, 173, 177)

[245] K. LIN AND L.-S. YOUNG, *Shear-induced chaos*, Nonlinearity, 21 (2008), pp. 899–922. (Cited on pp. 75, 81, 143)

[246] Z. LIN, J.-L. THIFFEAULT, AND C.R. DOERING, *Optimal stirring strategies for passive scalar mixing*, J. Fluid Mech., 675 (2011), pp. 465–476. (Cited on pp. 28, 181, 198, 211)

[247] D. LIPINSKI AND K. MOHSENI, *A ridge tracking algorithm and error estimate for efficient computation of Lagrangian coherent structures*, Chaos, 20 (2010), 017504. (Cited on p. 20)

[248] Z. LIU, Y. HUANG, Y. JIN, AND Y. CHEN, *Mixing intensification by chaotic advection inside droplets for controled nanoparticle preparation*, Microfluid. Nanofluid., 9 (2010), pp. 773–786. (Cited on pp. 5, 14, 15, 27, 28)

[249] H.E. LOMELI AND R. RAMIREZ-ROS, *Separatrix splitting in 3D volume preserving maps*, SIAM J. Appl. Dyn. Sys., 7 (2008), pp. 1527–1557. (Cited on pp. 5, 14, 210)

[250] E. LORENZ, *Determinstic nonperiodic flow*, J. Atmos. Sci., 20 (1963), pp. 130–141. (Cited on pp. 6, 8, 13, 208, 209)

[251] M. LOZIER, L. PRATT, A. ROGERSON, AND P. MILLER, *Exchange geometry revealed by float trajectories in Gulf Stream*, J. Phys. Oceanogr., 27 (1997), pp. 2327–2341. (Cited on p. 11)

[252] S. LOZIER AND D. BERCOVICI, *Particle exchange in an unstable jet*, J. Phys. Oceanogr., 22 (1992), pp. 1506–1516. (Cited on p. 9)

[253] C. LU, R. GONG, AND J. LUO, *Analysis of stagnation points for a pumping well in recharge areas*, J. Hydrology, 373 (2009), pp. 442–452. (Cited on pp. 45, 198)

[254] K. LU AND Q. WANG, *Chaotic behavior in differential equations driven by a Brownian motion*, J. Differential Equations, 251 (2011), pp. 2853—2895. (Cited on p. 202)

[255] E. LUNASIN, Z. LIN, A. NOVIKOV, A. MAZZUCATO, AND C. DOERING, *Optimal mixing and optimal stirring for fixed energy, fixed power, or fixed palenstrophy flows*, J. Math. Phys., 53 (2012), 115611. (Cited on p. 198)

[256] M. LUNEVA, C. CLAYSON, AND M. DUBOVIKOV, *Effects of mesoscale eddies in the active mixed layer: Test of the parametrisation in eddy resolving simulations*, Geophys. Astrophys. Fluid Dyn., 109 (2015), pp. 281–310. (Cited on p. 12)

[257] T. MA AND E. BOLLT, *Differential geometry perspective of shape coherence and curvature evolution by finite-time nonhyperbolic splitting*, SIAM J. Appl. Dyn. Sys., 13 (2014), pp. 1106–1136. (Cited on pp. 24, 93)

[258] R.M. MACKAY AND J.D. MEISS, *Relation between quantum and classical thresholds for multiphoton ionization of excited atoms*, Phys. Rev. A, 37 (1988), pp. 4702–4706. (Cited on pp. 104, 127, 128)

[259] R.M. MACKAY, J.D. MEISS, AND I.C. PERCIVAL, *Transport in Hamiltonian systems*, Phys. D, 13 (1984), pp. 55–81. (Cited on p. 104)

[260] R.S. MACKAY, *Transport in 3-D volume-preserving flows*, J. Nonlin. Sci., 4 (1994), pp. 329–354. (Cited on pp. 5, 14)

[261] Y. MAEKAWA, *On the invisdic limit problem of the vorticity equations for viscous incompressible flows in the half-plane*, Commun. Pure Appl. Math., 67 (2014), pp. 1045–1128. (Cited on pp. 159, 162)

[262] O. MALASPINAS, N. FIETIER, AND M. DEVILLE, *Lattice Boltzmann method for the simulation of viscoelastic fluid flows*, J. Non-Newtonian Fluid Mech., 165 (2010), pp. 1637–1653. (Cited on p. 200)

[263] N. MALHOTRA, I. MEZIĆ, AND S. WIGGINS, *Patchiness: A new diagnostic for Lagrangian trajectory analysis in time-dependent fluid flows*, Int. J. Bifurc. Chaos, 8 (1998), pp. 1073–1094. (Cited on p. 23)

[264] N. MALHOTRA AND S. WIGGINS, *Geometric structures, lobe dynamics, and Lagrangian transport in flows with aperiodic time-dependence, with applications to Rossby wave flow*, J. Nonlin. Sci., 8 (1998), pp. 401–456. (Cited on pp. 39, 50, 108, 135)

[265] A.M. MANCHO, D. SMALL, S. WIGGINS, AND K. IDE, *Computation of stable and unstable manifolds of hyperbolic trajectories in two-dimensional, aperiodically time-dependent vector fields*, Phys. D, 182 (2003), pp. 188–222. (Cited on pp. 23, 32, 45)

[266] A.M. MANCHO, S. WIGGINS, J. CURBELO, AND C. MENDOZA, *Lagrangian descriptors: A method for revealing phase space structures of general time dependent dynamical systems*, Commun. Nonlin. Sci. Numer. Simul., 18 (2013), pp. 3530–3557. (Cited on p. 23)

[267] G. MATHEW, I. MEZIĆ, S. GRIVOPOULOS, U. VAIDYA, AND L. PETZOLD, *Optimal control of mixing in Stokes fluid flow*, J. Fluid Mech., 580 (2007), pp. 261–281. (Cited on pp. 28, 181, 198)

[268] G. MATHEW, I. MEZIĆ, AND L. PETZOLD, *A multiscale measure for mixing*, Phys. D, 211 (2005), pp. 23–46. (Cited on p. 28)

[269] M. MATHUR, G. HALLER, T. PEACOCK, J. RUPPERT-FELSOT, AND H. SWINNEY, *Uncovering the Lagrangian skeleton of turbulence*, Phys. Rev. Lett., 98 (2007), 144502. (Cited on p. 20)

[270] F. MCGRATH, *Nonstationary plane flow of viscous and ideal fluids*, Arch. Ration. Mech. Anal., 27 (1968), pp. 328–348. (Cited on pp. 159, 162)

[271] V.K. MELNIKOV, *On the stability of the centre for time-periodic perturbations*, Trans. Moscow Math. Soc., 12 (1963), pp. 1–56. (Cited on pp. 36, 38, 40)

[272] P. MEUNIER, P. HUCK, AND C. NOBILL, *Experimental measurement of the Melnikov function*, Phys. Fluids, 27 (2015), 077103. (Cited on pp. 120, 177, 198)

[273] S. MEYERS, *Cross-frontal mixing in a meandering jet*, J. Phys. Oceanogr., 24 (1994), pp. 1641–1646. (Cited on pp. 9, 10)

[274] I. MEZIĆ, *Chaotic advection in bounded Navier-Stokes flows*, J. Fluid Mech., 431 (2001), pp. 347–370. (Cited on pp. 5, 14)

[275] I. MEZIĆ, S. LOIRE, V. FONOBEROV, AND P. HOGAN, *A new mixing diagnostic and Gulf oil spill movement*, Science, 330 (2010), pp. 486–489. (Cited on pp. 10, 12, 23)

[276] P. MILLER, C. JONES, A. ROGERSON, AND L. PRATT, *Quantifying transport in numerically generated velocity fields*, Phys. D, 110 (1997), pp. 105–122. (Cited on p. 10)

[277] P.D. MILLER, L.J. PRATT, K.R. HELFRICH, AND C.K.R.T. JONES, *Chaotic transport of mass and potential vorticity for an island recirculation*, J. Phys. Oceanogr., 32 (2002), pp. 80–102. (Cited on pp. 10, 112, 127, 128, 161)

[278] P. MIRON, J. VETEL, AND A. GARON, *On the use of the finite-time Lyapunov exponent to reveal complex flow physics in the wake of a mechanical valve*, Exp. Fluids, 55 (2014), p. 1814. (Cited on p. 20)

[279] N.A. MISHCHUK, T. HELDAL, T. VOLDEN, J. AUERSWALD, AND H. KNAPP, *Microfluidic pump based on the phenomenon of electroosmosis of the second kind*, Microfluid. Nanofluid., 11 (2011), pp. 675–684. (Cited on pp. 14, 186, 187, 196)

[280] B. MORRIS AND J. GRIZZLE, *Hybrid invariant manifolds in systems with impulse effects with application to periodic locomotion in bipedal robots*, IEEE Trans. Automat. Contol, 54 (2009), pp. 1751–1764. (Cited on p. 75)

[281] B.A. MOSOVSKY AND J.D. MEISS, *Transport in transitory dynamical systems*, SIAM J. Appl. Dyn. Sys., 10 (2011), pp. 35–65. (Cited on pp. 5, 14, 73, 93, 94)

[282] ———, *Transport in transitory, three-dimensional, Liouville flows*, SIAM J. Appl. Dyn. Sys., 11 (2012), pp. 1785–1816. (Cited on p. 94)

[283] M. MURADOGLU AND H.A. STONE, *Mixing in a drop moving through a serpentine channel: A computational study*, Phys. Fluids, 17 (2005), p. 073305. (Cited on pp. 15, 28)

[284] W.Y. NG, S. GOH, Y.C. LAM, C. YANG, AND I. RODRÍGUEZ, *DC-biased AC-electroosmotic and AC-electrothermal flow mixing in microchannels*, Lab Chip, (2009), pp. 802–809. (Cited on pp. 173, 177)

[285] K. NGAN AND T. SHEPHERD, *Chaotic mixing and transport in Rossby-wave critical layers*, J. Fluid Mech., 334 (1997), pp. 315–351. (Cited on p. 10)

[286] N.-T. NGUYEN AND Z. WU, *Micromixers—a review*, J. Micromech. Microeng., 15 (2005), pp. R1–R16. (Cited on p. 14)

[287] X. NIU AND Y.-K. LEE, *Efficient spatial-temporal chaotic mixing in microchannels*, J. Micromech. Microeng., 13 (2003), pp. 454–462. (Cited on pp. 5, 13, 15, 20, 173, 177)

[288] X. NIU, L. LIU, W. WEN, AND P. SHENG, *Active microfluidic mixer chip*, Appl. Phys. Lett., 88 (2006), 153508. (Cited on pp. 5, 13, 14, 15, 27, 28, 155, 173, 177, 179, 187)

[289] ——, *Hybrid approach to high-frequency microfluidic mixing*, Phys. Rev. Lett., 97 (2006), 044501. (Cited on pp. 5, 13, 14, 15, 20, 187)

[290] G. NORGARD AND P.-T. BREMER, *Second derivative ridges are straight lines and the implications for computing Lagrangian coherent structures*, Phys. D, 241 (2012), pp. 1475–1476. (Cited on p. 20)

[291] S. OBERTI, A. NEILD, AND T.W. NG, *Microfluidic mixing under low frequency vibration*, Lab Chip, 9 (2009), pp. 1435–1438. (Cited on pp. 173, 177)

[292] H. OKA, *Singular perturbations of autonomous ordinary differential equations and heteroclinic bufurcations*, in Dynamical Systems, Santiago de Chile 1990, Pitman Research Notes in Mathematics 285, R. Bamon, R. Labarca, J. Lewowicz, and J. Palis, eds., John Wiley & Sons, New York, 1993, pp. 159–194. (Cited on p. 39)

[293] Y. OKABE, Y. CHEN, R. PUROHIT, R. CORN, AND A. LEE, *Piezoelectrically driven vertical cavity acoustic transducers for the convective transport and rapid detection of DNA and protein binding to DNA microarrays with SPR imaging—a parametric study*, Biosensors Bioelectronics, 35 (2012), pp. 37–43. (Cited on pp. 15, 27)

[294] F. OKKELS AND P. TABELING, *Spatiotemporal resonances in mixing in open viscous flows*, Phys. Rev. Lett., 92 (2004), 038301. (Cited on p. 15)

[295] M. OLASCOAGA, *Isolation on the West Florida Shelf with implications for red tides and polutant dispersal in the Gulf of Mexico*, Nonlin. Proc. Geophys., 17 (2010), pp. 685–696. (Cited on pp. 11, 12)

[296] M. OLASCOAGA AND G. HALLER, *Forecasting sudden changes in environmental pollution patterns*, Proc. Natl. Acad. Sci. USA, 109 (2012), pp. 4738–4743. (Cited on pp. 12, 24)

[297] K. ONU, F. HUHN, AND G. HALLER, *LCS Tool: An algorithmic introduction to Lagrangian Coherent Structures*, J. Comp. Sci., 7 (2015), pp. 26–36. (Cited on p. 24)

[298] A. O'ROURKE AND G. VALLIS, *Jet interaction and the influence of minimum phase speed bound on the propagation of eddies*, J. Atmos. Sci., 70 (2013), pp. 2614–2628. (Cited on p. 134)

[299] V. OSELEDETS, *Multiplicative ergodic theorem: Characteristic Lyapunov exponents of dynamical systems*, Trudy MMO, 19 (1968), pp. 179–210. (Cited on pp. 16, 67)

[300] G. OSIPOV, A. KOZLOV, AND V. SHALFEEV, *Impulse control of chaos in continuous systems*, Phys. Lett. A, 247 (1998), pp. 119–128. (Cited on p. 75)

[301] E. OTT, C. GREBOGI, AND J. A. YORKE, *Controlling chaos*, Phys. Rev. Lett., 64 (1990), pp. 1196–1199. (Cited on pp. 197, 204, 209, 212)

[302] J.M. OTTINO, *The Kinematics of Mixing: Stretching, Chaos, and Transport*, Cambridge University Press, Cambridge, UK, 1989. (Cited on p. 13)

[303] N. OUELLETTE, *On the dynamical role of coherent structures in turbulence on the dynamical role of coherent structures in turbulence*, C. R. Physique, 13 (2012), pp. 866–877. (Cited on p. 20)

[304] K. PADBERG, T. HAUFF, F. JENKO, AND O. JUNGE, *Lagrangian structures and transport in turbulent magnetized plasmas*, New J. Phys., 9 (2007), p. 400. (Cited on p. 198)

[305] K.J. PALMER, *Exponential dichotomies and transversal homoclinic points*, J. Differential Equations, 55 (1984), pp. 225–256. (Cited on pp. 30, 39, 42, 117, 162, 163)

[306] L. PAN, *Stable invariant manifolds with impulses and growth rates*, Adv. Differential Equations, 2015 (2015), p. 221. (Cited on pp. 75, 81, 143)

[307] J. PARK, K. SEO, AND T. KWON, *A chaotic micromixer using obstruction-pairs*, J. Micromech. Microeng., 20 (2010), 015023. (Cited on pp. 5, 13, 20, 27)

[308] M. PARSA AND F. HORMOZI, *Experimental and CFD modeling of fluid mixing in sinusoidal microchannels with different phase shift between side walls*, J. Micromech. Microeng., 24 (2014), 065018. (Cited on pp. 5, 13, 14, 15, 28)

[309] T. PEACOCK AND G. HALLER, *Lagrangian coherent structures: The hidden skeleton of fluid flow*, Phys. Today, 66 (2013), pp. 41–47. (Cited on p. 8)

[310] J. PEDLOSKY, *Geophysical Fluids Dynamics*, Springer, New York, 1979. (Cited on pp. 11, 27, 134, 161)

[311] S. PENG, Y. QIAN, R. LUMPKIN, Y. DU, D. WANG, AND P. LI, *Characteristics of the near-surface currents in the Indian Ocean as deduced from satellite-tracked surface drifters. Part 1: Pseudo-Eulerian statistics*, J. Phys. Oceanogr., 45 (2015), pp. 441–458. (Cited on p. 11)

[312] ———, *Characteristics of the near-surface currents in the Indian Ocean as deduced from satellite-tracked surface drifters. Part 2: Lagrangian statistics*, J. Phys. Oceanogr., 45 (2015), pp. 459–477. (Cited on p. 11)

[313] V. PETROV, E. MIHALIUK, S.K. SCOTT, AND K. SHOWALTER, *Stabilizing and characterizing unstable states in high-dimensional systems from time-series*, Phys. Rev. E, 51 (1995), pp. 3988–3996. (Cited on p. 204)

[314] R. PIERREHUMBERT, *Chaotic mixing of tracer and vorticity by modulated travelling Rossby waves*, Geophys. Astrophys. Fluid Dyn., 58 (1991), pp. 285–319. (Cited on pp. 4, 5, 9, 10, 12, 100, 134, 161, 168, 198)

[315] R. PIERREHUMBERT AND H. YANG, *Global chaotic mixing in isentropic surfaces*, J. Atmospheric Sci., 50 (1993), pp. 2462–2480. (Cited on p. 20)

[316] A. POJE AND G. HALLER, *Geometry of cross-stream mixing in a double-gyre ocean model*, J. Phys. Oceanogr., 29 (1999), pp. 1649–1665. (Cited on pp. 4, 10)

[317] A. POJE, G. HALLER, AND I. MEZIĆ, *The geometry and statistics of mixing in aperiodic flows*, Phys. Fluids, 11 (1999), pp. 2963–2968. (Cited on p. 23)

[318] A. POJE, M. TONER, A. KIRWAN, AND C. JONES, *Drifter launch strategies based on Lagrangian templates*, J. Phys. Oceanogr., 32 (2002), pp. 1855–1869. (Cited on p. 10)

[319] A. POLYANIN AND A. MANZHIROV, *Handbook of Integral Equations*, CRC Press, Boca Raton, FL, 1998. (Cited on p. 89)

[320] C. POSTLETHWAITE AND M. SILBER, *Stabilizing unstable periodic orbits in the Lorenz equations using time-delayed feedback control*, Phys. Rev. E, 76 (2007), 056214. (Cited on p. 204)

[321] S. PRANTS, M. BUDYANSKY, V. PONOMAREV, AND M. ULEYSKY, *Lagrangian study of transport and mixing in a mesoscale eddy street*, Ocean Model., 38 (2011), pp. 114–125. (Cited on pp. 10, 20)

[322] S. PRANTS, M. BUDYANSKY, M. ULEYSKY, AND G. ZASLAVSKY, *Chaotic mixing and transport in a meandering jet flow*, Chaos, 16 (2006), 033117. (Cited on pp. 10, 24)

[323] L. PRATT, *Meandering and eddy detachment according to a simple (looking) path equation*, J. Phys. Oceanogr., 18 (1988), pp. 1627–1640. (Cited on pp. 4, 5, 100, 161, 168)

[324] L. PRATT, M. LOZIER, AND N. BELIAKOVA, *Parcel trajectories in quasigeostrophic jets · neutral modes*, J. Phys. Oceanogr., 25 (1995), pp. 1451–1466. (Cited on pp. 10, 12, 134)

[325] K.S. PRIYANKARA, S. BALASURIYA, AND E. BOLLT, *Quantifying the role of folding in nonautonomous flows: The unsteady double gyre*, submitted, 2016. (Cited on p. 65)

[326] K. PYRAGAS, *Continuous control of chaos by self-controlling feedback*, Phys. Lett. A, 170 (1992), pp. 421–428. (Cited on pp. 197, 204)

[327] K. PYRAGAS, V. PYRAGAS, I.Z. KISS, AND J.L. HUDSON, *Adaptive control of unknown unstable steady states of dynamical systems*, Phys. Rev. E, 70 (2004), 026215. (Cited on p. 204)

[328] R. QI AND S. XU, *Applications of Lagrangian coherent structures to expression of invariant manifolds in astrodynamics*, Astrophys. Space Sci., 351 (2014), pp. 125–133. (Cited on p. 20)

[329] S. RABEN, S. ROSS, AND P. VLACHOS, *Computation of finite-time Lyapunov exponents from time-resolved particle image velocimetry data*, Experiments Fluids, 55 (2014), p. 1638. (Cited on p. 20)

[330] T. RADKO, *On the generation of large-scale structures in a homogeneous eddy field*, J. Fluid Mech., 668 (2011), pp. 76–99. (Cited on pp. 62, 184)

[331] D. RANDALL, *An Introduction to the Global Circulation of the Atmosphere*, Princeton University Press, Princeton, NJ, 2015. (Cited on p. 134)

[332] M. RASMUSSEN, *Finite-time attractivity and bifurcation for nonautonomous differential equations*, Diff. Equat. Dyn. Sys, 18 (2010), pp. 57–78. (Cited on p. 32)

[333] N. REIS, A.A. VICENTE, J.A. TEIXEIRA, AND M.R. MACKLEY, *Residence times and mixing of a novel continuous oscillatory flow screening reactor*, Chem. Eng. Sci., 59 (2004), pp. 4967–4974. (Cited on pp. 173, 177)

[334] P. RESTO, E. BERTHIER, D. BEEBE, AND J. WILLIAMS, *An inertia enhanced passive pumping mechanism for fluid flow in microfluidic devices*, Lab Chip, 12 (2012), pp. 2221–2228. (Cited on pp. 14, 186, 187, 196)

[335] D. ROBINSON AND D. ROCKWELL, *Construction of three-dimensional images of flow structure via particle tracking techniques*, Experiments Fluids, 14 (1993), pp. 257–270. (Cited on pp. 5, 14)

[336] A.J.S. RODRIGO, R.C.R. RODRIGUES, N.F.C. FORMIGA, AND J.P.B. MOTA, *Mixing enhancement by frequency-selective chaotic advection in a 3-D time-periodic Stokes flow*, Chem. Eng. Comm., 193 (2006), pp. 743–753. (Cited on pp. 15, 173, 177)

[337] S.M. RODRIGUE AND E.V. ESCHENAZI, *Lobe transport analysis of the Kelvin-Stuart cat's eyes driven flow*, Chaos, 20 (2010), 013101. (Cited on pp. 108, 168)

[338] I. ROGACHEVSKII AND N. KLEEORIN, *Intermittency and anomalous scaling for magnetic fluctuations*, Phys. Rev. E, 56 (1997), pp. 417–426. (Cited on pp. 6, 8)

[339] A. ROGERSON, P. MILLER, L. PRATT, AND C. JONES, *Lagrangian motion and fluid exchange in a barotropic meandering jet*, J. Phys. Oceanogr., 29 (1999), pp. 2635–2655. (Cited on p. 10)

[340] J. ROHRS, A. SPERREVIK, K. CHRISTENSEN, G. BROSTROM, AND O. BREIVIK, *Comparison of HF radar measurements with Eulerian and Lagrangian surface currents*, Ocean Dyn., 65 (2015), pp. 679–690. (Cited on p. 11)

[341] V. ROM-KEDAR, A. LEONARD, AND S. WIGGINS, *An analytical study of transport, mixing and chaos in an unsteady vortical flow*, J. Fluid Mech., 214 (1990), pp. 347–394. (Cited on pp. 38, 40, 61, 107, 108, 113, 122, 127, 128, 129, 130, 131, 175)

[342] V. ROM-KEDAR AND A.C. POJE, *Universal properties of chaotic transport in the presence of diffusion*, Phys. Fluids, 11 (1999), pp. 2044–2057. (Cited on pp. 15, 129, 130, 175, 177, 178, 193)

[343] V. ROM-KEDAR AND S. WIGGINS, *Transport in two-dimensional maps*, Arch. Ration. Mech. Anal., 109 (1990), pp. 239–298. (Cited on pp. 104, 107, 127, 168)

[344] S. ROSS, M. TANAKA, AND C. SENATORE, *Detecting dynamical boundaries from kinematic data in biomechanics*, Chaos, 20 (2010), 017507. (Cited on p. 198)

[345] T. ROZI AND Y. FUKUMOTO, *The most unstable perturbation of wave-packet form inside Hill's vortex*, J. Phys. Soc. Japan, 69 (2000), pp. 2700–2701. (Cited on pp. 5, 14)

[346] B. RUTHERFORD, G. DANGELMAYR, J. PERSING, W. SCHUBERT, AND M. MONTGOMERY, *Advective mixing in a nondivergent barotropic hurricane model*, Atmos. Chem. Phys., 10 (2005), pp. 475–497. (Cited on pp. 4, 10)

[347] W. RYBCZYNSKI, *Über die fortschreitende bewegung einer flüssigen kugel in einem zähen medium*, Bull. Acad. Sci. Cracovi A, (1911), pp. 40–46. (Cited on pp. 5, 14)

[348] I. RYPINA, M. BROWN, F. BERON-VERA, H. KOCAK, M. OLASCOAGA, AND I. UDOVYDCHENKOV, *On the Lagrangian dynamics of atmospheric zonal jets and the permeability of the stratospheric polar vortex*, J. Atmos. Sci., 64 (2007), pp. 3595–3610. (Cited on pp. 10, 12, 134)

[349] I.I. RYPINA, M.G. BROWN, AND H. KOÇAK, *Transport in an idealized three-gyre system with applications to the Adriatic Sea*, J. Phys. Oceanogr., 39 (2009), pp. 675–690. (Cited on p. 198)

[350] I. RYPINA, L. PRATT, J. PULLEN, J. LEVIN, AND A. GORDON, *Chaotic advection in an archipelago*, J. Phys. Oceanogr., 40 (2010), pp. 1988–2006. (Cited on p. 10)

[351] I. RYPINA, S. SCOTT, L. PRATT, AND M. BROWN, *Investigating the connection between complexity of isolated trajectories and Lagrangian coherent structures*, Nonlin. Proc. Geophys., 18 (2011), pp. 977–987. (Cited on pp. 10, 23)

[352] R. SAMELSON, *Fluid exchange across a meandering jet*, J. Phys. Oceanogr., 22 (1992), pp. 431–440. (Cited on pp. 4, 9, 10, 12, 134, 161)

[353] ———, *Lagrangian motion, coherent structures, and lines of persistent material strain*, Annu. Rev. Marine Sci., 5 (2013), pp. 137–163. (Cited on pp. 9, 20)

[354] R. SAMELSON AND S. WIGGINS, *Lagrangian transport in geophysical jets and waves*, Interdisciplinary Applied Mathematics, Springer, New York, 2006. (Cited on pp. 9, 10, 11, 36, 38, 129, 130, 131, 134, 161)

[355] B. SANDSTEDE, S. BALASURIYA, C.K.R.T. JONES, AND P.D. MILLER, *Melnikov theory for finite-time vector fields*, Nonlinearity, 13 (2000), pp. 1357–1377. (Cited on pp. 4, 5, 10, 12, 27, 39, 100, 117, 161, 162, 163, 164, 165)

[356] T. SAPSIS AND G. HALLER, *Clustering criterion for inertial particles in two-dimensional time-periodic and three-dimensional steady flows*, Chaos, 20 (2010), 017515. (Cited on pp. 5, 14, 210)

[357] F. SARRAZIN, T. BONOMETTI, L. PRAT, C. GOURDON, AND J. MAGNAUDET, *Hydrodynamic structures of droplets engineered in rectangular micro-channels*, Microfluid. Nanofluid., 5 (2008), pp. 131–137. (Cited on p. 15)

[358] F. SARRAZIN, K. LOUBIÈRE, L. PRAT, C. GOURDON, T. BONOMETTI, AND J. MAGNAUDET, *Experimental and numerical study of droplet hydrodynamics in microchannels*, Am. Inst. Chem. Eng. J., 52 (2006), pp. 4061–4070. (Cited on p. 15)

[359] B. SCHINDLER, R. PEIKERT, R. FUCHS, AND H. THEISL, *Ridge concepts for the visualization of Lagrangian coherent structures*, in Topological Methods in Data Analysis and Visualization II, R. Peikert, H. Hauser, H. Carr, and R. Fuchs, eds., Springer, New York, 2012, pp. 221–236. (Cited on p. 20)

[360] D. G. SCHMALE AND S. D. ROSS, *Highways in the sky: Scales of atmospheric transport of plant pathogens*, Annu. Rev. Phytopath., 53 (2015), pp. 591–611. (Cited on p. 12)

[361] C. SEIS, *Maximal mixing by incompressible fluid flows*, Nonlinearity, 26 (2013), pp. 3279–3289. (Cited on p. 198)

[362] C. SENATORE AND S.D. ROSS, *Detection and characterization of transport barriers in complex flows via ridge extraction of the finite time Lyapunov exponent field*, Int. J. Num. Meth. Eng., 86 (2011), pp. 1163–1174. (Cited on p. 19)

[363] K.D. SEO AND D.S. KIM, *Microfluidic synthesis of thermo-responsive poly(n-isopropylacrylamide)-poly(ethylene glycol) diacreylate microhydrogels as chemo-embolic microspheres*, J. Micromech. Microeng., 24 (2014), 085001. (Cited on pp. 5, 14, 15)

[364] S.C. SHADDEN, F. LEKIEN, AND J.E. MARSDEN, *Definition and properties of Lagrangian coherent structures from finite-time Lyapunov exponents in two-dimensional aperiodic flows*, Phys. D, 212 (2005), pp. 271–304. (Cited on pp. 4, 8, 10, 12, 20, 21, 61, 64, 126)

[365] A. SHARMA AND B. MCKEON, *On coherent structure in wall turbulence*, J. Fluid Mech., 728 (2013), pp. 196–238. (Cited on p. 168)

[366] J. SHEN, K. LU, AND W. ZHANG, *Heteroclinic chaotic behavior driven by a Brownian motion*, J. Differential Equations, 255 (2013), pp. 4185–4225. (Cited on p. 39)

[367] S.M. SHIN, I.S. KANG, AND Y.-K. CHO, *Mixing enhancement by using electrokinetic instability under time-periodic electric field*, J. Micromech. Microeng., 15 (2005), pp. 455–462. (Cited on pp. 15, 173, 177)

[368] B.I. SHRAIMAN, *Diffusive transport in a Rayleigh-Bénard convection cell*, Phys. Rev. A, 36 (1987), pp. 261–267. (Cited on p. 184)

[369] E. SHUCKBURGH AND P. HAYNES, *Diagnosing transport and mixing using a tracer-based coordinate system*, Phys. Fluids, 15 (2003), pp. 3342–3357. (Cited on p. 10)

[370] M. SHUSSER AND D. WEIHS, *Compound drops as spherical shell vortices*, Fluid Dyn. Res., 42 (2010), 025502. (Cited on pp. 5, 14, 210)

[371] J. SIEBER, O.E. OMEL'CHENKO, AND M. WOLFRUM, *Controlling unstable chaos: Stabilizing chimera states by feedback*, Phys. Rev. Lett., 112 (2014), 054102. (Cited on p. 204)

[372] J. SIVASAMY, Z. CHE, T.N. WONG, N.-T. NGUYEN, AND L. YOBAS, *A simple method for evaluating and predicting chaotic advection in microfluidic slugs*, Chem. Eng. Sci., 65 (2010), pp. 5382–5391. (Cited on pp. 13, 15, 28)

[373] T.H. SOLOMON AND J.P. GOLLUB, *Chaotic particle transport in time-dependent Rayleigh-Bénard convection*, Phys. Rev. A, 38 (1988), pp. 6280–6286. (Cited on p. 184)

[374] T.H. SOLOMON, S. TOMAS, AND J.L. WARNER, *Role of lobes in chaotic mixing of miscible and immiscible impurities*, Phys. Rev. Lett., 77 (1996), pp. 2682–2685. (Cited on pp. 173, 177)

[375] H. SONG, M. BRINGER, J.D. TICE, C.J. GERDTS, AND R.F. ISMAGILOV, *Experimental test of scaling of mixing by chaotic advection in droplets moving through microfluidic channels*, Appl. Phys. Lett., 83 (2003), pp. 4664–4666. (Cited on pp. 5, 14, 15, 27)

[376] H SONG, Z. CAI, H. NOH, AND D. BENNETT, *Chaotic mixing in microchannels via low frequency switching transverse electroosmotic flow generated on integrated microelectrodes*, Lab Chip, 10 (2010), pp. 734–740. (Cited on pp. 5, 13, 14, 15, 28, 123, 173, 177, 187)

[377] H. SONG, Z. CAI, H.M. NOH, AND D.J. BENNETT, *Chaotic mixing in microchannels via low frequency switching transverse electroosmotic flow generated on integrated microelectrodes*, Lab Chip, 10 (2010), pp. 734–740. (Cited on p. 177)

[378] H. SONG, J.D. TICE, AND R.F. ISMAGILOV, *A microfluidic system for controlling reaction networks in time*, Angew. Chem. Int. Ed., 42 (2003), pp. 768–772. (Cited on pp. 5, 14)

[379] J. SOULAGES, M.S.N. OLIVEIRA, P.C. SOUZA, M.A. ALVES, AND G.H. MCKINLEY, *Investigating the stability of viscoelastic stagnation flows in T-shaped microchannels*, J. Non-Newtonian Fluid Mech., 163 (2009), pp. 9–24. (Cited on p. 198)

[380] H.A. STONE, A. NADIM, AND S. STROGATZ, *Chaotic streamlines inside drops immersed in steady Stokes flows*, J. Fluid Mech., 232 (1991), pp. 629–646. (Cited on pp. 5, 14, 15, 210)

[381] Z.B. STONE AND H.A. STONE, *Imaging and quantifying mixing in a model droplet micromixer*, Phys. Fluids, 17 (2005), 063103. (Cited on pp. 5, 13, 14, 15, 28, 210)

[382] R.S. STRICHARTZ, *A guide to distribution theory and Fourier transforms*, Studies in Advanced Mathematics, CRC Press, Boca Raton, FL, 1994. (Cited on p. 177)

[383] A. STROOCK, S. DERTINGER, A. ADJARI, I. MEZIĆ, H. STONE, AND G. WHITESIDES, *Chaotic mixer for microchannels*, Science, 295 (2002), pp. 647–651. (Cited on pp. 5, 13, 28)

[384] J. STUART, *On finite-amplitude oscillations in laminar mixing layers*, J. Fluid Mech., 29 (1967), pp. 417–440. (Cited on pp. 168, 169)

[385] H. SUGIOKA, *Chaotic mixer using electro-osmosis at finite Péclet number*, Phys. Rev. E, 81 (2010), 036306. (Cited on pp. 173, 177)

[386] X. SUN, K. TANG, R.D. SMITH, AND R.T. KELLY, *Controlled dispensing and mixing of pico- and nanoliter volumes using on-demand droplet-based microfluidics*, Microfluid. Nanofluid., 15 (2013), pp. 117–126. (Cited on pp. 5, 14, 15)

[387] H. SUZUKI, C.-M. HO, AND N. KASAGI, *A chaotic mixer for magnetic bead-based micro cell sorter*, J. Microelectromech. Sys., 13 (2004), pp. 779–790. (Cited on pp. 15, 173, 177)

[388] A.J. SZERI, W.J. MILLIKEN, AND L.G. LEAL, *Rigid particles suspended in time-dependent flows—irregular versus regular motion, disorder versus order*, J. Fluid Mech., 237 (2006), pp. 33–56. (Cited on p. 200)

[389] P. TABELING, *Recent progress in the physics of microfluidics and related biotechnological applications*, Current Opinion Biotech., 25 (2014), pp. 129–134. (Cited on p. 13)

[390] P. Tabeling, M. Chabart, A. Dodge, C. Jullien, and F. Okkels, *Chaotic mixing in cross-channel micromixers*, Phil. Trans. R. Soc. Lond. A, 362 (2004), pp. 987–1000. (Cited on pp. 5, 13, 14, 155, 179, 186, 196)

[391] P. Tallapragada and S. Ross, *A set oriented definition of finite-time Lyapunov exponents and coherent sets*, Commun. Nonlin. Sci. Numer. Simul., 18 (2013), pp. 1106–1126. (Cited on pp. 61, 126)

[392] P. Tallapragada, S.D. Ross, and D.G. Schmale, *Lagrangian coherent structures are associated with fluctuations in airborne microbial populations*, Chaos, 21 (2011), 033122. (Cited on p. 12)

[393] E. Tamaseviciute, G. Maikolaitis, S. Bumeliene, and A. Tamaseviciute, *Stabilizing saddles*, Phys. Rev. E, 88 (2013), 060901(R). (Cited on pp. 204, 212)

[394] M.K. Tan, L.Y. Yeo, and J.R. Friend, *Rapid fluid flow and mixing induced in microchannels using surface acoustic waves*, Europhys. Lett., 87 (2009), 47003. (Cited on pp. 173, 177)

[395] W. Tang, P. Chan, and G. Haller, *Accurate extraction of Lagrangian coherent structures over finite domains with application to flight data analysis over Hong Kong International Airport*, Chaos, 20 (2010), 017502. (Cited on p. 20)

[396] W. Tang, P.W. Chan, and G. Haller, *Lagrangian coherent structure analysis of terminal winds detected by Lidar. Part I: Turbulence structures*, J. Appl. Meteor. Climat., 50 (2011), pp. 325–338. (Cited on p. 24)

[397] ———, *Lagrangian coherent structure analysis of terminal winds detected by Lidar. Part II: Structure evolution and comparison with flight data*, J. Appl. Meteor. Climat., 50 (2011), pp. 2167–2183. (Cited on p. 24)

[398] G.I. Taylor, *The formation of emulsions in definable fields of flow*, Proc. R. Soc. Lond. A, 146 (1934), pp. 501–523. (Cited on p. 200)

[399] H. Teramoto, G. Haller, and T. Komatsuzaki, *Detecting invariant manifolds as stationary LCSs in autonomous dynamical systems*, Chaos, 23 (2013), 043107. (Cited on p. 23)

[400] J.-L. Thiffeault, *Using multiscale norms to quantify mixing and transport*, Nonlinearity, 84 (2012), pp. R1–R44. (Cited on pp. 14, 27, 28)

[401] Y.-P. Tian, *Controlling chaos using invariant manifolds*, Int. J. Control, 72 (1999), pp. 258–266. (Cited on p. 204)

[402] W. Tucker, *The Lorenz attractor exists*, C. R. Acad. Sci. Paris, Série I, 328 (1999), pp. 1197–1202. (Cited on p. 209)

[403] K.-Y. Tung, C.-C. Li, and J.-T. Yang, *Mixing and hydrodynamic analysis of a droplet in a planar serpentine micromixer*, Microfluid. Nanofluid., 7 (2009), pp. 545–557. (Cited on pp. 5, 14, 15, 28)

[404] Y. Ueda, S. Yoshida, H. Stewart, and J. Thompson, *Basin explosions and escape phenomena in the twin-well Duffing oscillator: Compound global bifurcations organizing behaviour*, Phil. Trans. R. Soc. Lond. A, 332 (1990), pp. 169–186. (Cited on pp. 198, 202)

[405] O.A. van Herwaarden, *Spread of pollutant by dispersive groundwater flow*, SIAM J. Appl. Math., 54 (1994), pp. 26–41. (Cited on p. 198)

[406] E. VAN SEBILLE, M. ENGLAND, AND G. FROYLAND, *Origin and evolution of the ocean garbage patches derived from surface drifter data*, Environ. Res. Lett., 7 (2012), 044040. (Cited on p. 12)

[407] A. VIKHANSKY, *Control of stretching rate in time-periodic chaotic flows*, Phys. Fluids, 14 (2002), pp. 2752–2756. (Cited on pp. 173, 177)

[408] ——, *Enhancement of laminar mixing by optimal control methods*, Chem. Eng. Sci., 57 (2002), pp. 2719–2725. (Cited on pp. 181, 198)

[409] M. VOLPERT, C. MEINHART, I. MEZIĆ, AND M. DAHLEH, *An actively controlled micromixer*, in Proc. Microelectromechanical Systems (MEMS), Nashville, TN, 1999, Vol. 1, American Society of Mechanical Engineers, New York, 1999, pp. 483–487. (Cited on pp. 13, 14, 186, 187)

[410] G. VOTH, G. HALLER, AND J. GOLLUB, *Experimental measurements of stretching fields in fluid mixing*, Phys. Rev. Lett., 88 (2002), 254501. (Cited on p. 20)

[411] C.-H. WANG AND G.-B. LEE, *Automatic bio-sampling chips integrated with micro-pumps and micro-valves for disease detection*, Biosensors Bioelectronics, 21 (2005), pp. 419–425. (Cited on pp. 14, 186, 187, 196)

[412] G. WANG, F. YANG, AND W. ZHAO, *There can be turbulence in microfluidics at low Reynolds number*, Lab Chip, 14 (2014), pp. 1452–1458. (Cited on pp. 5, 13)

[413] S.S. WANG, Z.J. JIAO, X.Y. HUANG, C. YANG, AND N.T. NGUYEN, *Acoustically induced bubbles in a microfluidic channel for mixing enhancement*, Microfluid. Nanofluid., 6 (2009), pp. 847–852. (Cited on pp. 15, 173, 177)

[414] Y. WANG, J. ZHE, B.T.F. CHUNG, AND P. DUTTA, *A rapid magnetic particle driven micromixer*, Microfluid. Nanofluid., 4 (2008), pp. 375–389. (Cited on pp. 15, 173, 177)

[415] M. WECHSELBERGER, *Extending Melnikov theory to invariant manifolds on non-compact domains*, Dyn. Sys., 17 (2002), pp. 215–233. (Cited on p. 39)

[416] L. WEIGUO AND S. ZUHE, *A constructive proof of existence and uniqueness of 2π-periodic solutions to Duffing's equation*, Nonlin. Anal., 42 (2000), pp. 1209–1220. (Cited on p. 45)

[417] J. WEISS AND E. KNOBLOCK, *Mass transport and mixing by modulated traveling waves*, Phys. Rev. A, 40 (1989), pp. 2579–2588. (Cited on pp. 5, 9, 12, 39, 50, 100, 131, 134, 135, 168)

[418] S.T. WERELEY AND S. GUI, *A correlation-based central difference image correction (CDIC) method and application in a four-roll mill flow PIV measurement*, Exp. Fluids, 34 (2003), pp. 42–51. (Cited on p. 200)

[419] J.A. WHITEHEAD, *Topographic control of oceanic flows in deep passages and straits*, Rev. Geophys., 36 (1998), pp. 423–440. (Cited on p. 198)

[420] G. WHITESIDES, *The origins and the future of microfluidics*, Nature, 442 (2006), pp. 368–373. (Cited on pp. 5, 13)

[421] S. WIGGINS, *Chaotic Transport in Dynamical Systems*, Springer-Verlag, New York, 1992. (Cited on pp. 9, 36, 38, 39, 40, 45, 49, 50, 61, 104, 107, 108, 113, 116, 119, 122, 127, 128, 129, 130, 131)

[422] ——, *The dynamical systems approach to Lagrangian transport in oceanic flows*, Annu. Rev. Fluid Mech., 37 (2005), pp. 295–328. (Cited on pp. 4, 9, 10)

[423] J.R. WINTERSMITH, L. ZOU, A.J. BERNOFF, J.C. ALEXANDER, J.A. MANN, E.E. KOOIJMAN, AND E.K. MANN, *Determination of interphase line tension in Langmuir films*, Phys. Rev. E, 75 (2007), 061605. (Cited on p. 200)

[424] P. WOIAS, *Micropumps—summarizing the first two decades*, Proc. SPIE, 4560 (2001), pp. 39–52. (Cited on pp. 14, 186, 187, 196)

[425] C.-Y. WU AND R.-T. TSAI, *Fluid mixing via multidirectional vortices in converging-diverging meandering microchannels with semi-elliptical side walls*, Chem. Eng. J., 217 (2013), pp. 320–328. (Cited on pp. 5, 13, 14, 15)

[426] F. WU AND D. VAINCHTEIN, *On localized mixing in action–action–angle flows*, Commun. Nonlin. Sci. Numer. Simulat., 19 (2014), pp. 67–73. (Cited on pp. 5, 14, 15)

[427] W. XU AND S.J. MULLER, *Exploring both sequence detection and restriction endonuclease cleavage kinetics by recognition site via single-molecule microfluidic trapping*, Lab Chip, 11 (2011), pp. 435–442. (Cited on p. 198)

[428] K. YAGASAKI, *Invariant manifolds and control of hyperbolic trajectories on infinite- or finite-time intervals*, Dyn. Sys., 23 (2008), pp. 309–331. (Cited on pp. 197, 212)

[429] ——, *Application of the subharmonic Melnikov method to piecewise-smooth systems*, Discrete Cont. Dyn. Sys. A, 33 (2013), pp. 2189–2209. (Cited on p. 76)

[430] D. YAN, C. YANG, J. MIAO, Y. LAM, AND X. HUANG, *Enhancement of electrokinetically driven microfluidic T-mixer using frequency modulated electric field and channel geometry effects*, Electrophoresis, 30 (2009), pp. 3144–3152. (Cited on pp. 123, 173, 177)

[431] J.M. YANG AND Y.Z. XU, *Coalescence of two viscoelastic droplets connected by a string*, Phys. Fluids, 20 (2008), 043101. (Cited on p. 200)

[432] I. YASHAYAEV, D. SEIDOV, AND E. DEMIROV, *A new collective view of oceanography of the Arctic and North Atlantic basins*, Prog. Oceanogr., 132 (2015), pp. 1–21. (Cited on p. 11)

[433] Y. YI, *A generalized integral manifold theorem*, J. Differential Equations, 102 (1993), pp. 153–187. (Cited on p. 41)

[434] ——, *Stability of integral manifold and orbital attraction of quasi-periodic motion*, J. Differential Equations, 103 (1993), pp. 278–322. (Cited on p. 41)

[435] X. YU, G. CHEN, Y. XIA, Y. SONG, AND Z. CAO, *An invariant-manifold-based method for chaos control*, IEEE Trans. Circuits Syst. I Fundam. Theory, 48 (2001), pp. 930–937. (Cited on pp. 204, 209)

[436] H. ZANG, M. HAN, AND D. XIAO, *On Melnikov of a homoclinic loop through a nilpotent saddle for planar near-Hamiltonian systems*, J. Differential Equations, 245 (2008), pp. 1086–1111. (Cited on p. 39)

[437] H. ZHANG, M. PRATER, AND T. ROSSBY, *Isopycnal Lagrangian statistics from the North Atlantic Current RAFOS float observations*, J. Geophys. Res. Oceans, 106 (2001), pp. 13817–13836. (Cited on pp. 11, 14, 187)

[438] X. ZHOU, X. ZHOU, AND B. ZHENG, *A pneumatic valve controlled microdevice for bioanalysis*, Biomicrofluidics, 7 (2013), 054116. (Cited on pp. 14, 15, 186, 187, 196)

Index